PHYSIOLOGY OF SMOOTH MUSCLE

P.146

Physiology of Smooth Muscle

Edited by

Edith Bülbring
Professor Emeritus of Pharmacology
University of Oxford
Oxford, England

M. F. Shuba
Professor of Physiology
Ukranian Academy of Sciences
Kiev, USSR

Raven Press ▪ New York

Made in the United States of America

International Standard Book Number 0–89004–051–6
Library of Congress Catalog Card Number 75–14566

ISBN outside North and South America only:
0–7204–7556–2

Preface

The Symposium on the Physiology of Smooth Muscles, held in Kiev in October 1974, offered a unique opportunity for an exchange of information from many countries otherwise not easily accessible. When I lightly undertook to edit the publication, I did not realize that the language barrier, so easily overcome in personal discussions, would become formidable when a clear presentation of experimental results is required for print.

Although I have had generous help from several participants, notably Dr. T. B. Bolton, Dr. Alison Brading, and Dr. G. Gabella, I must take the entire responsibility for any errors in interpretation that may have occurred, for consultation with the authors proved to be difficult once we had separated after the pleasant meeting. With this reservation in mind, the reader will find a wealth of interesting information on the physiology of many different smooth muscle types and a great variety in the experimental approach. Many new observations were reported.

In recent years, much progress has been made in our understanding of the cellular mechanisms that control ion distribution and ion movements and, in voltage-clamp experiments, ionic currents underlying membrane excitation have been identified. On the basis of such fundamental observations, the origin of rhythmic spontaneous activity and the nature of excitation–contraction–coupling is being investigated. Special features in the fine structure of smooth muscle cells and in the relationships between muscle cells and nerve terminals are recognized. By simultaneous recording of muscle responses and of neuronal activity in intrinsic nerve plexus, the nervous pathways in the peristaltic reflex activity of the gut are now being mapped. Finally, the characteristic mechanisms of action of neurotransmitters and of other drugs can now be specified.

The wide scope of this volume reflects the intense activity in smooth muscle research all over the world, and such a compilation is bound to be most stimulating for future work.

Edith Bülbring
(*April 1975*)

Contents

Ionic Composition, Ion Transport, and Transmembrane Ion Currents in Smooth Muscle

Origin and Characteristics of Spontaneous Activity in Smooth Muscle

**The Nature of Electromechanical Coupling
and Mechanisms of Smooth Muscle Activation**

Structure of Smooth Muscles and Neuromuscular Relations: Nervous Influence on Muscle Function

Mechanism of Action of Acetylcholine, Catecholamines, and Other Drugs on Smooth Muscle

Physiology of Smooth Muscle, edited by
E. Bülbring and M. F. Shuba.
Raven Press, New York © 1976.

Mechanisms Involved in Sodium Exchange in Smooth Muscle: Observations and Speculations

Alison F. Brading

University Department of Pharmacology, Oxford OX1 3QT, England

STEADY-STATE SODIUM FLUXES

The washout of tracer sodium from pieces of longitudinal smooth muscle from the taenia coli of the guinea pig equilibrated in normal Krebs solution, proceeds with a very rapid time course, so that after 15 min there is less than 1% of the tracer left in the muscle. It has been shown (Brading and Jones, 1969) that the major portion of the sodium, and more than could be freely dissolved in the extracellular fluid, behaves as if limited only by diffusion in the tissue. It now seems very likely that a large amount of sodium is held in some way in association with fixed negative charges on the cell surfaces. This was first postulated by Goodford (1970), and recently Widdicombe (1974; *personal communication*) has shown that 5mM La^{3+} can displace some 10 mmoles/kg fresh weight of sodium from the tissue, and that in the presence of La^{3+} the fast-exchanging Na is considerably reduced in size. The presence of this large amount of rapidly exchanging Na makes the study of truly transmembrane Na fluxes difficult to pursue; although the object of the work is to discover the factors involved in Na distribution in the normal tissue, it remains true that the only reasonable way of studying transmembrane fluxes is in tissues with high intracellular Na.

Such high-Na tissues can be prepared by exposing the muscle pieces to K^+-free solution at 35° to 37°C, when they rapidly gain Na and lose K (Axelsson and Holmberg, 1971; Casteels, Droogmans, and Hendrickx, 1971; Brading, 1973). The tissues reach a steady state, and have a membrane potential of about 7 to 10 mV, which is probably the same value as E_{Na} and E_{Cl} (Casteels, 1971). In such tissues the Na/K pump is not operative, and this simplifies the study of the fluxes. The properties of the Na/K pump were described recently by Brading and Widdicombe (1974).

Washout of tracer sodium from high-Na tissues also follows a complex time course, and the instantaneous rate constant of the efflux has not reached a steady value even after 40 to 50 min of washout. Figure 1 shows the decline in tissue counts during the washout and the instantaneous rate constant of the efflux from normal tissues washed out in Krebs solution, and high-Na tissues into K^+-free Krebs solution. The declining rate constant that

1

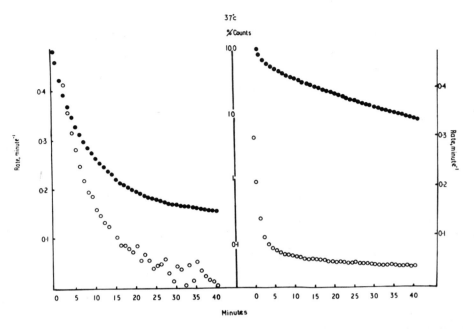

FIG. 1. Steady-state effluxes of ^{24}Na from normal tissues into Krebs solution (left) and high-Na tissues into K$^+$-free Krebs solution (right). (●) Percentage counts in the tissue (*central ordinate*); (○) instantaneous rate constant of the efflux (*outer ordinates*).

occurs in the high-Na tissues even after the washout of the rapidly exchanging component suggests either that the cells are heterogeneous with respect to Na permeability, or that there is some cellular compartmentalization of sodium.

Na EXCHANGE DIFFUSION

Casteels et al. (1973) have demonstrated that the rate of loss of Na from high-Na tissues is reduced when external Na is replaced by choline or lithium. They have suggested that a ouabain-insensitive Na-exchange diffusion process is operative in this tissue. I have studied this process in more detail (Brading, 1975), and some of the results obtained have led me to consider a model in which a large proportion of the Na/Na exchange is between the external solution and an internal compartment of limited volume. The implications of this model are also discussed in this chapter.

The Na/Na exchange has been studied using Mg^{2+} to replace Na$^+$ in the external solution and in the absence of external Ca^{2+}. Mg has proved to be the most satisfactory substitute for Na for periods of exposure longer than a few minutes, but other Na replacements have been used, and the results are very similar to those obtained with Mg as a substitute.

The easiest type of experiment to investigate the properties of the exchange mechanism is simply to load a tissue with ^{24}Na in K-free solution, and then to follow the rate of loss of Na into a $MgCl_2$-containing Na-free solution. The effects of introducing other cations can then be investigated under various conditions. Figure 2 shows such an experiment in the presence of 10^{-5} M ouabain. Incorporation of 10 mM Na into the washout solution causes a rapid increase in the rate of loss of Na, whereas the incorporation of 10 mM K has only a small effect on the rate of loss.

From experiments such as these the Na/Na exchange can be shown to be ouabain-insensitive, to have a fairly low affinity for external Na ions (the

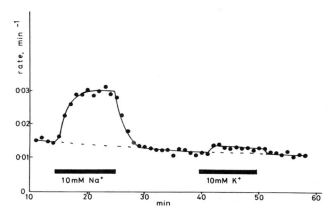

FIG. 2. The effect of introducing 10 mM Na$^+$ and 10 mM K$^+$ on the rate of loss of ^{24}Na into a Na-free $MgCl_2$ solution in the presence of 10^{-5} M ouabain.

results suggesting a K_e for binding of about 60 mM). The sites have an affinity for other alkali metal cations (Fig. 3) such that the affinity for Na > Li > K > Rb > K > Cs; this is series X of the Eisenman series (Eisenman, 1962). The exchange is very temperature-sensitive, with a Q_{10} of 6 between 15° and 25°C, and can be partially inhibited by some sulfhydryl reagents (p-chloromercuribenzoate and ethacrynic acid; Brading, 1975) Simultaneous measurement of loss of ^{22}Na and gain of ^{24}Na during a 10-min exposure to various concentrations of external sodium, suggest that the stoichiometry of the exchange process is 1:1. All these properties are consistent with a classical Na-exchange diffusion as first proposed by Ussing (1947), and are very similar to the properties of Na/Na exchange in the high-Na beef erythrocyte (Motais, 1973; Motais and Sola, 1973).

There are several results, however, that are not immediately consistent with a classical Na/Na exchange. One result is that the exchange is markedly reduced, but not abolished by levels of metabolic inhibition that deplete the

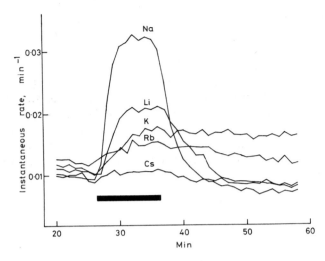

FIG. 3. The effect of alkali metal cations on the rate of loss of ^{24}Na into $MgCl_2$ washout solution. During the time indicated by the bar, the washout solution contained 46 mM alkali metal cation replacing some of the magnesium. All solutions contained ouabain 2×10^{-5} M, except the K^+- and Rb^+-containing solutions which contained 10^{-3} M ouabain. The effectiveness in stimulating the ^{24}Na efflux was of the order Na > Li > K > Rb > Cs.

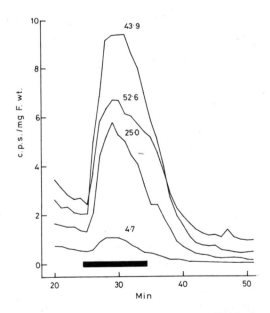

FIG. 4. The effects of altering the total cell Na on the ability of Na (126 mM) to stimulate ^{24}Na efflux. The counts per second per milligram fresh weight of tissue leaving the tissue in each minute have been plotted against time. For 10 min, as indicated by the bar, 126 mM Na was present in the washout solution and for the rest of the time the $MgCl_2$ washout solution was used. The figures indicate the sodium content of the tissue (millimoles per kilogram wet weight) at the time of introduction of the extracellular Na. Note that the stimulated efflux was greater when the Na content was 44 mmoles/kg wet wt than when it was 53 mmoles/kg wet wt. In the latter case, the tissue contained no potassium.

tissue of ATP. A second result is the relationship between total cell Na and the ^{24}Na loss stimulated by inclusion of 126 mM Na for 10 min in the external medium. This relationship is shown in Fig. 4, in which it can be seen that the greatest stimulation of the Na efflux does not occur at the highest level of cell Na. In this experiment the cells were allowed to exchange Na for K for limited periods of time after loading and K depleting. They were then washed in the Na-free solution before reintroducing Na into the medium. It is apparent that the highest stimulation of the efflux occurs when tissue Na content is high, but internal K is also present. The third unexpected observation is the behavior of the rate of loss of Na when Na is removed and then readmitted to the tissue during the efflux. Figure 5 shows the effect of replacing Na by Mg for 5, 10, and 20 min. The interesting thing to note is the "overshoot" that is observed in the rate constant when Na is readmitted. After 5-min exposure to MgCl$_2$ solution, the rate constant returns to its previous value with little overshoot, but with 10- or 20-min exposure, on reintroduction of Na the rate of loss overshoots its original level, and then returns toward it.

Consideration of these anomalous results has led to the adoption of a model for the compartmentalization of internal Na, which is discussed in the following section.

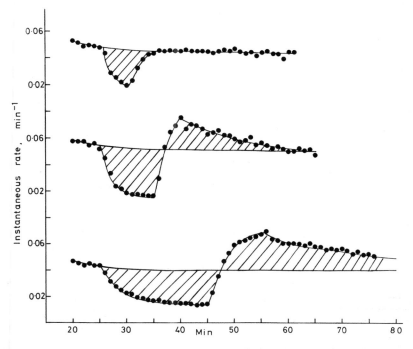

FIG. 5. The effect of removing Na on the rate of loss of ^{24}Na from high-Na tissues. The tissues were exposed to MgCl$_2$ washout solution for 5, 10, and 20 min. Note the effect on the overshoot.

COMPARTMENTALIZATION OF INTERNAL SODIUM

The model is shown in Fig. 6; the left is a schematic representation of the arrangement of cell compartments, and the right shows a possible interpretation of the model in terms of the structure of the cells of the taenia coli. Compartment 0 is the extracellular fluid, and compartments 1 and 2 are cell compartments. At steady state, fluxes of ions between compartments are such that $K_{02} = K_{20} > K_{12} = K_{21} > K_{10} = K_{01}$. These fluxes depend upon several factors. For example, K_{12} depends on the concentration of Na in compartment 1 (or the specific activity if tracer Na is being considered), the surface area of membrane separating compartments 1 and 2, and a con-

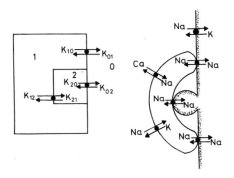

FIG. 6. (Left) A schematic representation of the arrangement of cell compartments. (Right) A possible interpretation of the model, showing an element of sarcoplasmic reticulum, and the plasma membrane with a membrane vesicle, and the basement membrane.

stant depending on the mechanisms involved in the transport of Na from 1 to 2. If the assumption is made that K_{20} and K_{02} involve only Na movement on the Na/Na exchange diffusion carrier, then the model can be shown to fit the results in the following ways.

Explanation of the "Overshoot"

If one considers the washout of tracer Na, as tracer is removed from the extracellular spaces the fluxes K_{02} and K_{01} rapidly approach zero.

Now because the model suggests that K_{20} is greater than K_{12}, the specific activity of the sodium in compartment 2 will decline more rapidly than will the specific activity in compartment 1, and this will lead to a progressive decline in the rate constant of washout of ^{24}Na from the whole tissue. Eventually this rate constant should reach a steady value when the loss of tracer from compartment 2 depends on the rate of entry from compartment 1. At this time the rate constant of loss from the whole tissue will depend on K_{10} and K_{12}. In practice, the rate constant still does decline, and although it has not reached a stable level after 40 to 50 min, it appears to be approaching a steady value. If Na is now removed from the extracellular fluid, K_{20}

will be zero, and the efflux from the tissue will now depend totally on K_{10} and should drop to a low steady value. During this time K_{12} will be greater than K_{21}, and the specific activity in compartment 2 will increase until it eventually equals that in compartment 1. When Na is reintroduced, K_{20} will be switched on and will be higher than when it was switched off. The total rate constant of loss of ^{24}Na from the whole tissue should overshoot its original level, and then decline as the specific activity in compartment 2 is once more lowered. Therefore the model will explain the results observed on the overshoot.

Relationship between Stimulated Efflux and Cell Sodium

As has been stated, the efflux stimulated by a 10-min exposure to 126 mM Na can be greater when there is some intracellular K than when there is none. This could be explained if the membrane separating compartments 1 and 2 possesses a Na/K pump. If this is the case, it could be imagined that the Na content in compartment 2 could be greater with the correct amount of K and Na present, than in the absence of K, when the pump would not be able to accumulate sodium in compartment 2.

Effect of Metabolic Inhibitors

If it is postulated that the effect of metabolic inhibition is to disconnect compartment 2 from the plasma membrane, then it is not necessary to assume that the Na/Na exchange is directly dependent on metabolic energy. The fraction of the exchange that is still present in metabolically inhibited tissue would then presumably represent an exchange between compartment 1 and the outside, as part of K_{10}.

POSSIBLE ANATOMICAL CORRELATES OF THE MODEL

The most likely structure to represent compartment 2 is the sarcoplasmic reticulum (SR), which has recently been demonstrated in the taenia, as well as many other smooth muscles, by Gabella (1971, 1973), and estimated by Devine, Somlyo, and Somlyo (1973) to account for about 2% of the cell volume. Gabella has shown that the SR comes in close contact with the plasma membrane, and also is closely associated with the membrane vesicles. Figure 6 gives an interpretation of the model, suggesting that the exchange of Na between this compartment and the extracellular fluid occurs at both sites.

The model now becomes analogous to the SR and T tubular system in cardiac and striated muscle. It is interesting in this context to note that Keynes and Steinhardt (1968) have suggested that a Na/Na exchange mechanism exists between the SR and the T tubule, and Rogus and Zierler (1973) have also suggested that the majority of the cell sodium is located in the SR, and exchanges with the external fluid more rapidly through the triadic junction with the T tubule than it does through the sarcoplasm.

SPECULATIONS

Such a model can lead to many speculations about the situation in normal muscle. What is the role of Na/Na exchange in the normal tissue? How is such a system involved in excitation–contraction coupling? First, one should reiterate the evidence. The results are most easily explained if there is a relatively small cellular compartment exchanging Na across the plasma membrane by a Na/Na exchange mechanism, and with the rest of the sarcoplasm with a rate coefficient of exchange greater than occurs between the sarcoplasm and the external medium across the plasma membrane. The rate coefficient of exchange between the sarcoplasm and the compartment can be greater in the presence of some cell potassium than in its absence, presumably because the membrane contains a Na/K pump. In order to fully explain the observations, the rate coefficient of exchange from the sarcoplasm across the compartment membrane has to be greater than across the plasma membrane, but this need not imply any difference in the membranes, and may simply reflect the greater surface area of the compartment membrane.

It is interesting to see how far the possibility that the SR membrane has exactly the same properties as the plasma membrane will go toward explaining some of the phenomena observed in smooth muscle, and especially in excitation–contraction coupling. It must be emphasized here that this section is entirely speculative.

First, such a compartment would accumulate Na through the Na/K pump. The amount of Na accumulated would presumably be limited by the K concentration inside the SR (which would be reduced by the pump), the Na concentration in the sarcoplasm which might be rather low, and the rate of diffusion of K into the SR. The accumulation of Na might be helped by the Na/Na exchange mechanism, as this has been shown to have some affinity for K at the outside site, and it probably has a similar affinity at the inside site (as I have observed in one experiment that the loss of ^{42}K from high-K solutions into $MgCl_2$ solution is stimulated more by adding external Na than it is by adding external K). Hence this mechanism could possibly exchange K for Na if the concentrations at each side are suitable.

There will be a gradient for sodium between the SR and the sarcoplasm, and it is possible that such a gradient could help accumulation of Ca through a Na/Ca exchange mechanism of the kind postulated by Baker, Blaustein, Hodgkin, and Steinhardt (1969), using the potential energy of Na moving down its gradient to move Ca up its gradient into the SR. If Ca is accumulated only up to 2 mM in the SR, and if it is released, this would raise the Ca ion concentration in the sarcoplasm by 4×10^{-5} M—presumably enough to initiate contraction.

There will also be a K gradient between the sarcoplasm and the SR, which would give rise to a potential between the two, and there will be a different potential (smaller) between the SR and the extracellular fluid where the two membranes are in contact. The plasma membrane is known to possess a potential dependent Ca permeability mechanism, which allows inflow of Ca into the cell during depolarization. If this mechanism also occurred in the SR membrane, Ca could be released into the sarcoplasm by depolarization of its membrane. When the plasma membrane is depolarized in high K, the tissues contract, then slowly relax. It is possible that the potential-dependent Ca permeability is inactivated by maintained depolarization. Under these conditions, the membrane of the SR may be able to repolarize and to continue to accumulate Ca, which then might be available for release by drugs activating receptors, possibly localized at the junctions between the SR and the plasma membrane.

REFERENCES

Axelsson, J., and Holmberg, B. (1971): The effects of K$^+$-free solution on tension development in the smooth muscle taenia coli from the guinea pig. *Acta Physiol. Scand.,* 82:322–332.

Baker, P. F., Blaustein, M. P., Hodgkin, A. L., and Steinhardt, R. A. (1969): The influence of calcium on sodium efflux in squid axons. *J. Physiol. (Lond.),* 200:431–458.

Brading, A. F. (1973): Ion distribution and ion movements in smooth muscle. *Philos. Trans. R. Soc. Lond. (Biol. Sci.),* 265:35–46.

Brading, A. F. (1975): Na/Na exchange in the smooth muscle of the guinea-pig taenia coli. *J. Physiol. (Lond.) (In press.)*

Brading, A. F., and Jones, A. W. (1969): Distribution and kinetics of CoEDTA in smooth muscle and its use as an extracellular marker. *J. Physiol. (Lond.),* 200:387–401.

Brading, A. F., and Widdicombe, J. H. (1974): An estimate of sodium/potassium pump activity and the number of pump sites in the smooth muscle of the guinea-pig taenia coli, using ^{3}[H] ouabain. *J. Physiol. (Lond.),* 238:235–249.

Casteels, R. (1971): The distribution of chloride ions in the smooth muscle cells of the guinea-pig taenia coli. *J. Physiol. (Lond.),* 214:225–244.

Casteels, R., Droogmans, G., and Hendrickx, H. (1971): Membrane potential of smooth muscle cells in K-free solution. *J. Physiol. (Lond.),* 217:281–295.

Casteels, R., Droogmans, G., and Hendrickx, H. (1973): Effect of sodium and sodium-substitutes on the active ion transport and on the membrane potential of smooth muscle cells. *J. Physiol. (Lond.),* 228:733–748.

Devine, C. E., Somlyo, A. V., and Somlyo A. P. (1973): Sarcoplasmic reticulum and mitochondria as cation accumulating sites in smooth muscle. *Philos. Trans. R. Soc. Lond. (Biol. Sci.),* 265:17–23.

Eisenman, G. (1962): Cation selective glass electrodes and their mode of operation. *Biophys. J.,* 2:259–323.

Gabella, G. (1971): Relationship between sarcoplasmic reticulum and caveolae intracellulares in the intestinal smooth muscle. *J. Physiol. (Lond.),* 216:42P.

Gabella, G. (1973): Fine structure of smooth muscle. *Philos. Trans. R. Soc. Lond. (Biol. Sci.),* 265:7–16.

Goodford, P. J. (1970): Ionic interactions in smooth muscle. In: *Smooth Muscle* edited by E. Bülbring, A. F. Brading, A. W. Jones, and T. Tomita. Arnold, London.

Keynes, R. D., and Steinhardt, R. A. (1968): Components of the sodium efflux in frog muscle. *J. Physiol. (Lond.)*, 198:581–599.

Motais, R. (1973): Sodium movements in high-sodium beef red cells: properties of a ouabain-insensitive exchange diffusion. *J. Physiol. (Lond.)*, 233:395–422.

Motais, R., and Sola, F. (1973): Characteristics of a sulphydryl group essential for sodium exchange diffusion in beef erythrocytes. *J. Physiol. (Lond.)*, 233:423–438.

Rogus, E., and Zierler, K. L. (1973): Sodium and water contents of sarcoplasm and sarcoplasmic reticulum in rat skeletal muscle: Effects of anisotonic media, ouabain and external sodium. *J. Physiol. (Lond.)*, 233:227–270.

Ussing, H. H. (1947): Interpretation of the exchange of radio sodium in isolated muscle. *Nature*, 160:262.

Widdicombe, J. H. (1974): The effect of lanthanum on ion content and movement in the guinea-pig's taenia coli. *J. Physiol. (Lond.)*, 241:106–107P.

Physiology of Smooth Muscle, edited by
E. Bülbring and M. F. Shuba.
Raven Press, New York © 1976.

Membrane Potential and Ion Transport in Smooth Muscle Cells

G. Droogmans and R. Casteels

Laboratorium voor Fysiologie Universiteit Leuven 3000 Leuven, Belgium

The nonequilibrium distribution of permeable ions under steady-state conditions requires that the passive fluxes of these ions down their electrochemical gradient be compensated by fluxes in the opposite direction. If the total current carried by the latter fluxes is different from zero, it produces a direct contribution to the membrane potential. The amplitude of this electrogenic component depends on both the current i_p and on the membrane resistance R_m. The total membrane potential E_m can be described by the relation

$$E_m = E_{\text{diff}} + R_m \cdot i_p$$

where E_{diff} is the diffusion potential, which can be estimated by the constant-field equation

$$E_{\text{diff}} = \frac{RT}{F} \ln \frac{\sum_i P_i{}^+ C_i{}^+ (\text{out}) + \sum_i P_i{}^- C_i{}^- (\text{in})}{\sum_i P_i{}^+ C_i{}^+ (\text{in}) + \sum_i P_i{}^- C_i{}^- (\text{out})}$$

The hypothesis that the active ion transport in taenia coli cells is electrogenic has been proposed by Burnstock (1) and Bülbring (2) to explain the hyperpolarization produced by epinephrine. This review presents more evidence for an electrogenic component of the membrane potential; some of the characteristics of the Na–K pump are discussed. Some of the results discussed below have been published previously (3–6), and are discussed in the following section.

ELECTROGENIC EFFECT OF THE ACTIVE TRANSPORT MECHANISM

In order to demonstrate unequivocally the electrogenic effect of the ion pump it is necessary to investigate this mechanism under conditions in which this effect can be supposed to be maximal, i.e., when the pumping rate is high and/or when the membrane resistance is high.

The pumping rate can be increased by increasing $(Na)_i$. This is obtained by exposing the tissues to a K-free solution. After 4 hr the intracellular K is replaced by Na, and the cells are depolarized to about -10 mV. Na and Cl ions are then passively distributed.

The membrane resistance can be increased by replacing Cl by a non-permeant anion (e.g., propionate, benzenesulfonate, isethionate). This procedure not only eliminates the contribution of Cl ions to the membrane conductance, but also reduces appreciably the K permeability (7) and the Na permeability, as shown below. Moreover, Ohashi (8) has shown that this substitution causes an approximate doubling of the membrane resistance. If the Na-rich, K-depleted cells are again exposed to a solution containing

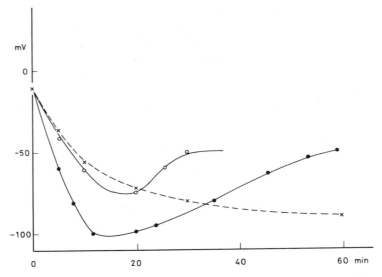

FIG. 1. The changes of E_K in Cl- and isethionate-containing solution (x), and the changes of the membrane potential in Cl (O) and isethionate solution (●) of K-depleted cells during K accumulation from a 5.9 mM K solution. The values of E_K have been calculated from the changes in $(K)_i$ by the Nernst equation. Ordinate: potential in millivolts; abscissa: time in minutes.

5.9 mM K, the intracellular Na is rapidly exchanged for K. The rate of this exchange is not significantly affected by the substitution of Cl by a non-permeant anion. During this extrusion of Na and accumulation of K, the cell membrane hyperpolarizes transiently, and the membrane potential temporarily becomes more negative than the K-equilibrium potential, calculated from the analytically determined values of $(K)_i$ (Fig. 1). This observation is a generally accepted criterion for an electrogenic pump, because under most circumstances E_k is more negative than E_{diff}. This hyperpolarization is more pronounced when Cl has been replaced by a nonpermeant anion. This finding is consistent with an electrogenic pump producing a higher voltage drop over a higher membrane resistance. Moreover, it contradicts the hypothesis that the pump is electroneutral, and that the observed electrogenic effect is an artifact caused by pericellular depletion of K by the pump activity.

The maximum difference of $|E_m - E_k|$ in Cl solution amounts to only about 12 mV. The rather small value of this difference can be explained by the decrease in membrane resistance that occurs during the pump stimulation, and which is largely due to a change in P_K (Fig. 2). Because of the initial increase of P_K, the diffusion potential closely follows E_K, and the electrogenic component remains small, causing E_m to be very close to E_K. It is only

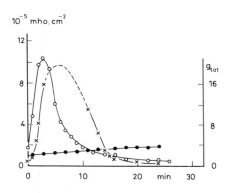

FIG. 2. Changes of the total membrane conductance (x, relative values), of the K conductance (O, mho · cm^{-2}) and of the Na conductance (●, mho · cm^{-2}) during K reaccumulation by K-depleted cells. The relative values of membrane conductance have been calculated from the measured electrotonic potentials, and the dashed line corresponds to the period during which the electrotonic potential was not different from the stimulus artifact. The K and Na conductances have been calculated from the fluxes of Na and K and the measured membrane potential.

when P_K decreases again, that the electrogenic effect is more pronounced and E_m becomes more negative than E_K. Finally, because of the reduction in $(Na)_i$, the decrease in pumping rate again reduces the electrogenic effect, which explains the transient nature of the observed hyperpolarization.

ION DISTRIBUTION AND TRACER FLUXES UNDER STEADY-STATE CONDITIONS IN CL-KREBS SOLUTION

In studying the characteristics of the pumping mechanism, the data of Casteels (9) have been analyzed using the constant-field equations and the Ussing flux-ratio equation. These data and the results of this analysis, which are based on a number of assumptions fully described by Casteels (9), have been summarized in Table 1. Calculation of the diffusion potential gives a value of −37 mV, which is 15 to 20 mV less negative than the measured value of E_m of about −55 mV. It can be observed that the sum of the ionic currents carried by Na, K, and Cl is different from zero, and amounts to 0.8 $\mu A \cdot cm^{-2}$. Because under steady-state conditions the passive fluxes are compensated by equal fluxes in the opposite direction, it can be concluded that active transport mechanisms generate a current of −0.8 $\mu A \cdot cm^{-2}$. With a value of the membrane resistance of 25 k$\Omega \cdot cm^2$ (10), it can be calculated that the pump current produces a potential drop of −20 mV across this resistance, giving a total membrane potential of −57 mV, which is close to the experimental value, and explains the difference between membrane potential and diffusion potential.

TABLE 1

	Cl-Krebs			Cl-free Krebs	
	Na	K	Cl	Na	K
Extracellular concentration (mM · liter^{-1})	137	5.9	134	137	5.9
Intracellular concentration (mM · liter^{-1})	13	164	58	10	170
Equilibrium potential (mV)	+ 62	− 89	− 22	+ 69	− 89
Unidirectional flux (pmoles · cm^{-2} · sec^{-1})	6.0	5.4	9.4	3.5	2.8
Permeability (10^{-8} cm · sec^{-1})	1.8	11	6.7	1.2	4.9
Net passive flux (pmoles · cm^{-2} · sec^{-1})	+ 6	− 4	− 6	+ 3.5	− 2.3
Ionic current (μA · cm^{-2})	+ 0.6	− 0.4	+ 0.6	+ 0.34	− 0.22
Total ionic current (μA · cm^{-2})	—	+ 0.8	—	+ 0.12	+ 0.12
E_{diff} (mV)	—	− 37	—	− 42	− 42
R_m (kΩ · cm^2)	—	25	—	50	50
E_m (measured, mV)	—	− 55	—	− 50	− 50
E_m (calculated, mV)	—	− 57	—	− 48	− 48

Analytical data and flux values for the different ions in taenia coli cells under steady-state conditions in Cl-Krebs and isethionate Krebs solution. The permeability coefficients, the passive and active fluxes, diffusion potentials, and electrogenic potentials have been calculated for these two experimental conditions.

A further conclusion from these data is the fact that the ratio of Na : K pumped equals 3 : 2, a value that is found for several other tissues.

ION DISTRIBUTION AND TRACER FLUXES IN CL-FREE MEDIUM

From the data of Casteels (9) it can be concluded that the Cl-equilibrium potential is significantly less negative than the membrane potential. On the basis of the Ussing flux-ratio equation, it can then be calculated that a net passive outward flux of 6 pmoles · cm^{-2} · sec^{-1} of Cl ions flows. In order to maintain the intracellular chloride concentration at its steady-state level, an influx of equal magnitude is needed. It has therefore been assumed in the above analysis that Cl contributes to the electrogenic component of the membrane potential. The amplitude of the Cl contribution is considerable and amounts to about 15 mV, whereas the contribution of the Na–K pump is only 5 mV. A similar value for the contribution of the Na current to the electrogenic component has been given by Brading and Widdicombe (11). It should be emphasized that the above analysis is independent of the mechanism by which Cl ions are transported against their electrochemical gradient, whether this is an active Cl pump, coupled or not to the Na–K pump, or whether the Cl influx is coupled to the passive Na influx. Moreover, if Cl is included in the calculation of the diffusion potential, it is also necessary to include the Cl current in the calculation of the electrogenic component of the membrane potential. There are some experimental findings consistent with the coupling of the Cl influx to the Na–K pump—inhibition of this pump either by ouabain or by exposing the cells to a K-free solution causes a decrease of $(Cl)_i$ (7,12).

We have also investigated the ion distribution and ion fluxes in Cl-free solutions, substituting Cl with a nonpermeant anion. It has been found by Casteels (7) that after 1-hr exposure to a Cl-free solution, the tissues are depleted of Cl without marked changes in Na and K content, and only a slight depolarization to about −50 mV. Under these conditions, it has been found that the ^{42}K efflux proceeds at a much slower rate (a reduction of about 40 to 50% in the rate of efflux). A similar effect has also been observed on the rate of ^{22}Na efflux (Fig. 3). In a chloride-free Krebs solution the intra-

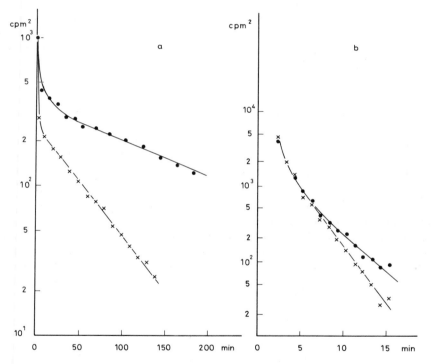

FIG. 3. Effect of nonpermeant anions on the rate of efflux of ^{42}K (a) and ^{22}Na (b). The dots represent the efflux in Cl-Krebs solution, the crosses in isethionate Krebs solution. Ordinate: the rate of efflux (cpm²) on a logarithmic scale; abscissa: time in minutes on a linear scale.

cellular Na content calculated from the ^{22}Na efflux data amounts to 4. ± 0.3 m$_M$/kg wet wt ($N = 4$), whereas the value calculated from ^{22}Na influx measurements, using the method described by Casteels (9), equals 5.4 ± 0.4 m$_M$/kg wet wt ($N = 5$). These values are significantly lower than are those in Cl-Krebs solution (6.2 ± 0.3, $N = 10$) of Casteels (9). In contrast, no significant changes in the total K content, determined analytically, could be observed: 79 ± 2 $(N = 6)$ in Cl-Krebs solution as compared to 81 ± 1 $(N = 6)$ in isethionate Krebs solution. Moreover, no significant effects could be detected on either the ^{14}C-sorbitol space or on the dry/wet weight ratio. The values of intracellular concentrations, and of unidirectional fluxes calculated from the above data in isethionate Krebs solution, as well as the values of permeabilities and net passive fluxes have been summarized in Table 1. Calculation of the diffusion potential gives a value of -42 mV, and the calculated active pump current equals 0.12 μA $\cdot$ cm^{-2}. The ratio of Na:K pumped equals $3.5/2.3 = 3/2$ as in Cl solution. In order to calculate the electrogenic component of E_m an estimate of the membrane resistance must be made. As Ohashi (8) observed, a doubling of the membrane resistance when Cl was substituted by a nonpermeant anion in hypertonic solution, a value of 50 kΩ $\cdot$ cm^2 seems a possible estimate. This gives us an electrogenic component of the membrane potential of -6 mV, and a calculated resting potential of -48 mV. The pumping rate in Cl-free solution is about 40% lower than in Cl-Krebs solution. This is consistent with the decrease in intracellular Na, accompanying the substitution of Cl by a nonpermeant anion.

FURTHER CHARACTERISTICS OF THE NA–K PUMP IN TAENIA COLI CELLS

The effect of different external K concentrations on the rate of Na extrusion has been investigated by stimulating the Na–K pump in K-depleted cells by means of different $(K)_o$. The initial rate of decrease in $(Na)_i$ after readmitting K in the external solution may be taken as a measure of the rate of Na extrusion, because initially $E_m = E_{Na}$. Therefore, we have calculated this initial rate of decrease from the quadratic regression line through the experimental points giving the intracellular Na concentration as a function of the time of exposure to the K-containing solution. These values are represented in Fig. 4a as a function of $(K)_o$. The estimated maximal value for the Na efflux is 22.2 pmoles $\cdot$ cm^{-2} $\cdot$ sec^{-1} and the value of $(K)_o$, which causes half-

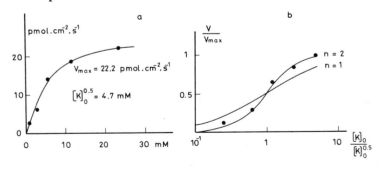

FIG. 4. Initial rate of Na extrusion as a function of $(K)_0$ (a) and comparison of the normalized data with the theoretical dose-response curves of a first- and a second-order system (b). (a) The rate of Na extrusion (pmoles $\cdot$ cm^{-2} $\cdot$ sec^{-1}) is plotted versus $(K)_o$ (mM), both on a linear scale. (b) The rate of Na extrusion/maximal rate of Na extrusion on a linear scale is plotted versus the ratio $(K)_o/(K)_o^{0.5}$ on a logarithmic scale. The solid lines are the corresponding theoretical curves for a first- and a second-order system.

maximal activation $(K)_o^{0.5}$ is 4.7 mM. This latter value is higher than the value observed for the isolated Na–K ATPase (13), but we will present arguments in favor of the hypothesis that the activation curve is shifted to the right by a competitive inhibition between external K and Na. The plot of the ratio Na efflux/maximal Na efflux versus the logarithm of $(K)_o/(K)_o^{0.5}$ (Fig. 4b) shows that the experimental points are best-fitted by a second-order system, which is consistent with a carrier that has to be occupied by two K ions in order to activate the ion transport. This number is the same as the number of K ions transported per cycle as found in the above analysis, which suggests that the activating K ions are also the transported ones.

The inhibitory effect of external Na ions on the pump has been demonstrated by following the reaccumulation of K-depleted cells in solutions in which half of the external Na has been replaced by choline or Li (Fig. 5). It can be observed that the K uptake at the same $(K)_o$ proceeds slower in

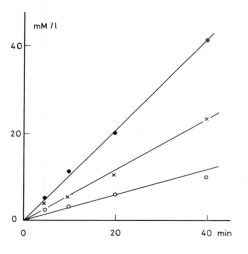

FIG. 5. K-uptake (in mM/liter cell water) by K-depleted cells from a solution containing 0.59 mM K and 71 mM Na–71 mM Li (●), 71 mM Na–71 mM choline (x), or 142 mM Na (○) as a function of exposure time (in minutes).

the solutions containing 142 mM Na than in the solutions containing 71 mM Na.

The discrepancy between the Li- and the choline-containing solutions can be explained by a stimulating action of Li ions on the ion transport. The active uptake of Li ions has been demonstrated by following the uptake of Li during exposure of K-depleted cells to a 71 mM Li-containing solution. Moreover, this uptake can be inhibited by 10^{-5} M ouabain (5).

A further interesting characteristic is demonstrated when the influence of the substitution of Na by Li on the ^{22}Na efflux is investigated. Whereas in the absence of ouabain there is no effect when Na is replaced by Li, a decrease of about 50% in the rate of ^{22}Na efflux is observed in the presence of ouabain. A similar reduction in the rate of ^{22}Na efflux could also be observed when Na is replaced by choline. Under these conditions ouabain has no effect on the reduction in efflux rate. This result can be interpreted as a ouabain-insensitive Na–Na exchange mechanism. It cannot be excluded that also under physiological conditions some Na–Na exchange occurs, but so far no experimental evidence could be brought forward.

CONCLUSIONS

In taenia coli cells there is a large contribution by the active ion transport to the membrane potential. The Na–K pump contributes about 5 mV, as seen in Cl-free solution, whereas in the presence of Cl ions, which contribute about 15 mV, a total of 20 mV is due to electrogenic ion transport. However, the nature of the Cl influx as well as the possible role of its contribution to the membrane potential need further detailed investigation.

The other characteristics of the Na–K exchange are similar to those observed in many other cells, i.e., this mechanism is activated by $(Na)_i$ and $(K)_o$, while it is inhibited by $(Na)_o$. The ratio of Na:K pumped equals 3:2. In K-depleted cells about 50% of the ^{22}Na efflux is due to a ouabain-insensitive Na–Na exchange.

REFERENCES

1. Burnstock, G. (1958): *J. Physiol. (Lond.)*, 143:183–194.
2. Bülbring, E. (1962): *Physiol. Rev.*, 42,Suppl. 5:160–178.
3. Casteels, R., Droogmans, G., and Hendrickx, H. (1971): *J. Physiol. (Lond.)*, 217:281–295.
4. Casteels, R., Droogmans, G., and Hendrickx, H. (1971): *J. Physiol. (Lond.)*, 217:297–313.
5. Casteels, R., Droogmans, G., and Hendrickx, H. (1973): *J. Physiol. (Lond.)*, 228:733–748.
6. Casteels, R., Droogmans, G., and Hendrickx, H. (1973): *Philos. Trans. R. Soc. Lond. [Biol. Sci.]*, 265:47–56.
7. Casteels, R. (1971): *J. Physiol. (Lond.)*, 214:225–243.
8. Ohashi, H. (1970): *J. Physiol. (Lond.)*, 210:405–419.
9. Casteels, R. (1969): *J. Physiol. (Lond.)*, 205:193–208.
10. Tomita, T. (1966): *J. Theoret. Biol.*, 12:216–229.
11. Brading, A., and Widdicombe, J. (1974): *J. Physiol. (Lond.)*, 238:235–249.
12. Casteels, R. (1966): *J. Physiol. (Lond.)*, 184:131–142.
13. Skou, J. C. (1965): *Physiol. Rev.*, 45:596–617.

Physiology of Smooth Muscle, edited by
E. Bülbring and M. F. Shuba.
Raven Press, New York © 1976.

Ionic Composition and Ion Exchange in Vascular Smooth Muscle*

G. Siegel, H. Roedel, J. Nolte, H. W. Hofer, and O. Bertsche

Institute of Physiology, Biophysical Research Group, Free University of Berlin, D-1000 Berlin 33; Department of Medicine, University of Freiburg, D-7800 Freiburg Br.; Department of Biology, University of Konstanz, D-7750 Konstanz, Germany

Vascular smooth muscle represents one of the terminal members in the long chain of neurovegetative regulation. An integration of nervous, humoral, renal, and local influences occurs here. In this chapter, we will discuss one aspect of blood vessel regulation—the control by local, myogenic factors. In order to gain insight into this control, we must consider the basic membrane physiology of the smooth muscle cells in the vessel wall.

Smooth muscle organs can be classified into two groups—one, a spike-producing, more phasic group, the other a non-spike-producing, more tonic group (1). The cells exhibit a basic rhythm in the membrane potential, the so-called second or minute rhythm, which manifests itself by "slow waves" (2,3). During the depolarization phases, spike potentials may or may not be superimposed. Examples of spiking smooth muscle are the intestine (4) and the V. portae of the blood vessels (5,6). In contrast, most of the large and middle arteries belong to the tonic class, which only produces slow waves in the membrane potential and graded depolarization (1,7). Investigations on such structures provide a good opportunity for studying the ionic basis of the spontaneous rhythm of the slow waves. As we have a special interest in this problem, we have chosen the A. carotis communis of the dog for our experiments.

METHODS

Experimental Preparations and Solutions

The common carotid arteries were surgically removed from anesthetized dogs in 3 to 4 min and were equilibrated in a modified Krebs solution (8) having the following composition: Na^+ 151.16; K^+ 4.69; Ca^{2+} 2.52; Mg^{2+} 0.11; Cl^- 140.79; HCO_3^- 16.31; $H_2PO_4^-$ 1.38; dextrose 7.77 mM (37^0C; pH 7.2 to 7.4). The external Na^+ concentration of the Krebs solution was altered simultaneously with that of K^+, so that the sum $[K^+]_0 + [Na^+]_0 =$

---

* Dedicated to Professor Dr. O. H. Gauer on his 65th birthday.

155.85 mM remained constant. The following gradations of external K^+ concentration were used: 0; 0.47; 1.17; 2.35; 4.69; 11.73; 23.45; 46.90; and 93.80 mM. The solutions were equilibrated with a 95% O_2–5% CO_2 gas mixture.

Water Content, Extracellular Space, and Ion Concentrations

The tissue wet weight was determined after the preparations had been squeezed between two pieces of filter paper by a 5-kp weight for 1 min. The extracellular space of the preparations was measured with the aid of an electron microscopic–morphometric technique, which has been described in detail in earlier publications (9,10). The inulin space determination was performed after 2 hr of loading with ^{14}C-inulin ($a = 9.8$ mCi/mmole; $c \approx 0.02$ mmoles/liter); the inulin efflux was measured at intervals over a 3-hr period in 5-ml vials (preparation volume < 50 μl) containing modified Krebs solution (10,11).

The total ion concentrations of the isolated tissue were determined using an atomic absorption spectrometer (Perkin-Elmer model 403). After determination of wet and dry weight, the preparation was placed in an ice-cold quartz Erlenmeyer flask containing pure oxygen and Pt catalyst and burned ash-free. The gaseous products were dissolved in a defined volume of 1.3% HNO_3. Atomic absorption measurements were made directly from this fluid for Na and K and for Cl by measurement of Ag after quantitative precipitation with $AgNO_3$. The total method error is $< \pm 5\%$.

Electrical Recording

Intracellular recordings of membrane potential were made with glass microelectrodes filled with 3 M KCl. The resistances ranged from 10 to 60 MΩ and the tip potentials from -5 to -60 mV. The electrodes were shielded to the tips; in all other cases, conventional recording techniques were used.

All intracellular recordings in *normal* Krebs solution were corrected for tip and junction potentials by a method described earlier (12). At least 30 penetrations per carotid artery were executed and averaged. The membrane potentials given in Table 2 were derived from 19 different dog carotids. Carotids with membrane potentials between -30 mV and -70 mV were selected for the final averaging. The membrane potentials in the test solutions with various external K^+ concentrations were registered by achieving a stable penetration in normal Krebs solution and then switching to the desired test solution. Time was allowed for the membrane potential to reach a new, stable value.

Na^+–K^+ Exchange Measurements

The preparations were loaded to saturation for several hours in a radioactive bath containing ^{24}Na and ^{42}K. Washout was then performed in inactive

Krebs solution. A NaI-crystal (3 in. × 3 in.) detector was used, which permitted adequate discrimination of the γ-lines of ^{24}Na at 1.37 MeV and of ^{42}K at 1.52 MeV by means of a multichannel analyzer. The γ-spectra of the washout fluid were measured at intervals of 18 sec down to 1 sec (cf. 11,13, 14).

Calculations with a Dynamic Membrane Model (15–17)

With the voltage-clamp technique, the membrane current ΣI, which is composed of specific ionic components I

$$I = z \cdot F \cdot M \tag{1}$$

is regulated by a negative feedback amplifier such that the membrane potential remains at a constant given value V_c.

Each net ion flux M can be described as the difference between two unidirectional fluxes, efflux and influx

$$M = P'a'c' - P''a''c'' \tag{2}$$

where c' is the intracellular and c'' the extracellular concentration of the ion species in question, and a' and a'' are factors that account for the influence of the membrane potential on the transmembrane ion movement:

$$a' = \frac{z \cdot F \cdot V_c/RT}{1 - \exp(-z \cdot F \cdot V_c/RT)} \qquad a'' = \frac{z \cdot F \cdot V_c/RT}{\exp(z \cdot F \cdot V_c/RT) - 1} \tag{3}$$

In general, the permeabilities P' and P'' of each ion species are supposed to be different for efflux and influx. The active transport processes are expressed in this difference. Conceiving the membrane as an energy barrier for ionic transport, all permeabilities can be described by the activation energies A' and A''

$$p'^{,\,''} \sim e^{-\frac{A',''}{RT}} \tag{4}$$

The activation energy A' (A'') must be attained by an ion that passes outward (inward) through the membrane. Because active processes that show a definite direction-specificity also participate in transmembrane ion transport, the average activation energies for outwardly and inwardly directed ion transport are generally different.

The activation energy of an ion in the membrane phase understandably depends on the conformation of the membrane lipoproteins. This, on the other hand, depends on the electrical field strength across the membrane. Therefore, the activation energy is a function of the clamp potential. Above that, the lipoprotein conformations, which correspond to specific activation energies (A', A''), change only protractedly after a voltage jump from resting to clamped potential. The membrane system rather relaxes into a new equilibrium. As precise knowledge of the membrane relaxation processes is lacking, the time- and voltage dependence of the activation energies can only be formulated empirically:

$$\tau \cdot \dot{A}'^{,''}(t) + A'^{,''}(t) = S'^{,''}(x_c)$$

with $S'^{,''}(x_c) = A_m'^{,''}[1 - L(x_c - x_r'^{,''})]/2$ (5)

$$L(x_c) = cth\ x_c - \frac{1}{x_c} \quad \text{(Langvin function)}$$

$$x_c = \sigma \cdot V_c \cdot z \cdot F/RT \qquad x_r'^{,''} = \sigma \cdot V_r'^{,''} \cdot z \cdot F/RT$$

Under resting conditions,

$$A'^{,''}(t = 0) = S'^{,''}(x_o) \quad \text{with} \quad x_o + \sigma V_o \cdot z \cdot F/RT$$

so $A_m'^{,''}$ can be eliminated in Eq. (5).

In a dynamic model the mutual hindrance of the ions in the membrane phase has to be included, all the more when assuming that the ion passage only takes place in closely circumscribed areas of the membrane. With help of Maxwell-Boltzmann statistics, for each ion species and for each of the two transport directions one can give analytical expressions which make possible a quantitative description of the ion interaction.

$$w'^{,''} = \frac{\gamma \cdot \exp[-(A'^{,''} - \eta'^{,''})/RT]}{\Sigma'\Sigma''\gamma \cdot \exp[-(A'^{,''} - \eta'^{,''})/RT]} \tag{7}$$

The differences between the activation energies and the electrochemical potentials appear in the exponents. The values of γ are empirically determined, ion-specific weight factors. The summation in the denominator of the interaction term is over all ion species as well as over both transport directions.

Now, the proportionality expressions for the permeabilities (4) can be expanded by the interaction term

$$P'^{,''} \sim (w'^{,''})^{n-1} \cdot \exp(-A'^{,''}/RT) \tag{8}$$

The proportionality factor f is empirically determined. This analytical connection shows the dependence of the permeability on the activation energy, which in return is time-dependent [Eq. (5)]. Thus, with Eqs. (1) and (2) the time-dependence of the ion fluxes can be ascertained.

The participation of active processes in the transmembrane transport of an ionic species is manifested analytically in the relation of their unidirectional fluxes.

$$\frac{\Phi'}{\Phi''} = \frac{P'a'c'}{P''a''c''} = e^{n \cdot \frac{zF}{RT}(V_c - E - K)}$$

(9)

with $E = (RT/zF) \ln \frac{c''}{c'}$ and $K = (A' - A'')/zF$

A nonzero value of the driving force K is reason to assume active transport processes. To separate an active transport component from the net ion flux, one should assume that this component is unidirectional. Then with help of the given flux relation, its percentage of the equidirectional flux can be determined.

$$\left(1 - e^{-n\frac{|A' - A''|}{RT}}\right) \cdot 100\%$$ (10)

RESULTS AND DISCUSSION

Extracellular Space and Ion Concentrations

The extracellular fluid space was determined using ^{14}C-inulin and the fast components of $^{24}Na^+$ and $^{42}K^+$ tracer washout curves. From the inulin measurements it amounted to 38.2% (Table 1). The influence of a 10-fold increase

TABLE 1. Computer analysis of the ^{14}C-inulin efflux curves in the dog carotid artery upon variation of external K^+ concentration

Krebs solution	A_1 [%]	A_2 [%]	ΣA [%]	τ_1 [s]	τ_2 [s]
Normal (I)	20.8	17.4	38.2	272	2175
Hyperkalemic (II)	16.9	13.6	30.5	228	2055
Normal (III)	25.0	14.8	39.8	338	2307

The amplitudes A_1 and A_2 and the time constants τ_1 and τ_2 are obtained with a double exponential approximation of the inulin efflux curves according to a least-square fit. I: Equilibrated 3 hr in normal Krebs solution, II: Equilibrated 3 hr in normal Krebs solution, then equilibrated 2 hr in hyperkalemic Krebs solution ($[K^+]_o = 46.9$ mM), III: Equilibrated 3 hr in normal Krebs solution, then equilibrated 2 hr in hyperkalemic Krebs solution, then 3 hr in normal Krebs solution (reversibility).

in external K^+ concentration on the extracellular fluid space and its washout kinetics was likewise tested. It is noteworthy that in hyperkalemic solution a decrease of the inulin space occurs, which is reversible in normal Krebs solution. This is seen clearly in the computer analysis of the double exponential inulin washout kinetics, which shows, moreover, that the exchange from the medial (A_1, τ_1) as well as from the adventitial compartment (A_2, τ_2) of the extracellular fluid space occurs more rapidly in the hyperkalemic solution. A complete compartmental analysis of the three-phase system (media, adventitia, bath), with a calculation of the exchange amounts and the kinetic coefficients, has been given in a more thorough paper (10).

As the total extracellular space of the vessel wall consists of at least two compartments, the extracellular water space and the volume of extracellular connective tissue fiber, and as only the first value can be measured by inulin, we have additionally applied an electron microscopic–morphometric technique for determining the extracellular space. According to this, the total extracellular space amounts to 57.4%, and the extracellular connective tissue portion is 19.2% (cf. Table 2). With the data of the compartmental analysis of the three-phase system, this space of 57.4% can be divided into 33.3% medial and 24.1% adventitial extracellular space (10,18). All of these values are required for calculating ionic concentrations.

Furthermore, we investigated the dependence of the inulin space on the external K^+ concentration (Fig. 1). When $[K^+]_o$ is varied in a wide range of 0 to 100mM, the inulin space changes in a characteristic way—The extracellular fluid space decreases for external K^+ concentrations below and above 2 mM. With a 20- to 30-min equilibration after a change of the K^+ solution, the inulin space varies at high external K^+ solutions according to the dotted line,

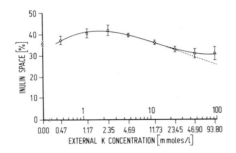

FIG. 1. Inulin space of the dog carotid artery in dependence on the external K^+ concentration. Dotted curve gives the size of the inulin space after 20 to 30 min equilibration in changed K^+ solution, the solid curve after a 2-hr equilibration following the change of the Krebs solution.

with a 2-hr equilibration following the change of the Krebs solution according to the solid line.

The Na, K, and Cl content of the carotid changes during the preparation and long incubation in Krebs solution. After preparation of the test object and superficial removal of the adventitial connective tissue, it can be seen that during the following equilibration, the Na and Cl concentrations show an initial rise, whereas the K concentration decreases. After about 2 hr the concentrations, and thus the ionic fluxes, have again reached a steady state (11). All measurements under the various test conditions were carried out in the steady state.

It has been claimed by numerous authors that the extracellular K^+ concentration is of decisive importance for the regulation of the vessel diameter, for example, in working skeletal muscle. That is why we have determined ionic concentrations and membrane potentials with different external K^+ solutions. For instance, in a hyperkalemic Krebs solution, the total K concentration increases and the total Na concentration decreases parallel to the external K^+ and Na^+ concentrations, respectively (Table 2). The intracellular concentration of a given ion species can be calculated if, in addition to the known ionic concentration of the extracellular fluid space, the concentration

TABLE 2. *Water content, extracellular space, content of connective tissue fiber, ionic concentrations, equilibrium potentials (E) and membrane potentials (V) of the carotid artery of the dog with variation of the external Na^+ and K^+ concentrations*

	Krebs solution		
	Normal (I)	Hyperkalemic (II)	Normal (III)
Water content	72.3	70.5	71.8%
Extracellular space[a]	57.4	45.6	59.4%
Inulin space[b]	38.2	30.8	41.6%
Intracellular water	34.1	39.7	30.3%
Connective tissue volume	19.2	14.9	17.8%
$[Na]_o$	151.2	108.9	151.2 mmoles/liter
$[Na]_t{}^c$	108.6	76.5	113.5 mmoles/kg wet wt
$[Na]_b{}^d$	29.6	33.6	28.3 mmoles/kg dry wt
$[Na]_{i,b}{}^e$	51.4	44.0	56.2 mmoles
$[Na]_i$	68.2	35.4	75.1 mmoles/liter f.w.[f]
E_{Na}	+ 21.3	+ 30.0	+ 18.7 mV
$[K]_o$	4.7	46.9	4.7 mmoles/liter
$[K]_t$	34.4	66.6	29.7 mmoles/kg wet wt
$[K]_b$	13.7	29.1	18.7 mmoles/kg dry wt
$[K]_i$	87.7	120.4	80.8 mmoles/liter f.w.
E_K	− 78.3	− 25.2	− 76.1 mV
$[Cl]_o$	140.8	140.8	140.8 mmoles/liter
$[Cl]_t$	77.7	70.9	79.8 mmoles/kg wet wt
$[Cl]_b$	(− 1.2)	(− 1.3)	(− 0.8) mmoles/kg dry wt
$[Cl]_i$	70.2	69.3	65.3 mmoles/liter f.w.
E_{Cl}	− 18.6	− 18.9	− 20.5 mV
V	− 43.3	− 31.6	− 42.0 mV

[a] Determined by electron microscopic–morphometric methods.

[b] Determined by ^{14}C-inulin as an extracellular marker.

[c] Total concentration of the ion species in question.

[d] Fraction of the ion species in question bound to connective tissue.

[e] Fraction of the ion species in question bound to intracellular structures.

[f] f.w. = fiber water.

in the extracellular connective tissue fiber space is known. The latter was determined in isolated adventitial connective tissue, assuming that the ionic binding capacities of adventitial and of medial connective tissue are equal. The fractions of the ion species bound to connective tissue fibers are: K 14, Na 30, Ca 30, Mg 4.5, and Cl 0 mmoles/kg dry weight. In hyperkalemic solution the binding capacity for K^+ ions is by no means at its limit; the bound Na fraction, on the other hand, remains unchanged. Cl^- ions are not bound due to the negative groups of the connective tissue structures and their matrix. Because all values of the different compartments—volume of connective tissue fibers, inulin and intracellular space, intracellular water—are now known, one can calculate the intracellular concentrations, and then the ionic equilibrium potentials according to the Nernst equation.

The marked increase of the internal K^+ concentration in hyperkalemic solution is evident. There are apparently two reasons for this.

1. The electrogenic, active inward K^+ transport depends on the extracellular K^+ concentration and is thus increased (19).
2. The driving force $(V - E_K)$ for the passive K^+ net flux has changed sign in comparison with normal Krebs solution and is now directed inward.

The strong increase of the internal K^+ concentration is mainly responsible for the observation that when $[K^+]_o$ is varied, the slope of the E_K curve amounts to only 53 mV instead of 61.5 mV per 10-fold change of $[K^+]_o$. A similar value was reported by Casteels (20) for the taenia coli of the guinea pig.

The total internal Na concentration in hyperkalemic solution is decreased analogous to the external concentration. The amount of intracellular free Na^+ ions can be well approximated if one assumes that on the basis of stabilized osmotic conditions $[Na^+]_o + [K^+]_o \approx [Na^+]_i + [K^+]_i$ (cf. 12). If one takes into account that the intracellular space is about 80% water and 20% solid substances, then one can separate a bound Na fraction from the total internal Na. As can be easily calculated from Table 2, the intracellular bound Na amounts to an average of 54% of the total internal Na. Similar values were reported for skeletal muscle and kidney from nuclear magnetic resonance experiments (21,22).

The generally high values for $[Na^+]_i$ lead to the assumption that in addition to the intracellular bound Na fraction given in Table 2, further Na^+ ions are to be found structurally bound. Thus, the calculation of E_{Na} using the concentrations can show uncertainties. Hinke (23) and Lev (24) reported for Na an activity coefficient with a strong intracellular deviation from 1, which fits in well with this assumption. Using the expanded Debye-Hückel boundary law (25), we calculated for vascular smooth muscle an activity coefficient of 0.7 for univalent ions, of 0.3 for divalent ions outside and inside the membrane. This boundary law can be successfully applied to electrolytic solutions up to 0.1 M. With solutions up to 0.2 M (Krebs solution $\approx$ 0.16 M), it leads to usable estimates. However, the calculated intracellular values of the activity coefficients must be considered as too high, as the adsorption of electrolytes on large molecular substances is not taken into account in the calculation.

The Cl equilibrium potential E_{Cl} is over a wide range independent of the external Na^+ and K^+ concentrations. As a result of the relatively high intracellular Cl^- concentration E_{Cl} deviates strongly from the membrane potential V, so a purely passive Cl^- ion distribution does not exist. That is why the membrane resting potential is also influenced by Cl^- ions. The possibility that the calculations of E_{Cl} and E_K are in error seems very unlikely as Keynes (26) and Casteels (20) reported nearly the same intra- and extracellular activity coefficients for these ion species.

Influence of K^+ Ions and pH on Membrane Potential

The membrane potential is relatively independent of $[K^+]_o$, which is demonstrated by intracellular recordings following variation of the external K^+

concentration (Fig. 2). The flat course of the membrane potential curve of vascular smooth muscle is especially striking considering the dependence of the membrane potential of other structures on $[K^+]_o$. In Fig. 2 one can see for comparison the analogous experimental results of Adrian (27) for skeletal muscle, Haas et al. (12) for heart muscle, Huxley and Stämpfli (28) for nerve, and Casteels (20) for smooth muscle of the taenia coli. It becomes immediately evident that the membrane potential of vascular smooth muscle changes least of all with $[K^+]_o$. Below and above 2mM $[K^+]_o$ depolarization occurs, which, mechanically speaking, would signify increased muscle tone or constriction. There is indirect evidence for this. When one recalls the dependence of the extracellular fluid space on the external K^+ concentration, illustrated in Fig. 1, it becomes immediately apparent that membrane potential and extracellular fluid space take the same course. Obviously the contractile tone of vascular smooth muscle is coupled very closely to the membrane potential. Direct evidence is brought by the fact that the curve for the development of mechanical tension in dependence on $[K^+]_o$ found by Konold

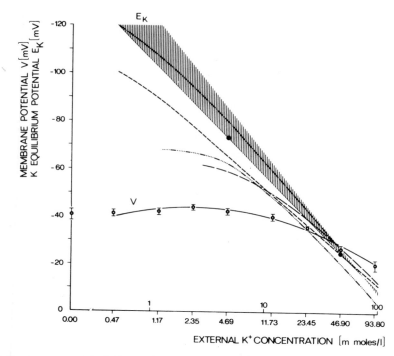

FIG. 2. Membrane potential V ($\bigcirc$) and K^+ equilibrium potential E_K($\bullet$) in the dog carotid as a function of the external K^+ concentration $[K^+]_o$; semilogarithmic plot. $[K^+]_o$ and $[Na^+]_o$ vary in an opposite manner so that the sum is always 155.9 mM. The dashed–dotted curve shows the course of the membrane potential for skeletal muscle (27), the short-dashed curve for heart muscle (12), the multidotted–dashed curve for nerve (28), and the long-dashed curve for smooth muscle of taenia coli (20). The hatched area represents the range of E_K changes, bounded above by the theoretically calculated and below by the experimentally determined lines.

et al. (29) and Betz (30) in the facial artery and brain blood vessels runs almost parallel to our V curve. Mekata and Niu (31) have shown for the rabbit carotid artery that the slow, small-amplitude waves of the membrane potential are accompanied by synchronous mechanical oscillations. Even during the increase in mechanical tension under noradrenaline, a light depolarization can be recognized in the simultaneously registered membrane potentials of Keatinge (32) and Mekata et al. (31). These findings can be explained when one notes that the membrane potential listed in Table 2 lies with -43.3 mV, very near to the mechanical threshold (33). Wahlström (34) and Keatinge (32, 35) found very similar membrane potentials in the portal vein and carotid artery, respectively. A small change of the membrane potential in this range should therefore also cause a change in the mechanical tension. Other authors, however, claim that no direct correlation between membrane potential and the development of mechanical tension exists under norepinephrine and other pharmaceuticals (1,36,37).

The dependence of the K equilibrium potential E_K on the extracellular K^+ concentration is represented in Fig. 2 by the hatched area. The upper limit of the area is the theoretically calculated line with a slope of 61.5 mV per 10-fold alteration of $[K^+]_o$; the lower limit is the experimentally determined line with a slope of 53 mV per 10-fold variation of the extracellular K^+ concentration. An explanation for this deviation can be provided by the observation of a strong increase of the internal K^+ concentration in a high K^+ solution. Just as with heart muscle (12), the V and E_K curves intersect with high extracellular K^+ concentration, approximately 35 mM. For vascular smooth muscle this indicates that in high K^+ solution the driving force $(V - E_K)$ for the passive K net current is directed into the cell interior and is thus reversed.

In further experiments the smooth muscle cells were subjected to simulated acidosis. A hyperpolarization always resulted (mean: -4.3 mV following transition from pH 7.3 to 6.8), whereas alkalosis led to depolarization of the membrane potential (mean: $+3.4$ mV following transition from pH 7.3 to 7.8). The former case corresponds to a dilatation of the vessels, the latter to constriction. From this it becomes clear that the actual state of metabolism of the tissue surrounding the vascular muscles undoubtedly has a great influence on the regulation of vessel caliber.

At the conclusion of this section, we wish to discuss the problem of potential generation. The theory of an approximated K diffusion potential had to be discarded because the membrane potential distinctly deviates from the K equilibrium potential in a wide range of the external K^+ concentration. Thus, it was necessary to consider the participation of other ion species in generating the membrane potential. Taking the Na equilibrium potential of $+21.3$ mV and the K equilibrium potential of -78.3 mV together with the membrane potential of -43.3 mV with normal extracellular ion concentrations, one could suppose that the membrane potential of this preparation is a mixed Na–K diffusion potential, in which the K conductance slightly outweighs that of Na. V lies about 15 mV negative to the halfway-point between the two equilibrium potentials. This assumption remains valid if the Na^+ ions, the intracellular concentration of which is high compared to other structures, are freely ionized to the extent shown in Table 2. According to the investigations of Hinke (23) and Lev (24) on nerve and skeletal muscle as well as our calculation by means of the Debye-Hückel theory mentioned previously, we must suppose that the activity coefficient of the intracellular Na^+ ions is 0.7 at the most. Therefore, E_{Na} should be even more positive. Up to now this activity coefficient has not been determined experimentally in vascular smooth muscles. But, based on the investigations mentioned, the theory of a Na–K mixed potential loses some of its probability. This is emphasized after flux experiments on blood vessels showed a limited Na permeability in comparison with the K permeability (34,38).

The Cl equilibrium potential also turns out to be far more positive (-18.6 mV) than in other structures, which can be attributed to a relatively high intracellular Cl^- concentration. It is conceivable that the membrane potential is held in a depolarized state at -40 mV by a high Cl permeability. This depolarizing effect would be supported by the Na^+ ions. The hypothesis of a mixed K–Cl diffusion potential receives support by the latest investigations by Wahlström (34), who found with the aid of flux measurements a P_{Cl} close to P_K in vascular smooth muscle.

Effect of Electrogenic, Active Transport on Membrane Potential

Finally, let us turn to a third way of considering this problem of potential generation. If we recall the course we previously demonstrated for active Na and K transport in cardiac muscle in dependence on the extracellular Na^+ and K^+ concentrations (19), and if we examine the same dependence for the membrane potential of vascular smooth muscle, we come to the surprising finding that the Na transport curve and the membrane potential take the same course (Fig. 3, upper). On the abscissa the external K^+ concentration is plotted, on the ordinate the active Na and K transport in the Na^+ efflux or K^+ influx, respectively, as flux deficit under 2,4-dimitrophenol ($c_{DNP} = 2$ mM). Apart from mentioning that both transport curves are not to be brought to congruence, so that an electroneutral pump must be excluded, the Na transport curve strongly calls to mind the course of the membrane potential in Fig. 2. Preliminary experiments on vascular smooth muscle, while making opposing changes in $[K^+]_o$ and $[Na^+]_o$, have shown that its active transport rate resembles that of cardiac muscle.

The Na–K transport curves show that the carrier system is in no case saturated with K^+ ions in the given range of external K^+ concentration, whereas it demonstrates a saturation behavior with Na^+ ions. The double reciprocal plot of the K transport curve is indicative of weak negative cooperativity (39). Such kinetics are possible when one realizes that the mem-

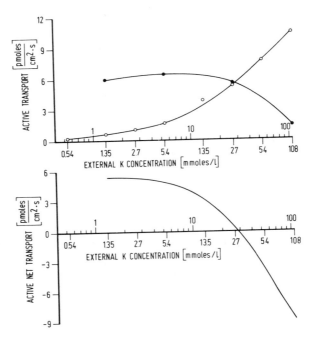

FIG. 3. *Upper part:* K and Na transport in frog atrium in dependence on the K^+ and Na^+ content of the external solution. $[K^+]_o$ and $[Na^+]_o$ vary oppositely such that their sum remains 118.2 mM. (●) Na transport (corresponding to the Na efflux deficit under DNP); (○) K transport (corresponding to the K influx deficit under DNP). *Lower:* Stationary, active net transport in dependence on the external K^+ concentration. The curve was obtained by subtraction of the active K^+ inward transport from the active Na^+ outward transport shown in the upper half of the figure.

brane ATPase is an allosteric enzyme complex (40,41) with two active sites, one for Na^+, the other for K^+.

In conclusion, it must be mentioned that the Na–K transport curves can also be considered as stationary, active current–voltage (concentration) curves, because the membrane potential of the cells varies monotonically with external K^+ concentration. In the lower part of Fig. 3 the active net transport is plotted. It shows a saturation behavior with a sigmoid course and gives again some information on the enzyme kinetics of the active transport mechanism. Up to 30 mM $[K^+]_o$ the electrogenic transport is directed outward (mainly Na^+ transport), above this value inward (mainly K^+ transport). K^+ and Na^+ transport are equal in *one* range of the external K^+ and Na^+ concentration, in which the crossover of the V and E_K curves in Fig. 2 also lies. This means that the membrane potential of vascular smooth muscle cells can be considered as generated partly by an electrogenic, active net outward transport of ions (mainly Na^+) in the range of normal external ion concentrations up to 30 mM $[K^+]_o$. By even higher external K^+ concentrations the membrane potential is depolarized by an electrogenic, active net inward transport of ions (mainly K^+). The active, electrogenic net ion transport causes then a steepening of the membrane potential curve depending on external K^+ and Na^+ concentrations.

These findings are further corroborated by application of a physical-chemical membrane model to vascular smooth muscle (15–17). This model, which includes active transport processes, has been shortly described in the section on *Methods*. The separation of active transport components from the Na^+ efflux and K^+ influx was derived. The following measured data from vascular smooth muscle was used for the simulation of a voltage-clamp experiment on a digital computer: $[Na^+]_o = 151.2$; $[Na^+]_i = 68.2$; $[K^+]_o = 4.7$; $[K^+]_i = 87.7$ mM; $V_o = -43.3$ mV; $\Phi_{Na} = 6.73$; $\Phi_K = 2.53$ pmoles/cm^2/sec; $n = 1.9$; $C_m = 3$ μF/cm^2 (42); $t = 37.0^\circ$C. The following parameters, which were chosen in accordance with heart muscle membrane characteristics (15,16), are taken into the calculations: $\tau_{Na} = 0.01$; $\tau_K = 0.1$ sec; $\sigma_{Na} = \sigma_K = 1.3$; $V'_{T,Na} = -59.4$; $V''_{T,Na} = -40.0$; $V'_{T,K} = -40.0$; $V''_{T,K} = -54.1$ mV; $\gamma_{Na} = 1$; $\gamma_K = 5$; $f_{Na} = f_K = 10^{-4}$ cm/sec. After being fed these data on vascular smooth muscle, the dynamic membrane model produces for a normal Krebs solution an active Na^+ outward transport which exceeds all the other unidirectional fluxes in this membrane several times over. At the resting potential the active Na^+ outward transport comprises 0.63 μA/cm^2, the active K^+ inward transport only -0.22 μA/cm^2. These time-independent active transport components, whose electrogenic character is obvious, can be read from Fig. 4 for the resting potential. This figure shows the *active, dynamic* Na and K current-voltage relations at different times after setting a voltage clamp. Before the time-dependent membrane currents were calculated for the various clamping steps, the input parameters for the model were chosen such that the time-dependent total ionic current curve at a clamp potential of 0 mV was optimally simulated as it was obtained by Anderson (43) from uterine smooth muscle. Comparable voltage-clamp experiments for vascular smooth muscle have not yet appeared in the literature.

The upper part of Fig. 4 shows the active, dynamic Na^+ current-voltage relations for a potential range from -80 mV to $+60$mV after four clamping periods. Starting from the resting potential of the vascular smooth muscle cells ($V_o = -43.3$ mV), the active Na^+ outward transport increases steeply toward depolarization and saturates in the positive potential range with longer clamping periods. The active, electrogenic Na^+ current is therefore oriented against the potential change, and attempts to hinder it (cf. 16). The time-dependent Na^+ transport curves fall in the direction of hyperpolarization to small current values. All Na^+ curves cross at a point that gives the amount of active Na^+ outward transport at the resting potential. This Na^+ current is therefore time-independent, as it must be. The same considerations hold for the active K^+ inward transport at the resting potential, as is shown in the lower part of Fig. 4. Upon consideration of the active, dynamic K^+ current-voltage relations, it strikes one immediately that the K^+ transport curves fall

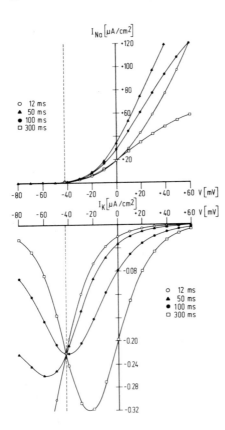

FIG. 4. Dynamic current–voltage relations of the active Na+ outward transport (upper half) and K+ inward transport (lower half) after four different clamping periods. The transport curves were calculated with a physicochemical membrane model using experimentally determined membrane parameters from vascular smooth muscle. Note that the active Na+ and K+ transport is time-independent at the resting potential.

off steeply in the direction of membrane depolarization as well as in that of membrane hyperpolarization. Clearly a "potential window", in which K+ ions are optimally pumped into the cells, exists for the active K+ inward transport. This potential range lies between −60 mV and −20 mV. A similar potential window, with an optimum at −40 mV was found by Dudel et al. (44) for the passive Na+ inward current in heart muscle. If active K+ inward transport and passive Na+ inward current are both carrier-mediated, then the potential-dependent conformational changes in the carrier (45) will influence Na and K transport in the same way.

One final thought should be pursued further. After long clamping periods, when the relaxation processes in the membrane are nearly completed, the dynamic Na+–K+ current–voltage relations much approach the stationary, active ones. To test the correctness of this prediction, we compare the upper part of Fig. 3 with Fig. 4. For K+ ions one must imagine the stationary active characteristic in Fig. 3 to be turned down the x axis (negative K+ inward transport) in order to be able to compare the K+ current after 300 msec of clamp duration with that of Fig. 4. Because the various external K+ concentrations in Fig. 3 cover a potential range between -80 and -20 mV from left to right (12,19), the active K^+ transport curve for a 300-msec clamping period—in the lower part of Fig. 4—must be taken for comparison in the same potential range. Both K^+ transport characteristics take a similar course. In the case of Na^+ ions the stationary active Na^+ transport curve in Fig. 3 must be turned around the y axis, because the external Na^+ and K^+ concentrations vary in an opposite manner. Then the stationary, active Na^+ characteristic is once again nearly in accordance with the dynamic relation 300 msec after setting the clamp in the upper part of Fig. 4.

The advantage of the simulation of membrane characteristics of the vascular smooth muscle cells is that with the described model one can obtain the active, time- and voltage-dependent Na^+ and K^+ transport curves which have until now avoided experimental determination.

Oscillations of Transmembrane Na⁺ and K⁺ Fluxes

In order to explore the supposed coupling between metabolism and membrane potential via an electrogenic ion transport we carried out simultaneous ^{24}Na–^{42}K flux measurements. The ionic flux measurements were performed at least 3 hr after preparation of the tissue (11,13). The Na^+ and K^+ *influxes* have a multiple exponential time course and point to a compartmentalization. After 1 minute, 50% of the total K and 70% of the total Na are exchanged. In the stationary equilibrium the Na^+ influx = Na^+ efflux was fixed at 6.73 pmoles/cm²/sec and the K^+ influx = K^+ efflux at 2.53 pmoles/cm²/sec. Using the data of Table 2 for the ionic equilibrium potentials and the membrane potential it is possible by the Ussing-Teorell flux relation (46,47) to separate out an active component of 6.13 pmoles/cm²/sec in the Na^+ efflux and 2.32 pmoles/cm²/sec in the K^+ influx. A comparison with the active transport components obtained through application of the dynamic membrane model gives very good agreement. Since active Na^+ and K^+ transport deviate strongly from each other, one concludes the existence of an active, electrogenic net ion transport across the membrane of vascular smooth muscle cells even in resting conditions. The membrane potential is thereby dependent on the metabolism.

A typical example of a simultaneous ^{24}Na–^{42}K *efflux* experiment using indirect methods in normal Krebs solution is shown in Fig. 5. Time is given on the abscissa, the instantaneous exchange quotient on the ordinate. Every increase of the instantaneous exchange quotient corresponds to an increase of flux of equal magnitude; every decrease to a reduction of flux. In the later, flat part of the washout curves, the recorded flux fluctuations are much larger than in the initial, steep course. This indicates that they are related to the cellular exchange. In order to investigate this phenomenon more closely, the experimental flux curves were fitted by ideal washout curves with the help of a digital computer using a triple exponential approximation. The ideal time

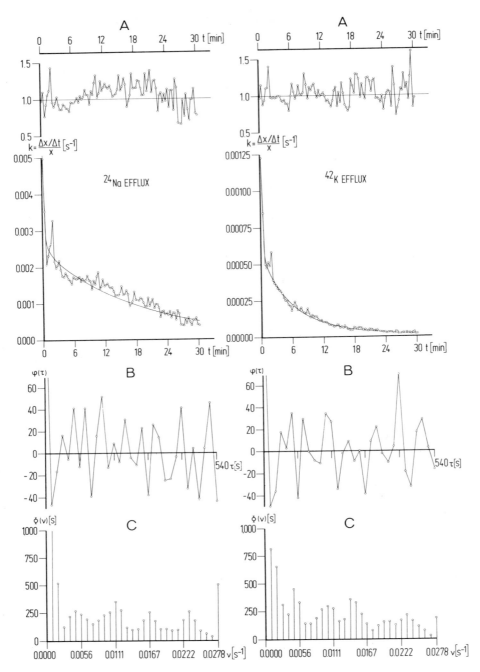

FIG. 5. $^{24}Na^+$ and $^{42}K^+$ efflux determined simultaneously in the carotid artery of the dog according to the indirect method. The graphs show the time course of the instantaneous exchange quotient $k(t)$ at intervals of 18 sec in a single experiment. The thin-traced line represents the optimal triple-exponential function found by computer fitting. (A) Relative deviation of the measured values from the functional values of the optimal curve. (B) Autocorrelation functions of the percent deviations given in part A. Oscillations are already clearly recognizable. (C) Power spectra of the autocorrelation functions. The frequencies of the oscillations shown in B may be directly observed on the abscissa of the power spectra.

course of the Na$^+$ and K$^+$ efflux is given by the thinly drawn curves. The time course of the deviations of the measured values from the functional values of the fitted curves is clarified in the insets labeled A. Increase and decrease rates for the Na$^+$ and K$^+$ effluxes of 10 to 60% are obtained.

Fig. 5 (B,C) result from calculation of the autocorrelation functions and the power spectra using the deviation. The mathematical procedure for the construction of the autocorrelation functions and the power spectra was described in detail in a previous report (14). Flux oscillations are already clearly seen in the autocorrelations (cf. Fig. 5B). The power spectrum (Fourier analysis of the autocorrelation function; cf. Fig. 5C) enables an exact analysis of the frequencies of significant flux oscillations. The amplitude

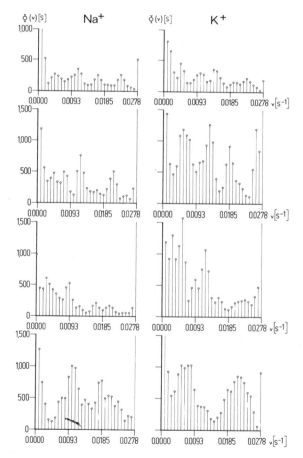

FIG. 6. Power spectra from four separate efflux experiments with simultaneous measurement of ^{24}Na$^+$ and ^{42}K$^+$ in the dog carotid artery according to the indirect method. The mathematical evaluation is the same as illustrated in Fig. 5. Distinct periodicities of the Na$^+$ and K$^+$ efflux are recognizable. Some frequency ranges were preferred by the individual preparations.

of each point of the function is a measure of the relative frequency of occurrence of the corresponding oscillation frequency. The background noise of the power spectrum of this experiment is about 0.08 sec, so a distinct rhythm exists in the Na^+ and K^+ efflux. From part C of Fig. 5 oscillations with the *principal periods* of 90, 215, 45, and 60 sec for Na^+, and 215, 70, 100, and 45 sec for K^+ can be ranked according to probability.

In Fig. 6 the power spectra for Na^+ and K^+ efflux from 4 individual preparations are given one below the other as a function of frequency. It becomes immediately apparent that a clear periodicity exists for the Na^+ efflux as well as the K^+. The flux oscillations occur often at integral multiples of one frequency (cf. 48). It has been determined from a large number of experiments that several peaks in the power spectra lie one below the other, or at most a slight frequency shift occurs. This suggests that the individual preparations favor certain frequencies.

In all probability, fluctuations of the Na^+ efflux can be attributed to periodic changes of active transport as this component amounts to about 95% of the total Na^+ efflux in this tissue. That is why one can assume an oscillatory, electrogenic Na^+ pump. Concerning K^+, however, it is not certain if the oscillations originate in an electrogenic, oscillating K^+ inward pump or arise as a result of slow waves of the membrane potential being produced by an oscillating electrogenic Na^+ pump. From these findings we come to the general conclusion that a rhythmically active, electrogenic net ion pump may be responsible for the slow waves of the membrane potential of vascular smooth muscle.

SUMMARY

1. Through a combination of electron microscopic-morphometric (point counting method) and tracer physiological techniques (^{14}C-inulin method) it is possible to delineate a compartmentation of the extracellular space of vascular smooth muscle (extracellular fluid space and extracellular connective tissue volume). By measurement of the concentration of the most important univalent cations and anions in these compartments as well as in the entire tissue, one can arrive at the intracellular ionic content and can estimate the specific equilibrium potential. Because, in addition, the membrane potential was recorded intracellularly, the electrochemical potential $(V - E)$ for the transmembrane current of the ion species in question can also be estimated.

2. The inulin space of the carotid artery and the membrane potential of the vascular smooth muscle cells vary in parallel with the external K^+ concentration. Since the membrane potential varies relatively little with $[K^+]_o$, a tight electromechanical coupling can be postulated.

3. The curve that relates the membrane potential to $[K^+]_o$ shows a strong deviation from the K equilibrium potential and is rather flat, which underlines the importance of ion species other than K^+. The distribution of Cl^- and the

fact that E_{Cl} is less negative than the membrane potential appears to play a decisive role in passive potential generation. Na^+ ions have a hyperpolarizing influence on the membrane potential, largely through their active, electrogenic net outward transport throughout a wide range of the external K^+ and Na^+ concentration. Steady-state Na transport kinetics and the membrane potential show the same dependence on extracellular ion concentrations.

4. The application of a dynamic membrane model to vascular smooth muscle gives the time-dependent, active current–voltage relations. The dynamic, active Na^+ current–voltage characteristics show a sigmoid course which reaches saturation with membrane depolarization. The dynamic, active K^+ current–voltage curves exhibit an optimal functioning of the inwardly directed K^+ pump in the potential range -60 mV $<$ V < -20 mV. Active Na^+ and K^+ transport are time-independent at the resting potential.

5. Simultaneous, indirect ^{24}Na-^{42}K efflux experiments demonstrate significant oscillations in Na^+ and K^+ exchange across the membrane of vascular smooth muscle cells. The flux oscillations occur often at integral multiples of one frequency. A rhythmically active, electrogenic net ion transport seems to be responsible for the slow waves of the membrane potential.

6. Finally, we can state that the extracellular K^+ and H^+ ion concentrations, as well as the metabolic state of the vascular smooth muscle cells, must be considered to be among the local myogenic factors that decisively influence the regulation of the vessel caliber via their effect on the actual membrane potential.

ACKNOWLEDGMENTS

The authors thank Mrs. Ch. Fuhrmann for her excellent technical assistance and Mr. L. Seldon for his aid and advice at all stages in the preparation of this manuscript. We are grateful to Mrs. H. Vorthaler for her outstanding work in preparing the illustrations. We would also like to thank the staff of the Grossrechenzentrum für die Wissenschaft in Berlin for their cooperation and for generously making computer time available. This work was supported by the Deutsche Forschungsgemeinschaft (Si 182/1, 3) within its overall program "Structure and Function of Biological Membranes" and the Stiftung Volkswagenwerk (Az. 111 327).

REFERENCES

1. Somlyo, A. P., and Somlyo, A. V. (1968): *Pharmacol. Rev.*, 20:197–272.
2. Bülbring, E. (1961): *Pfluegers Arch.*, 273:1–17.
3. Golenhofen, K. (1970): In: *Smooth Muscle,* edited by E. Bülbring, A. F. Brading, A. W. Jones, and T. Tomita, pp. 316–342. Arnold, London.
4. Kuriyama, H. (1963): *J. Physiol. (Lond.),* 166:15–28.
5. Funaki, S. (1967): In: *Symposium on Electrical Activity and Innervation of Blood Vessels,* edited by W. R. Keatinge (*Bibl. Anat.*), Vol. 8, pp. 5–10. S. Karger, Basel, New York.

6. Golenhofen, K., Hermstein, N., and Lammel, E. (1973): *Microvasc. Res.*, 5:73–80.
7. Speden, R. N. (1970): In: *Smooth Muscle*, edited by E. Bülbring, A. F. Brading, A. W. Jones, and T. Tomita, pp. 558–588. Arnold, London.
8. Bülbring, E. (1953): *J. Physiol. (Lond.)*, 122:111–134.
9. Bertsche, O., and Siegel, G. (1974): *Pfluegers Arch.*, 347 (Suppl.):R14.
10. Siegel, G., Roedel, H., Bertsche, O., and Neumann, B. (1975): In: *Quantitative Analysis of Microstructures in Medicine, Biology and Materials Development*, edited by H. E. Exner, pp. 153–167. Dr. Riederer-Verlag GmbH, Stuttgart.
11. Siegel, G., Jäger, R., Nolte, J., Bertsche, O., Roedel, H., and Schröter, R. (1974): In: *Pathology of Cerebral Microcirculation*, edited by J. Cervós-Navarro, pp. 96–120. Walter de Gruyter, Berlin, New York.
12. Haas, H. G., Glitsch, H. G., Kern, R., Hantsch, F., and Siegel, G. (1966): *Pfluegers Arch.*, 288:43–64.
13. Koepchen, H. P., Siegel, G., and Warta, H. (1967): *Pfluegers Arch.*, 297:R63.
14. Siegel, G., Koepchen, H. P., and Roedel, H. (1972): In: *Vascular Smooth Muscle*, edited by E. Betz, pp. 3–6. Springer-Verlag, Berlin, Heidelberg, New York.
15. Roedel, H., and Siegel, G. (1972): In: *Passive Permeability of Cell Membranes. Biomembranes:* Vol. 3, edited by F. Kreuzer and J. F. G. Slegers, pp. 449–471. Plenum Press, New York, London.
16. Siegel, G., and Roedel, H. (1972): *Abstr. Contrib. Papers, IV. Int. Biophys. Congr.*, Vol. 3, pp. 322–323. Acad. Sci. USSR, Moscow.
17. Roedel, H., and Siegel, G. (1974): *Proc. Int. Union Physiol. Sci., XXVI Int. Congr.*, Vol. XI, p. 5. Thomson Press, New Delhi.
18. Bertsche, O., Siegel, G., and Roedel, H. (1974): *Proc. Int. Union Physiol. Sci., XXVI Int. Congr.*, Vol. XI, p. 18. Thomson Press, New Delhi.
19. Haas, H. G., Hantsch, F., Otter, H. P., and Siegel, G. (1967): *Pfluegers Arch.*, 294:144–168.
20. Casteels, R. (1970): In: *Smooth Muscle*, edited by E. Bülbring, A. F. Brading, A. W. Jones, and T. Tomita, pp. 70–99. Arnold, London.
21. Cope, F. W. (1967): *J. Gen. Physiol.*, 50:1353–1375.
22. Shporer, M., and Civan, M. M. (1972): *Biophys. J.*, 12:114–122.
23. Hinke, J. A. M. (1961): *J. Physiol. (Lond.)*, 156:314–335.
24. Lev, A. A. (1964): *Nature*, 201:1132–1134.
25. Barrow, G. M., editor (1972): *Physikalische Chemie, Teil III: Mischphasenthermodynamik, Elektrochemie, Reaktionskinetik*. Bohmann-Verlag and F. Vieweg-Verlag, Heidelberg, Wien, Braunschweig.
26. Keynes, R. D. (1963): *J. Physiol. (Lond.)*, 169:690–705.
27. Adrian, R. H. (1956): *J. Physiol. (Lond.)*, 133:631–658.
28. Huxley, A. F., and Stämpfli, R. (1951): *J. Physiol. (Lond.)*, 112:496–508.
29. Konold, P., Gebert, G., and Brecht, K. (1968): *Pfluegers Arch.*, 301:285–291.
30. Betz, E. (1975): In: *The Working Brain, Alfred Benzon Symposium*, Vol. VIII. Munksgaard, Copenhagen.
31. Mekata, F., and Niu, H. (1972): *J. Gen. Physiol.*, 59:92–102.
32. Keatinge, W. R. (1967): In: *Symposium on Electrical Activity and Innervation of Blood Vessels*, edited by W. R. Keatinge *(Bibl. anat.)*, Vol. 8, pp. 21–24. S. Karger, Basel, New York.
33. Lüttgau, H. Ch. (1965): In: *Fortschritte der Zoologie*, edited by H. Bauer, Vol. 17, pp. 272–312. Gustav Fischer Verlag, Stuttgart.
34. Wahlström, B. A. (1973): *Acta Physiol. Scand.*, 89:436–448.
35. Keatinge, W. R. (1968): *J. Physiol. (Lond.)*, 194:169–182.
36. Su, C., Bevan, J. A., and Ursillo, R. C. (1964): *Circ. Res.*, 15:20–27.
37. Holman, M. E. (1973): *Philos. Trans. R. Soc. Lond. [Biol. Sci.]*, 265:157–165.
38. Siegel, G., Koepchen, H. P., and Roedel, H. (1967): *Pfluegers Arch.*, 297:R64.
39. Koshland, D. E., Jr., Nemethy, G., and Filmer, D. (1966): *Biochemistry*, 5:365–385.
40. Squires, R. F. (1965): *Biochem. Biophys. Res. Commun.*, 19:28–32.
41. Robinson, J. D. (1967): *Biochemistry*, 6:3250–3258.
42. Tomita, T. (1966): *J. Theoret. Biol.*, 12:216–227.
43. Anderson, N. C., Jr. (1969): *J. Gen. Physiol.*, 54:145–165.

44. Dudel, J., Peper, K., Rüdel, R., and Trautwein, W. (1966): *Pfluegers Arch.,* 292:255–273.
45. Armstrong, C. M., and Bezanilla, F. (1973): *Nature,* 242:459–461.
46. Ussing, H. H. (1950): *Acta Physiol. Scand.,* 19:43–56.
47. Teorell, T. (1949): *Arch. Sci. Physiol.,* 3:205–219.
48. Golenhofen, K., and Loh, D. v. (1970): *Pfluegers Arch.,* 314:312–328.

Physiology of Smooth Muscle, edited by
E. Bülbring and M. F. Shuba.
Raven Press, New York © 1976.

The Spread of Current in Electrical Syncytia

Lloyd Barr and Eric Jakobsson

Department of Physiology and Biophysics, University of Illinois, Urbana, Illinois 61801

It is now well established that in nerve (1), epithelia (2), and in smooth (3) and cardiac muscle (4) there are conductance pathways that directly connect cell interiors without interposition of any region of extracellular space. This alone is not sufficient to prove that propagation of action potentials occurs between cells in these tissues by means of local circuits. However, this sufficiency has also been demonstrated in certain cases for nerve (5) and for smooth and cardiac muscle (6,7). The magnitude of the conductance between cells necessary to allow electrical transmission to occur is a complex function of cell geometries, the passive membrane parameters, and the kinetics of the excitable processes. Therefore, one must understand that in most usages the phrase "low-resistance intercellular pathways" properly means a conductance high enough for electrical transmission to occur and not necessarily a higher conductance than the plasma membrane.

There are few situations in which the investigators of smooth muscle physiology find themselves with the relatively simplest of the excitable systems. This unlikely event does obtain with regard to the site of electrical coupling between cell interiors. The nexus is the only morphological junction between smooth muscle cells, which has been a serious candidate for the intercellular connection. The situation has been quite a bit more complicated in other tissues. In epithelia, for example, the original definitions of the nexus (tight junction, etc.) apply to two different junctions—one has a hexagonal array of particles in the junctional membranes and the other has a filamentous substructure. A useful resolution of this has been suggested by McNutt and Weinstein (8) on the basis of the apparent difference of function of the two junctions. Because the junction with filaments in the plane of the junction is the one that blocks diffusion between extracellular compartments, they propose to restrict the term "tight junction" to this junction. The "tight junction" then occurs only in epithelia. Whether "tight junctions" electrically couple cells is unknown. Since the nexus was first defined for regions of membrane apposition in smooth muscle and these junctions have hexagonal arrays of membrane particles in the plane of the junction, McNutt and Weinstein (8) proposed to restrict the term "nexus" to junctions with hexagonal arrays. Nexuses by this definition have been found in plants and between cells of all tissue types in nearly all phyla above Protozoa. Although widely sought, there are no known cases of electrical coupling in the absence of nexuses. Formerly, it was thought that the septate junction provided electrical coupling in various invertebrate systems (2,9), especially between the salivary epithelial cells of *Chironomus thummi*. However, Berger and Uhrik (10) have recently demonstrated nexuses of an extent about equal to the area occupied by the septate junctions. Moreover, hypertonic solutions disrupt the nexuses and the electrical connections between the cells (10) but leave the septate junctions undisturbed (9). The relationships between nexuses and septate junctions in these experiments are similar to those found between nexuses and desmosomes in vertebrate heart muscle (6). The role of nexuses in electrical transmission in smooth muscle has been demonstrated by use of a sucrose gap (7) in which a low-conductance sucrose replaces the higher conductance extracellular electrolyte in a middle portion of a small bundle of smooth muscle. The sucrose gap strongly attenuates any extracellular action current and thereby blocks the propagation of action potentials. When "an artificial extracellular pathway" (an ordinary carbon resistor) is provided, action potentials jump past the gap. As the excitatory current is much smaller than *in vivo,* it is concluded that action potentials propagate with a considerable safety factor *in vivo.* When hypertonic sucrose (more than two times isosmolar) was used in the gap, the transgap resistance went up and action potentials would not jump in the presence of a gap-shunting resistor. The ability to jump returned when isotonic sucrose was again used in the gap. Correspondingly, it was found that hypertonicity reversibly disrupted the nexuses in similar small bundles. The conclusions from these experiments were (a) propagation in smooth muscle occurs by electrical transmission; and (b) nexuses are the sites of electrical coupling. Because hypertonicity can cause hyperpolarization and anode blockade, these experiments have been criticized as being inconclusive (11). However, this criticism is based on misunderstanding. Only the cells in the gap are exposed to the hypertonic sucrose, but they are neither excited nor excitable. They act only as an intercellular conductance path across the gap. The postgap cells, which are potentially excitable, are in normal electrolyte and hence cannot experience any hypertonic hyperpolarization. Similarly, criticisms related to the damaging effects of isotonic sucrose miss the point of these experiments, which is the reversibility of the blocking effects of hypertonic sucrose solutions. The proper conclusions from the gap-jumping experiments do not require the cells in the gap to be normal—quite the opposite. All that is required is that they lose their nexuses reversibly and decouple electrically. It has recently been shown using freeze-cleave techniques that separation of the membranes of the nexus by hypertonic solutions does not disturb the hexagonal array of particles in the constituent membranes (12). Measurements of the spread of current in various smooth muscles have convinced most investigators (3,11,13), but not all (14), that these tissues behave like electrical syncytia. One would expect that electrotonic spread in three dimensions will be quantitatively different from spread along a cable. Indeed, analysis of three-dimensional core-conducting tissue is necessary before it is possible to set limits for the use of cable equations to describe electrotonic spread in small bundles. In the following analysis, we will make severe simplifying assumptions that ignore the organization of most smooth muscle cells into bundles. On the basis of many precedents it seems that this approach can provide a foundation for further and more realistic modeling of the electrical behavior of tissues comprising many small cells interconnected by nexuses.

A vector formalism is well suited for the purpose of extending cable theory to a three-dimensional system. In the following section, we will apply this formalism to a special case of current being injected into one cell of such a preparation and withdrawn from the extracellular space just outside the same cell (Fig. 1). The impaled cell is many space constants from the edge of

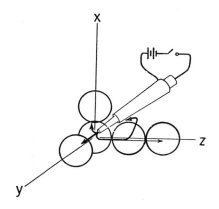

FIG. 1. Schematic drawing of a model experiment on a multicellular tissue with isotropic electrical properties and which would yield the same boundary conditions as are used in the isotropic case.

the tissue, so that we can assume the tissue to be infinite in extent and the current injection and withdrawal both to occur at the origin. We will use a vector formalism to analyze the distribution of steady-state transmembrane voltage and current in this tissue first under the assumption that the tissue is isotropic, and then under the more realistic assumption that resistance to current flow is different along the length of the cells from what it is across the cells.

ISOTROPIC CASE (STEADY STATE)

$$V_m = V_i - V_o; \; \bar{I}_i = -\bar{I}_o \tag{1}$$

where V_m is the transmembrane potential difference (mV); V_i is the intracellular potential (mV); and V_o is the extracellular potential (mV), both electrical potentials relative to arbitrarily distant point; $\bar{I}_i$ is the density of current flow in intracellular space (mA/cm^2); and $\bar{I}_o$ is the density of current flow in extracellular space (mA/cm^2).

All of the above are functions of r in spherical coordinates; V_m, V_o, V_i are scalars; $\vec{I}_o$, $\vec{I}_i$ are vectors. The right side of Eq. (1) comes from conservation of charge.

By Ohm's Law,

$$\vec{I}_o = -\frac{1}{R_o}\vec{\nabla}V_o \tag{2}$$

$$\vec{I}_i = -\frac{1}{R_i}\vec{\nabla}V_i \tag{3}$$

where R_o is resistivity of extracellular space ($\Omega \cdot$ cm) and R_i is resistivity of intracellular space ($\Omega \cdot$ cm). (These differ from bulk resistivities of the respective fluids because of the differences in fractional cross section of the two pathways in addition to other geometical factors.) $\vec{\nabla}$ is the gradient.

Combining Eqs. (1)–(3),

$$\vec{I}_o = [1/(R_o + R_i)]\vec{\nabla}V_m \tag{4}$$

If G_m is cell-membrane specific conductance (mmhos/cm^2) and A_m is density of membrane

$$\left(\frac{\text{cm}^2 \text{ surface area}}{\text{cm}^3 \text{ volume of tissue}}\right)$$

then

$$G_m A_m V_m = I_m = \vec{\nabla}\cdot\vec{I}_o \tag{5}$$

where I_m is transmembrane current density (mA/cm^3), a scalar function of r and $\vec{\nabla}\cdot\vec{I}_o$ is the divergence of $\vec{I}_o$.

The right-hand side of Eq. (5) follows from conservation of charge. Taking the divergence of Eq. (4) and substituting in Eq. (5),

$$G_m A_m V_m = [1/(R_o + R_i)]\nabla^2 V_m \tag{6}$$

Making the definition

$$\lambda^2 \equiv 1/[G_m A_m(R_o + R_i)] \tag{7}$$

and invoking spherical symmetry, Eq. (6) reduces to

$$d^2 V_m/dr^2 + 2/r\, d V_m/dr - (1/\lambda^2) V_m = 0 \tag{8}$$

which has the solution

$$V_m = (A/r)\, e^{-r/\lambda} + (R/r)\, e^{+r/\lambda} \tag{9}$$

where A is a finite constant and $B = 0$, since an infinite transmembrane potential at $r = \infty$ is physically inadmissible.

$$V_m = (A/r)\, e^{-r/\lambda} \tag{9'}$$

This solution has previously been found without use of the vector formalism (15). We may construct a hypothetical experiment with one current-injecting and two voltage-measuring electrodes which would determine several of the tissue parameters. If we inject current at $r = 0$ and measure V_m at $r = r_1$ and $r = r_2$, then take the ratio of Eq. (9) at r_1 and r_2

$$[V_m(r_1)/V_m(r_2)] = \frac{r_2}{r_1} e^{-(r_1 - r_2)/\lambda} \tag{10}$$

Rearranging Eq. (10),

$$\lambda = \frac{r_2 - r_1}{\ln[r_1 V_m (r_1)/r_2 V_m (r_2)]} \tag{11}$$

using the value of λ determined in Eq. (11) A can be calculated from the value of V_m at any value of r, such as r_1.

$$A = r_1 V_m (r_1) e^{r_1/\lambda} = \lambda e V_m(\lambda) \tag{12}$$

The membrane conductance can also be obtained as follows.

The current electrode injects a current I_T into the intracellular space and withdraws it from the extracellular space. Thus

$$I_T = \int_{\substack{\text{all} \\ \text{space}}} I_m \, dV \tag{13}$$

where dV is an element of volume.

Substituting Eqs. (5) and (9) in Eq. (13)

$$I_T = \int_{\substack{\text{all} \\ \text{space}}} G_m A_m V_m \, dV = G_m A_m A \int_{r=0}^{\infty} \frac{(e^{-r/\lambda})}{r} 4\pi r^2 dr$$

$$= 4\pi \, G_m A_m A \int^{\infty} r e^{-r/\lambda} \, dr$$

$$= 4\pi \, G_m A_m A \lambda^2 \int_0^{\infty} r e^{-r} \, dr$$

$$= 4\pi \, G_m A_m A \lambda^2$$

and rearranging Eq. (14)

$$G_m A_m = (I_T/4\pi A \lambda^2) \tag{14'}$$

Equations (7), (11), (12), and (14'), applied to our hypothetical but quite plausible measurements of $V_m(r_1)$, $V_m(r_2)$, and I_T, combine to determine $G_m A_m$ and $(R_o + R_i)$, the transmembrane and bulk resistive properties of the tissue, respectively. A_m, of course, can be estimated from electron micrographs.

THE CASE OF ANISOTROPY IN ONE DIMENSION (STEADY STATE)

In practice, we would not expect a tissue such as smooth muscle to have isotropic conductance properties, but rather to have different conductance along the length of cells and across cells. We therefore set up a cylindrical coordinate system with the z coordinate along the length of cells (Fig. 2).

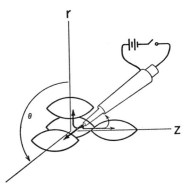

FIG. 2. Same as Fig. 1 except that the tissue is now anisotropic along the cell axes.

Equation (1) still holds with the provision that the quantities are now functions of r, θ, and z.

The equations corresponding to (2) and (3) are now different, however.

$$\vec{I}_o = -\frac{1}{R_{o,r}}\nabla_r V_o \hat{e}_r - 1/R_{o,z}\,\nabla_z\,V_o\hat{e}_z \tag{15}$$

$$\vec{I}_i = -\frac{1}{R_{i,r}}\nabla_r V_i \hat{e}_r - 1/R_{i,z}\,\nabla_z V_i\hat{e}_z \tag{16}$$

where V_r and V_z are the components of the gradient in the r and z directions, respectively; $\hat{e}$ and $\hat{e}_z$ are unit vectors in the respective directions, $R_{o,r}$ and $R_{o,z}$ are the resistivities of the extracellular space in the respective directions, and $R_{i,r}$ and $R_{i,z}$ are the resistivities of the intercellular space in the respective directions. Again, in general, these parameters will vary from the bulk resistivities by the areas of the pathways and unknown tortuosity factors. We leave out the gradient component in the θ direction on grounds of symmetry. Combining Eqs. (1), (15), and (16)

$$\vec{I}_o = \frac{1}{R_{o,r}+R_{i,r}}\nabla_r V_m \hat{e}_r + \frac{1}{R_{o,z}+R_{i,z}}\nabla_z V_m \hat{e}_z \tag{17}$$

Equation (5) is still valid for the anisotropic case, with the provision that I_m is now a function of both r and z. Taking the divergence of Eq. (17) and substituting in Eq. (5) gives

$$G_m A_m V_m = \frac{1}{R_{o,r}+R_{i,r}}\frac{1}{r}\frac{\partial}{\partial r}\left(r\frac{\partial V_m}{\partial r}\right) + \frac{1}{R_{o,z}+R_{i,z}}\frac{\partial^2 V_m}{\partial z^2} \tag{18}$$

If we make the definitions

$$\lambda_r^2 \equiv \frac{1}{G_m A_m (R_{o,r} + R_{i,r})} \qquad \lambda_z^2 \equiv \frac{1}{G_m A_m (R_{o,z} + R_{i,z})} \qquad (19)$$

then Eq. (18) becomes

$$\lambda_r^2 \left(\frac{1}{r} \frac{\partial V_m}{\partial r} + \frac{\partial^2 V_m}{\partial r^2} \right) + \lambda_z^2 \frac{\partial^2 V_m}{\partial z^2} - V_m = 0 \qquad (20)$$

It is possible to transform Eq. (20) from a partial to an ordinary differential equation by the transformation

$$\rho \equiv r^2 + \alpha z^2 \qquad \alpha \equiv \lambda_r^2 / \lambda_z^2 \qquad (21)$$

then the differential operators in Eq. (20) can be transformed as follows

$$\frac{\partial}{\partial r} = \frac{\partial}{\partial \rho} \frac{\partial \rho}{\partial r} = 2r \frac{\partial}{\partial \rho} \qquad (22)$$

$$\frac{\partial^2}{\partial r^2} = 2 \frac{\partial}{\partial \rho} + 2r \frac{\partial^2}{\partial r \partial \rho} = 2 \frac{\partial}{\partial \rho} + 4r^2 \frac{\partial^2}{\partial \rho^2} \qquad (23)$$

$$\frac{\partial}{\partial z} = 2\alpha z \frac{\partial}{\partial \rho} \qquad (24)$$

$$\frac{\partial^2}{\partial z^2} = 2\alpha \frac{\partial}{\partial z} + 2\alpha z \frac{\partial^2}{\partial z \partial \rho} = 2\alpha \frac{\partial}{\partial z} + 4\alpha^2 z^2 \frac{\partial^2}{\partial \rho^2} \qquad (25)$$

substituting Eqs. (21), (22), (24), and (25) into Eq. (20) gives

$$4\lambda_r \rho \frac{\partial^2 V_m}{\partial \rho^2} + 6\lambda_r^2 \frac{\partial V_m}{\partial \rho} - V_m = 0 \qquad (26)$$

which has the solution

$$V_m = \frac{C}{\sqrt{\rho}} e^{-\sqrt{\rho}/\lambda_r} + \frac{D}{\sqrt{\rho}} e^{+\sqrt{\rho}/\lambda_r} \qquad (27)$$

where C is a finite constant and $D = 0$, since an infinite transmembrane potential at $\rho = \infty$ is physically inadmissible. Thus so

$$V_m = \frac{C}{\sqrt{\rho}} e^{-\sqrt{\rho}/\lambda_r} = \frac{C}{\sqrt{r^2 + \alpha z^2}} e^{-\sqrt{r^2 + \alpha z^2}/\lambda_r} \qquad (27')$$

Clearly Eq. (27') is closely analogous to Eq. (9'). Now, in contrast to the isotropic case, our hypothetical experiment to determine tissue parameters will involve one current-injecting electrode and *three* (rather than two) voltage electrodes, since the description of the tissue involves an extra parameter α. In general, the three parameters in Eq. (27') (C, λ_r, and α) may be determined from measurements at any three or more positions in the preparation.

Equations (13) and (5) for the injected current are still valid for the anisotropic case. Thus

$$I_T = \int_{\substack{all \\ space}} G_m A_m V_m \, dV = G_m A_m C \int_{r=0}^{\infty} \int_{z=-\infty}^{\infty} 2\pi r dr dz \frac{e^{-\sqrt{r^2+\alpha z^2}/\lambda_r}}{\sqrt{r^2 + \alpha z^2}}$$

$$= \frac{G_m A_m C}{\alpha^{1/2}} \int_{r=0}^{\infty} \int_{z=-\infty}^{\infty} 2\pi r dr d(\alpha^{1/2} z) \frac{e^{-\sqrt{r^2+\alpha z^2}/\lambda_r}}{\sqrt{r^2 + \alpha z^2}}$$

$$= \frac{G_m A_m C}{\alpha^{1/2}} \int_{\substack{all \\ space}} dV \frac{e^{-r/\lambda_r}}{r} = \frac{4\pi G_m A_m C \lambda_r^2}{\alpha^{1/2}} = 4\pi G_m A_m C \lambda_r \lambda_z \qquad (28)$$

where r is defined as distance from the origin.
rearranging (28):

$$G_m A_m = \frac{I_T}{4\pi C \lambda_r \lambda_z} \qquad (29)$$

Thus, we now have in Eq. (21), (27′), and (29) the means to take the steady-state values of injected current at the origin and transmembrane potential at three other points in the tissue and determine the following physical characteristics of the anisotropic multicellular preparation.
1) $G_m A_m$, 2) $(R_{o,z} + R_{i,z})$ and 3) $(R_{i,r} + R_{o,r})$

The above theoretical considerations may or may not in themselves provide a quantitatively accurate representation of an actual tissue, because the assumption of infinite extent of the tissue is questionable. However, with the appropriate boundary conditions the vector extension of cable theory described above should give a good description of actual pieces of electrically syncytial tissue, such as bundles of smooth muscle.

REFERENCES

1. Bennett, M. V. L. (1966): *Ann. NY Acad. Sci.*, 137:509.
2. Loewenstein, W. R. (1966): *Ann. NY Acad. Sci.* 137:441.
3. Barr, L. (1963): *J. Theoret. Biol.* 4:73
4. Weidman, S. (1952): *J. Physiol. (Lond.)*, 118:348.
5. Furshpan, E. J., and Potter, D. D. (1959): *J. Physiol. (Lond.)*, 145:289.
6. Barr, L., Dewey, M. M., and Berger, W. (1965): *J. Gen. Physiol.*, 48:797.
7. Barr, L., Berger, W., and Dewey, M. M. (1968): *J. Gen. Physiol.*, 51:347.
8. McNutt, N. S., and Weinstein, R. S. (1973): *Progr. Biophys. Mol. Biol.*, 26:47.
9. Bullivant, S., and Loewenstein, W. R. (1968): *J. Cell. Biol.*, 37:621.
10. Berger, W. K., and Uhrik, B. (1972): *Z. Zellforsch. Mikrosk. Anat.*, 127:116.
11. Tomita, T. (1970): In: *Smooth Muscle*, edited by E. Bülbring, A. F. Brading, A. W. Jones, and T. Tomita, pp. 197–243. Williams & Wilkins, Baltimore, Maryland.
12. Goodenough, D. A., and Gilula, N. B. (1974): *J. Cell. Biol.*, 61:575.
13. Prosser, C. L. (1974): *Ann. Rev. Physiol.*, 36:503.
14. Sperelakis, N. and Tarr, M. (1965): *Am. J. Physiol.*, 208:737.
15. Barr, L. and Dewey, M. M. (1968): In: *Handbook of Physiology*, Section 6: *Alimentary Canal. IV. Motility*, edited by C. F. Code, pp. 1733–1742. Williams & Wilkins, Baltimore, Maryland.

Physiology of Smooth Muscle, edited by
E. Bülbring and M. F. Shuba.
Raven Press, New York © 1976.

Ionic Currents in an Intestinal Smooth Muscle

H. Inomata and C. Y. Kao

Department of Pharmacology, State University of New York, Downstate Medical Center, Brooklyn, New York, 11203

PREPARATION

The guinea pig taenia coli, which has been studied extensively with the microelectrode technique, flux measurements, and chemical analyses, has been subjected to a voltage-clamp study in a double sucrose-gap apparatus. The observations, although similar in nature to those obtained in the longitudinal myometrium of the pregnant rat uterus (Kao and McCullough, 1975), differ from the latter in some important details. To appreciate the nature of the ionic currents in the taenia coli, it is necessary to compare and contrast them with those in the myometrium.

The method of the double sucrose-gap voltage clamp has been described (Anderson, 1969; Kao and McCullough, 1975). The preparation used is an isolated portion of a taenia coli, measuring, on the average, about 0.25 mm wide and about 0.2 mm thick. The "node" formed by the two sucrose streams averaged about 0.05 mm. The total capacitance, as estimated from the rate of capacitative discharge produced by a small voltage step, averaged 0.4 μF, i.e., about three times as large as that found in the studies on the pregnant rat myometrium (Kao and McCullough, 1975). The number of individual cells was around 10^4, as contrasted to 10^2 in a node of the pregnant rat myometrium. In addition to the more numerous cells, the taenia coli cells may also be packed tighter together than the myometrial cells. Kinetically, the distinction is very obvious: whereas in the myometrium washing for 4 to 5 min with the isotonic sucrose solution was usually adequate to establish appropriate conditions for voltage clamping, in the taenia coli washing for 30 to 40 min was required. The slowness of the extracellular washout has important effects on some experimental results.

LATE CURRENT

As in other excitable tissues, the taenia, on depolarization, develops a transient inward current followed by a late outward current. The current–voltage relation of these currents also resembles those of other tissues. The late outward current is primarily carried by K^+, as shown by the following observations. When the taenia was depolarized in a medium with high $[K^+]_o$

and was held back to the original resting potential, depolarization, which originally elicited only outward current, could now produce inward currents within a range between the resting potentials in the natural state and the depolarized state. The reversal potential of the late current (E_K) has been estimated by repolarizing to different voltage levels. After correcting for capacitive and leakage currents, E_K was 15 to 20 mV more negative than the resting potential. The instantaneous current–voltage (I–V) relation for the tail current was linear. In these respects, the observations on the taenia are similar to those on the myometrium. Although the amplitude and the duration of the activating step influenced the slope conductance of the instantaneous I–V relation, E_K, taken as the zero-current voltage, remained fixed. These observations were made in media containing 5.9, 26, 59, and 149 mM of K^+. The similarity of these observations suggests that contributions of other ions to the late outward current is relatively small.

Holding the taenia at potentials requiring considerable holding current tended to change E_K, as in the pregnant myometrium. The duration of the holding current was important in that a minimum duration was needed to reach some sort of steady state for detectable changes in E_K to show up. Very long holding currents would cause enough passive redistribution to create new equilibrium states in which the formerly unphysiological holding potential would become that potential at which negligible holding current would be needed. In all cases, we allowed about 5 min, which is enough to produce changes. In the hyperpolarizing direction, requiring inward holding currents, the change in holding potential and the change in E_K were linearly related, with a slope very close to 1. In the depolarizing direction, which requires outward holding currents, however, the changes in E_K were greater than would be expected from a purely passive decrease in $[K^+]_i$. This deviation can be explained if $[K^+]_o$ immediately surrounding the taenia cells were somewhat higher than that in the medium, a view consistent with a diffusion lag from the extracellular space. In this respect the taenia differs importantly from the pregnant myometrium in which the relation between E_K and holding potential is linear with the same slope of 1 for both hyperpolarizing and depolarizing changes.

Tetraethylammonium (TEA) ion markedly reduced the late outward current, and its maximum chord conductance, $\bar{g}_K$. Action potentials were markedly prolonged by TEA. Mn^{2+} also reduced $\bar{g}_K$, and prolonged the action potential, suggesting that its action may not be specific against any early Ca^{2+} channels involved in the initial current.

EARLY CURRENT

The nature of the inward current in the taenia is considerably more complex than that in the pregnant myometrium. In the pregnant myometrium, in a Na^+-free medium, spikes were abolished in 10 min; and under voltage-clamp conditions the initial current, which had been inward, became outward. This reversal also occurred in the tail current following repolarization. When $[Na^+]_o$ was reduced to 50%, a shift in the reversal potential of the early current, E_a, by -17 mV was observed, a change consistent with expectations of a Na^+-selective membrane. In the taenia, however, similar treatment with a Na^+-free medium did not abolish spikes within a 10-min test period. On the contrary, the amplitude of the spike often increased slightly, possibly because of a reduced leakage conductance. Reduction of $[Na^+]_o$ to 50% did not produce any electrophysiological effects. When $[Ca^{2+}]_o$ was reduced to 25%, a shift of E_a by -14 mV occurred; when $[Ca^{2+}]_o$ was reduced to 10%, a shift of -23 mV occurred. These observations are qualitatively consistent with an idea that Ca^{2+} may serve as a charge carrier for the initial current.

However, increasing $[Ca^{2+}]_o$ to 9.5 mM failed to produce significant changes in the E_a. This failure can be interpreted as an obstacle to the idea of an early inward Ca^{2+} current unless some alternative resolution can be found. Because the taenia coli preparation used is multicellular, there must be some temporal dispersion of the various ionic currents originating from each individual cell because of the unavoidable series resistances in the extracellular fluid between cells. Under these conditions, the reversal potential of the early current, E_a, is influenced as much by the voltage threshold for activating the outward current as by the electrochemical driving forces on the ion responsible for the early inward current. With this consideration in mind, some taenia coli preparations were pretreated with TEA to reduce the outward current. In such TEA-treated preparations, increasing $[Ca^{2+}]_o$ to 9.5 mM caused a marked and significant shift of the reversal potential toward the positive. From all these observations, therefore, it can be concluded that the early current in the taenia coli is to be attributed more to Ca^{2+} influx than to Na^+ influx.

CATECHOLAMINE ACTIONS ON IONIC CURRENTS

When the guinea pig taenia coli was treated with either epinephrine or norepinephrine, a hyperpolarization developed. The membrane time constant, as observed in the double sucrose-gap current-clamp mode, was not affected in any way by the catecholamines, although the hyperpolarization might be as much as 10 to 15 mV. This observation, which is fundamentally different from those observations of Shuba and Klevets (1967) and Bülbring and Tomita (1969), has been studied in detail in the voltage-clamp mode. The slope conductance of the tail K^+ current was identical for untreated and catecholamine-treated nodes. The maximum chord conductances of the early current, $\bar{g}_a$, and of the late current, $\bar{g}_K$, the maximum inward current, and the leakage conductances were all unaffected. The reversal potential of the late current, E_K, after correction for capacitative and leakage currents, was about

15 mV more negative in the catecholamine-treated condition as compared with that in the control condition. From these analyses it is concluded that the catecholamine-induced hyperpolarization in the taenia, like that in the pregnant myometrium, is due to an increase in the electromotive force of the K^+ cell, and not to any changes in any K^+ conductances.

Possible mechanisms of the increase in the EMF of the K^+ cell have been investigated by testing the effects of catecholamines in media containing different $[K^+]_o$. In $[K^+]_o$ = 5.9, 26, 59, or 149 mM, the K^+ conductances were unaffected by the catecholamines, but when $[K^+]_o$ = 0.1 or 1.2 mM, $\bar{g}_K$ was significantly reduced by both epinephrine and norepinephrine. That a change in K^+ conductance can be observed in some conditions but not others is strong evidence that the absence of a change in conductance cannot be a technical artifact. From the experiments in media containing different $[K^+]_o$ it has also been concluded that catecholamines must also alter the selective cation permeability of the resting membrane in favor of K^+. Therefore, the catecholamine-induced hyperpolarization can be attributed to a combination of a change in the EMF of the K^+ cell, and a change in the selective cation permeability, but not to an increase in potassium conductance. The effect cannot be attributed to any significant alterations in either intracellular Na^+ or Ca^{2+}, because even when the catecholamine effect had remained in a steady state for some time, the reversal potential for the early current was unaltered. It is possible that the effect is coupled with an exchange with intracellular H^+, and that cyclic AMP is involved.

Although the detailed mechanism remains to be studied, the observations to date indicate the importance of metabolic activity in smooth muscles (which have very small individual cells) in affecting (within relatively short periods of time) fundamental electrophysiological properties, and consequent mechanical activities. In other words, in smooth muscles, alternatives to changes in ionic conductances may be equally important as means through which physiological properties are modulated by neurotransmitters and pharmacological agents.

REFERENCES

Anderson, N. C. (1969): Voltage clamp studies on uterine smooth muscle. *J. Gen. Physiol.*, 54:145–165.

Bülbring, E., and Tomita, T. (1969): Increase of membrane conductance by adrenaline in the smooth muscle of guinea-pig taenia coli. *Proc. R. Soc. Lond. [Biol.]*, 172:89–102.

Kao, C. Y., and McCullough, J. R. (1975): Ionic currents in the uterine smooth muscle, *J. Physiol. (Lond.)*, 246:1–36.

Shuba, M. F., and Klevets, M. Y. (1967): Ionic mechanisms of the inhibitory actions of adrenaline and noradrenaline on smooth muscle cells. *Fiziol. Zh.*, 13:1–11.

Physiology of Smooth Muscle, edited by
E. Bülbring and M. F. Shuba.
Raven Press, New York © 1976.

Interaction Between Pacemaker Electrical Behavior and Action Potential Mechanism in Uterine Smooth Muscle

Nels C. Anderson and Fidel Ramon

Departments of Physiology and Obstetrics and Gynecology, Duke University, Durham, North Carolina 22710

Spontaneously active smooth muscles have two basic components of electrical activity: slow, subthreshold, rhythmic fluctuation in membrane potential and action potentials. However, the amplitude, shape, frequency, and time course of subthreshold electrical behavior differs greatly among smooth muscles and even within a given type of smooth muscle (1–3). Prosser (3) has recently reviewed the classification of subthreshold electrical phenomena in spontaneously active smooth muscles, from which it seems clear that there are indeed distinctly different mechanisms associated with membrane potential oscillations in different types of smooth muscle. Furthermore, a given smooth muscle may have more than one mechanism for generating intrinsic membrane potential oscillations (1). The nature of these mechanisms, however, remains to be clearly defined. The relative contribution and interaction of electrogenic ion pumps (2) and slow voltage-dependent ionic conductance changes (1) during a spontaneous cycle of activity is at the center of the issue.

Tomita and Watanabe (1) have concluded that a basic metabolically driven pacemaker potential is present in all spontaneously active smooth muscles, which functions either to trigger a second voltage-dependent prepotential-type pacemaker or to initiate directly action potential activity and contraction.

Subthreshold electrical behavior in the myometrium may be classified into two basic components according to time course and function (4). One type occurs with a period varying between 10 sec and minutes and represents the pacemaker for contractile events. This slow pacemaker potential varies in amplitude from a few millivolts to 20 mV, depending on the condition of the preparation, hormonal environment, and method of recording (4; Fig. 1).

The second type of pacemaker potential appears in a narrow range of membrane potentials near the crest of the slow pacemaker potential where it functions as a prepotential-type pacemaker to trigger action potentials directly (4,5).

The question explored in this chapter concerns the functional relationship between the basic slow pacemaker potential and the state of activation of the ionic conductance mechanisms of the action potential.

MATERIALS, METHODS, AND CRITIQUE OF VOLTAGE-CLAMP

In all of these studies, strips of estrogen-dominated rat myometrium were used in the double sucrose-gap current-clamp and voltage-clamp system as described previously (6,7).

Efforts to investigate the ionic mechanisms of excitation in tissues of complex geometry (cardiac and smooth muscle), through double sucrose-gap voltage-clamp techniques, have yielded controversial results (8–10). Basically, the problem is to evaluate the effect of shunt, series, and access resistance with respect to the isopotential control of membrane contributing to net ionic current measurements.

Because the distribution and absolute magnitudes of these resistances within the preparation are unknown and may vary with time of exposure to sucrose, the rationale for justifying voltage-clamp procedures has rested on observations other than direct analysis of the contribution of tissue geometry to the observed results. As recently pointed out by Tarr and Trank (10), the rationale for accepting voltage-clamp data has frequently been the apparent good agreement of observed membrane currents with the predicted voltage-clamp results. However, theoretical considerations (9) have clearly demonstrated that major deviations from the command potential may exist in a "voltage-clamped" cable, and yet the waveform of net membrane current may not differ significantly from that predicted from a perfectly voltage-clamped preparation.

Tarr and Trank (10) have recently carried out an elegant series of experiments under voltage-clamp conditions in which the homogeneity of potential control of the nodal membrane was monitored with an intracellular microelectrode. Their results confirm the theoretical prediction that voltage control is poor during periods of transient change in membrane conductance and good during the steady holding potential and for small depolarizing steps.

The meaning of these results with respect to interpreting smooth muscle voltage-clamp data may be summarized as follows: (a) because of the likely event of nonisopotential control of nodal membrane during fast transient changes in membrane conductance, precise quantitative kinetic analysis of membrane currents seems unlikely; (b) experimental designs in which the voltage command is slow and of low amplitude optimizes the homogeneity of nodal membrane control; (c) the interpretation of voltage-clamp data on the fast transient inward current associated with action potential mechanisms must be viewed in qualitative terms and is subject to error in proportion to the degree of nonisopotential control of node membrane.

RESULTS

Current–Clamp Studies

The basic rhythm of excitation alternating with periods of rest is clearly seen in the myometrium (Fig. 1). Excitation cycles in the estrogen-dominated rat uterus occur within a period of a few seconds to minutes and are responsive to exogenous stimulants such as oxytocin (Fig. 1). The intrinsic pacemaker behavior of this excitation cycle is characterized by relatively large (5 to 20 mV), slow (10 to 60 sec), rhythmic, subthreshold fluctuations in the membrane potential which, during the crest of the depolarizing phase, are coupled to small threshold depolarizations which initiate bursts or trains of action potentials; i.e., each action potential within the burst is preceded by a prepotential depolarization which directly triggers the spike (4) (Fig. 1).

Action potential amplitude, maximal rate of rise and fall, and frequency have been shown to vary during the spontaneous burst (1,5,12). The frequency appears to be determined primarily by the rate of prepotential depolarization and by the level of membrane potential during the spontaneous slow pacemaker potential (5). It has long been recognized that there is a direct correlation between the mean level of membrane potential and the

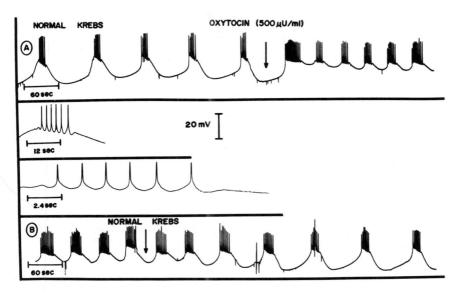

FIG. 1. A and B, continuous record of spontaneous electrical activity and response to oxytocin. Records with 12- and 2.4-sec calibration marks present the first spontaneous cycle in A at expanded time scale, and show in more detail the prepotential activity and spikes associated with the crest of the slow pacemaker potential.

frequency of action potential discharge and tension development (13). It is also clear from Fig. 1 that, although the slow pacemaker potential may have an amplitude of 20 mV, the threshold for spontaneous action potentials is restricted to a very narrow range of membrane voltage, of the order of a few millivolts, on the crest of the slow pacemaker potential. Slight hyperpolarization following an action potential may result in failure of the next prepotential to trigger an action potential. Characteristically this is followed by initiation of the repolarizing phase of the slow pacemaker potential without further prepotential activity (Fig. 1).

Decreased excitability of uterine smooth muscle to stimulating current pulses during the period between spontaneous bursts of action potentials has been clearly described by Casteels and Kuriyama (12) in the rat myometrium. This phenomenon of intrinsic modulation of membrane excitability is illustrated in Fig. 2. In this experiment, a constant current pulse was

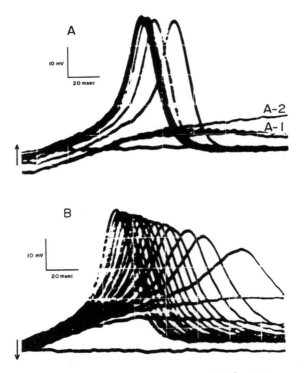

FIG. 2. Response of the node membrane to a constant current pulse (1/sec) during one cycle of the slow pacemaker potential. A: Response prior to the initiation of the depolarizing phase of the slow pacemaker potential, during the rising phase and at the crest. A-1 shows the electrotonic potential elicited during the inactive period between slow pacemaker potentials. A-2 shows increased electrotonic response just prior to initiation of the slow pacemaker potential. Succeeding action potentials appeared on the rapid rising phase of the slow pacemaker potential and assumed a stable amplitude and dV/dt at the crest of the slow pacemaker potential. B: Change in action potential shape and time course during the slow repolarizing phase of the slow pacemaker potential.

selected which failed to trigger a regenerative response during the period between spontaneous slow pacemaker potentials. Under these conditions, the stimulus elicited only electrotonic potentials between slow pacemaker potentials (A-1). As the period of the next spontaneous burst approached the magnitude of the electrotonic potential appeared to increase before any detectable depolarizing shift in the baseline (A-2). As the membrane potential entered the depolarizing phase of the slow pacemaker potential, the stimulus elicited action potentials of increasing amplitude and maximum dV/dt. In this experiment, the maximum depolarization of the slow pacemaker was approximately 5 mV from the resting potential between slow pacemakers.

The amplitude and maximum dV/dt of the action potential was transiently maintained during the crest of the slow pacemaker potential and then gradually decreased in amplitude and dV/dt as the membrane potential repolarized (Fig. 2B). At this point, the stimulus again elicited only non-regenerative electrotonic potentials until a few seconds before the next slow pacemaker potential.

If a stronger stimulating current pulse was applied in the period between spontaneous burst of action potentials, spikes could be triggered. However, under these conditions the intrinsic membrane potential oscillations are abolished.

Voltage-Clamp Experiments

The basic question to be explored in these voltage-clamp experiments concerns the functional relationship of the oscillating slow pacemaker behavior described above to the state of activation of the ionic conductance mechanisms of the action potential. We have long recognized in voltage-clamp experiments that the optimal holding potential for maximum inward current is frequently near the action potential threshold, although there is some variability in this observation. Furthermore, it is readily demonstrated that changes in the holding potential in either depolarizing or hyperpolarizing direction from this potential leads to inactivation of the transient inward current (Fig. 3). A hyperpolarizing inactivation curve similar to that in the uterus has been described for the transient inward current mechanism in *Aplysia* (19) and snail neurons (20). However, the interpretation differs: being attributed to the voltage dependence of the transient calcium current in *Aplysia* and to an early, transient outward potassium current in the snail neuron. In uterine smooth muscle, exposure to tetraethylammonium (TEA) does not increase the amplitude of the action potential or magnitude of the transient inward current, suggesting that an early transient increase in potassium conductance does not explain the hyperpolarizing inactivation (*unpublished observations*).

To evaluate further the relationship between oscillatory changes in the level

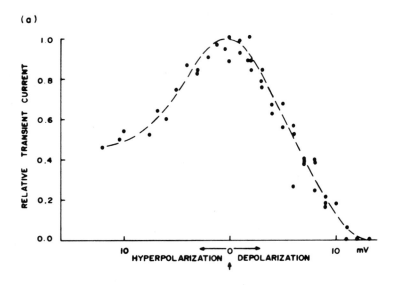

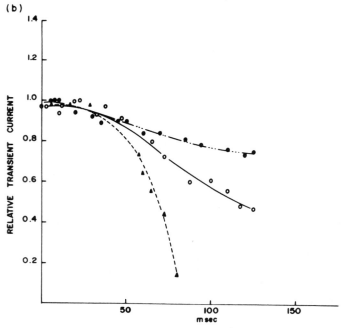

FIG. 3. Relationship between membrane potential and state of inactivation of the transient inward current mechanism. a: The data in this figure were obtained by either a standard double-pulse h^{∞}-type experiment, or by manually changing the level of the holding potential between test pulses. Both techniques give similar results. The zero value on the voltage axis is taken as the holding potential at which maximum inward current is obtained for a given test pulse. As described in the text, this value is taken as near spike threshold for preparations that have been driven electrically and near the crest of the slow pacemaker potential in spontaneously active preparations. b: Time dependence of the inactivation process as obtained from double-pulse experiments. In this case a conditioning pulse of variable length, followed by a standard depolarizing test pulse, showed inactivation of the time course indicated. (●) Hyperpolarization, 7 mV; (○) depolarization, 5 mV; (▲) depolarization, 7 mV.

of membrane potential and the state of activation of the action potential mechanism, the following experiments were carried out: under voltage-clamp conditions an oscillating voltage command signal, simulating the normal slow pacemaker in terms of amplitude and time course, was applied to the muscle. During the period of the cycle, brief (60 msec; 1 V/sec) depolarizing ramp voltage command signals were given at 2-sec intervals (Fig. 4). With the ramp-clamp technique as described by Palti and Adelman (15), it is possible to sample the entire transient or steady-state current–voltage relationship within a few milliseconds. Thus, having determined the parameters of a depolarizing ramp command signal which gives maximum negative slope conductances, it is possible to follow the relative state of activation of the transient current mechanism during the period of the simulated slow pacemaker potential. These experiments showed that slow, depolarizing potential changes, which drive the membrane potential toward the level of action potential threshold, produced activation of the regenerative inward current mechanism as judged by increased negative slope conductance and amplitude of inward transient current. In contrast, during the hyperpolarizing phase of the cycle, the peak transient current decreased.

DISCUSSION

The basic aim of this chapter is to investigate functional interaction of subthreshold slow pacemaker activity and action potential mechanisms in uterine smooth muscle.

Spontaneously active smooth muscles demonstrate subthreshold electrical behavior which functions to determine the period of contractions and to trigger action potentials. However, the variability of this behavior, even within a given type of smooth muscle, has been emphasized by several authors (1,2). Tomita and Watanabe (1) also carefully point out that the relative size of subthreshold oscillations and action potentials varies with the condition of the preparation and the method of recording.

In the present experiments, using the double sucrose-gap technique, two components of subthreshold membrane potential oscillations have been clearly observed. Relatively large (5 to 20 mV) slow to (10 to 60 sec) pacemaker potentials were demonstrated, which, during the crest of depolarization, appeared to trigger prepotential-type pacemakers which initiated action potential activity. This behavior is similar to that described by Kleinhaus and Kao (4) in the rabbit myometrium.

The basic hypothesis tested in these experiments was that the slow pacemaker potential has a unique role in controlling the intrinsic excitability of the smooth muscle cell, i.e., the state of activation of the action potential mechanism.

The observation by Casteels and Kuriyama (12) in the rat uterus clearly described the phenomenon of cyclic change in smooth muscle cell excitability.

However, their experiments tended to rule out the possibility that the action potential refractory period was the basis of this change in excitability.

The basic observation from voltage-clamp experiments, which suggests a functional role of slow pacemaker behavior in controlling action potential mechanisms, was that hyperpolarization of the membrane resulted in partial transient current inactivation (Figs. 3 and 4). These data suggest the possibility that the slow pacemaker potential forces the membrane potential to oscillate over a voltage range in which the ionic conductance mechanism of the action potential would be partially inactivated during the repolarizing phase and the inactivation gradually removed during the depolarizing phase.

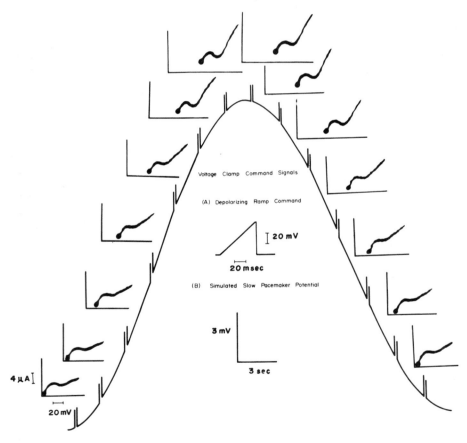

Voltage Clamp Command Signals

(A) Depolarizing Ramp Command

I 20 mV

20 msec

(B) Simulated Slow Pacemaker Potential

3 mV

3 sec

4 μA

20 mV

FIG. 4. Voltage-clamp experiment in which depolarizing ramp command signals were superimposed on an oscillating voltage command (simulated slow pacemaker potential). Current–voltage plots obtained from individual ramp commands are illustrated above each successive command (CRO in X–Y mode). Note the increased negative slope conductance associated with ramps near the crest of the simulated slow pacemaker potential.

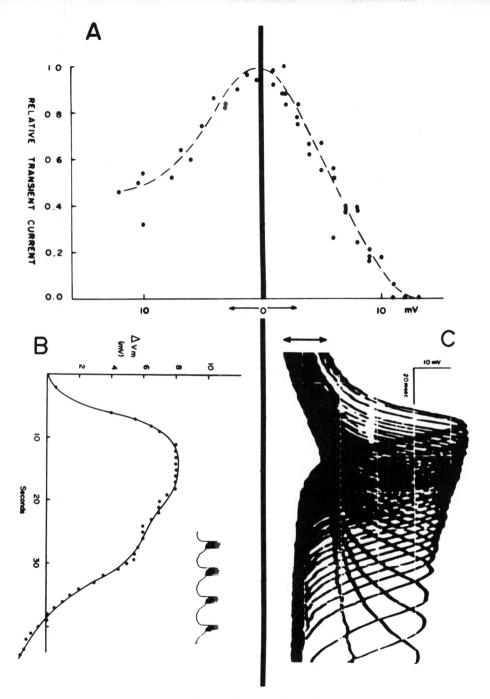

FIG. 5. A summary of the experiments discussed in text on the possible interaction of slow pacemaker potentials and action potentials. A: Voltage dependence of the transient inward current as described in Fig. 3. B: This curve was obtained by following the voltage change during a spontaneous slow pacemaker potential. Each point on the curve represents the membrane potential at that time during the slow pacemaker potential at which a depolarizing current pulse was applied. The shape of this slow pacemaker potential agrees well with that producing a spontaneous burst of action potentials (*inset*). C: The response to stimulating current pulses. In this case action potentials associated with the rising and falling phases of the slow pacemaker potential are superimposed and cannot be distinguished as in Fig. 2. This figure suggests that the slow pacemaker potential may contribute to the control of uterine smooth muscle excitability by influencing the state of activation of the action potential mechanism. The depolarizing inactivation curve would control the regenerative ionic event once the membrane potential passes threshold.

Hence, according to this scheme (Fig. 5), as the slow pacemaker potential approaches spike threshold, voltage-dependent prepotential pacemaker activity appears and triggers a spontaneous burst of action potentials. Repetitive spiking is maintained during the crest of the slow pacemaker potential, which is also the potential of maximum activation of the transient inward current mechanism.

The repolarizing phase of the slow pacemaker potential seems likely to involve the activation of an electrogenic ion pump (16). Furthermore, activation of this metabolically driven hyperpolarizing current source may be triggered by increased $[K^+]_o + [Na^+]_i$ owing to action potential activity and possibly to ionic conductances underlying pacemaker activity (17). According to this view, as electrogenic pump activity increases, the membrane repolarizes, driving the membrane potential below spontaneous spike threshold and at the same time inactivating the voltage-dependent prepotential-type pacemaker mechanism and partially inactivating the regenerative current mechanism of the action potential.

As briefly reviewed above and elsewhere (8), there is now substantial evidence, both theoretical (9) and experimental (10), that under voltage-clamp conditions the nodal membrane is not isopotential during fast transient changes in membrane conductance. Clearly, with the present voltage-clamp technology, one cannot unequivocally rule out the compromising effects of complex tissue geometry in these experimental results.

However, as recently pointed out by Morad and Goldman (18), "most investigators in the field are aware of these shortcomings and try to bypass or minimize them experimentally." Because the voltage clamp represents such a unique and powerful tool in our analysis of excitation mechanisms, the struggle to develop a totally defined system will surely have been worth the effort.

ACKNOWLEDGMENTS

This work was supported by NIH Grant HD-02742 and by General Research Support Grant RR 05405 from the General Research Branch, Division of Research Facilities and Resources, National Institutes of Health.

REFERENCES

1. Tomita, T., and Watanabe, H. (1973): *Philos. Trans. R. Soc. Lond.* [Biol. Sci.], 265:73.
2. Connor, J. A., Prosser, C. L., and Weems, W. A. (1974): *J. Physiol. (Lond.),* 240:671–701.
3. Prosser, C. L. (1974): In: *Proceedings of the Fourth International Symposium on Gastrointestinal Motility.*
4. Kleinhaus, A. L., and Kao, C. Y. (1969): *J. Gen. Physiol.,* 53:578.
5. Marshall, J. M. (1969): *Am. J. Physiol.,* 197:935.
6. Anderson, N. C. (1969): *J. Gen. Physiol.,* 54:145.
7. Anderson, N. C., Ramon, F., and Snyder, A. (1971): *J. Gen. Physiol.,* 58:322.
8. Johnson, E. A., and Lieberman, M. (1971): *Ann. Rev. Physiol.,* 33:479.

9. Kootsey, J. M., and Johnson, E. A. (1972): *Biophys. J.,* 12:1496.
10. Tarr, M., and Trank, J. W. (1974): *Biophys. J.* 14:627–643.
11. Ramon, F., Anderson, N. C., Joyner, R. W., and Moore, J. W. (1974): *Axon Voltage Clamp Simulation IV A Multicellular Preparation.*
12. Casteels, R., and Kuriyama, H. (1965): *J. Physiol. (Lond.),* 177:263.
13. Bülbring, E. (1962): *Physiol. Rev.,* 42:160.
14. Abe, Y. (1970): In: *Smooth Muscle,* edited by E. Bülbring, A. F. Brading, A. W. Jones, and T. Tomita. Williams and Wilkins, Baltimore, Md., p. 396.
15. Palti, Y., and Adelman, W. J. (1969): *J. Membrane Biol.,* 1:431–458.
16. Job, D. D. (1969): *Am. J. Physiol.,* 217:1534.
17. Bolton, T. B. (1973): *J. Physiol. (Lond.),* 228:693–712.
18. Morad, M., and Goldman, Y. (1973): *Prog. Biophys. Mol. Biol.,* 27:257–313.
19. Geduldig, D., and Gruener, R. (1970): *J. Physiol. (Lond.),* 211:217.
20. Neher, E. (1971): *J. Gen. Physiol.,* 58:36.

Physiology of Smooth Muscle, edited by
E. Bülbring and M. F. Shuba.
Raven Press, New York © 1976.

Transmembrane Ionic Currents in Smooth Muscle Cells of Ureter During Excitation

V. A. Bury and M. F. Shuba

Department of Nerve-Muscle Physiology, A. A. Bogomoletz Institute of Physiology, Kiev, USSR

Recently our ideas about the mechanisms generating the action potential of smooth muscle cells have been significantly enriched owing to the application of the voltage-clamp technique (1–10). Evidently, the solution of the problem to what degree Na^+ and Ca^{2+} ions participate in the generation of the action potential requires an individual approach for each kind of smooth muscle.

Therefore, in smooth muscle of taenia coli it has been suggested that the action potential is mainly due to an inward calcium current (5,11). In uterine smooth muscle the inward current during excitation is carried both by Ca and Na ions. But the quantitative contribution of each of these ions has not yet been finally established (1,2,3,6–9).

In our experiments we attempted to apply the voltage-clamp technique to analyze the mechanism that generates the action potential of ureter smooth muscle. Kobayashi (12,13) suggested that in cat ureter, Ca ions enter the fiber during the action potential in excess calcium solutions, whereas in a normal Krebs solution the dominating role in this process is played by Na ions. In the smooth muscle of the guinea pig ureter Kuriyama and Tomita (14) and Kochemasova (15) suggested that the plateau phase of the action potential is due to an increase in sodium conductance of the membrane, whereas the spike potentials on the plateau have a calcium origin.

Our voltage-clamp experiments (10,16–18) enabled us to gain some new support for this supposition.

METHODS

Experiments were performed on smooth muscle of isolated ureter of the guinea pig. Pieces of 1.2 to 1.5 cm length and about 400 μm diameter were mounted in a sucrose-gap chamber, similar to that described by Berger and Barr (19). In some experiments small longitudinal muscle strips from ureter (about 100 μm wide) were used. The width of the nodal area was 500 μm. The normal Krebs solution used in these experiments contained: NaCl, 133.0 mM; $NaHCO_3$, 16.3 mM; NaH_2PO_4, 1.38 mM; KCl, 5.0 mM; $CaCl_2$, 2.8 mM; $MgCl_2$, 0.1 mM; and glucose, 7.8 mM.

Sodium-free solution was prepared by replacing NaCl with sucrose. NaHCO₃ was also omitted and partly replaced with KHCO₃. The resulting composition of sodium-free solution was (mM): KHCO₃ 5.0; NaH₂PO₄ 1.4; CaCl₂ 2.8; MgCl₂ 0.2; glucose 7.8; and sucrose 295.8.

Calcium-free solution was made by replacing in Krebs solution CaCl₂ with equimolar NaCl.

In experiments with manganese, a Ringer-Locke solution was used containing (mM) NaCl 154.0; NaHCO₃ 1.8; KCl, 5.6; CaCl₂ 2.2; and glucose 5.6.

Tetrodotoxin (TTX) (Sankyo Co., Tokyo) was added to normal Krebs solutions to obtain a final concentration of 10^{-6} g/ml.

All solutions were maintained at 35°C.

For voltage clamping we used an ordinary circuit with differential feedback amplifier and operational amplifier for measurement of clamp current. In order to avoid the errors in voltage control due to the voltage drop in the solution and electrode resistances, the membrane potential was measured differentially. For this purpose an additional electrode was used, placed in the test compartment and connected to noninverting input of a high input impedance differential preamplifier.

Because many factors (such as diffusion potentials between neighboring solutions, incomplete prevention by the sucrose solution of short-circuiting between test and reference compartments, and the impossibility of estimating total membrane area involved in voltage clamping) limit quantitative analysis, we performed mainly a qualitative analysis of data gained by this method. We analyzed the changes of resting potential, membrane resistance, amplitude and time course of action potentials, transmembrane ionic currents, and equilibrium potential for inward currents.

As an illustration we present data obtained in one experiment showing the procedure regularly repeated in the whole series. All of the transmembrane current records and current–voltage relations are corrected for leakage current by summing currents from equal but opposite polarity clamp pulses. It is assumed here that leakage current is a linear and time-independent function of the membrane potential.

RESULTS

Experiments with Normal Krebs Solutions

Smooth muscle cells of guinea pig ureter do not exhibit spontaneous activity in normal Krebs solution. In response to extracellular stimulation they generate a complex action potential with a burst of discharges on a plateau phase (Fig. 1Ab).

Under voltage-clamp conditions a stepwise change of membrane potential resulted in an initial capacitive current followed by a slow transient inward current. Thereafter the current changed its direction and reached a nearly steady-state level (Fig. 1Ae). With larger voltage steps the slow inward current became larger and a fast component of transmembrane current in the shape of a train of damped oscillations was superimposed on the slow current (Fig. 1Af-h). The first peak of these oscillations always had an inward direction, followed by a fast outward current. Several alternate inward and outward currents followed. Finally, a slow inward current was followed by a slow outward current. Both the fast and the slow components of inward cur-

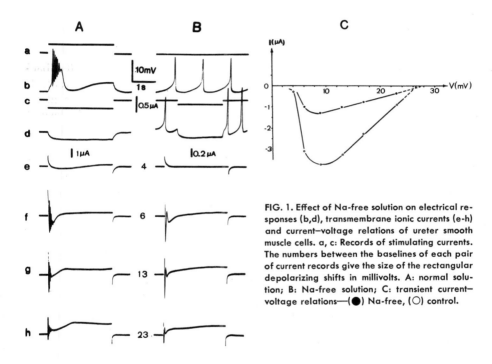

FIG. 1. Effect of Na-free solution on electrical responses (b,d), transmembrane ionic currents (e-h) and current–voltage relations of ureter smooth muscle cells. a, c: Records of stimulating currents. The numbers between the baselines of each pair of current records give the size of the rectangular depolarizing shifts in millivolts. A: normal solution; B: Na-free solution; C: transient current–voltage relations—(●) Na-free, (○) control.

rents are voltage-dependent and have different values of reversal potential.

As we can see from Fig. 1Ah, the slow current associated with the depolarization step of 23 mV was only outward, while the fast component current still had an inward direction. But we could not measure the reversal potential for the fast inward current directly because feedback currents associated with strong depolarization irreversibly damage the muscle cells.

The typical current–voltage relation plotted for the maximum fast inward current (for the first inward peak) is shown in Fig. 1C. This curve is similar to that of other excitable tissues with a region of negative resistance.

The reversal potential of the inward current can be estimated by extrapolation of the current–voltage relation before crossing with the voltage axis. Its magnitude is close to the amplitude of the action potential.

Effects of Sodium-free Solution

Immediately after changing Krebs solution for Na⁺-free solution a fast hyperpolarization of the muscle cells occurred, which was followed in 20 to 30 sec by a slow depolarization. After 5 to 10 min, the resting potential reached nearly steady-state level, which was 2 to 3 mV more positive than the initial one. These changes of membrane potential were associated with a significant increase of membrane resistance (about three times). We can see this from the increased amplitude of the electrotonic potential in Na-free solution (Fig. 1d). (Note different current scales for A and B in Fig. 1.)

As a rule, with the slow depolarization, spontaneous activity appeared in a pattern of single-action potentials without plateau, which can be inhibited by hyperpolarizing current. Action potentials, which were triggered by breaking the hyperpolarizing current, had a 3 to 5 mV larger amplitude than those in normal Krebs solution (Fig. 1Bd). The spontaneous activity continued for 15 to 20 min, but after cessation of spontaneous activity the muscle cells still remained excitable and for a long time action potentials could be evoked by extracellular stimulation.

In voltage-clamp conditions only the fast component of the transmembrane ionic current accompanied depolarizing shifts of the membrane potential (Fig. 1Be-h). Although the amplitude of action potentials was sometimes larger in Na-free solution than in normal Krebs solution, the transmembrane ionic currents in this condition were significantly decreased.

The current–voltage relations (Fig. 1C) show no essential changes in the reversal potential of the inward current in Na-free solution.

Effects of Tetrodotoxin

The experiments with TTX were undertaken to establish the existence of sodium channels in the membrane of ureter muscle cells specifically blocked by this toxin. The resting potential, membrane resistance, amplitude and time course of action potential, transmembrane ionic currents under voltage-clamp conditions, and the reversal potential for inward currents were measured. These experiments showed that TTX in concentration of 10^{-6} g/ml had no effect on any of the tested parameters. Figure 2 shows the

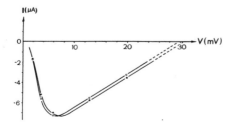

FIG. 2. Effect of TTX (10^{-6} g/ml) on transient current–voltage relations of ureter smooth muscle cells. (●) TTX; (○) control.

transient current–voltage relation for the fast inward currents plotted for one such experiment. Neither the amplitude of the inward current nor the reversal potential were affected by TTX.

Effects of Calcium-free Solutions

In Ca-free solution the membrane of ureter muscle cells became depolarized by 3 to 5 mV and the membrane resistance decreased by about 20%. Within the first minute in Ca-free solution the oscillations on the action potential plateau disappeared and the duration of the plateau became slightly shorter (Fig. 3Ba). Such altered action potentials can be triggered for 10 to 15 min in Ca-free solution.

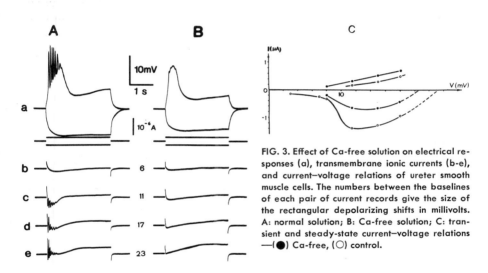

FIG. 3. Effect of Ca-free solution on electrical responses (a), transmembrane ionic currents (b-e), and current–voltage relations of ureter smooth muscle cells. The numbers between the baselines of each pair of current records give the size of the rectangular depolarizing shifts in millivolts. A: normal solution; B: Ca-free solution; C: transient and steady-state current–voltage relations —(●) Ca-free, (○) control.

Transmembrane ionic currents, measured during this period by voltage-clamping of the membrane, showed no fast component even with large depolarizing shifts of the membrane potential (Fig. 3Bb-e). Transient and steady-state current–voltage relations plotted for such an experiment are shown in Fig. 3C. Both the maximum amplitude of inward currents and the reversal potential decreased in Ca-free solution, while the steady-state outward current (upper curves) became larger. The threshold of inward current activation shifted toward more positive values of membrane potential.

Figure 4 shows simultaneous records of intracellularly recorded action potentials, triggered by extracellular depolarizing currents, and contractile activities of a ureter muscle strip. The disappearance of oscillations on the action potential plateau in Ca-free solution is associated with a significant decrease of the magnitude of contractions (Fig. 4b,c). As we can see from

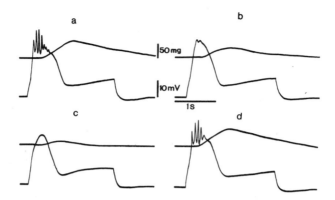

FIG. 4. Effect of Ca-free solution on electrical and contractile activity of ureter smooth muscle cells. a: Normal Krebs solution; b, c: 1st and 6th min of exposure to Ca-free solution; d: after 10 min washing with normal Krebs solution.

Fig. 4, this effect of Ca-free solution was fully reversible in normal Krebs solution.

Effects of Manganese

Effects of Mn²⁺ on the ureter smooth muscle cells depended on its concentration. In large concentrations Mn²⁺ irreversibly abolished action potentials. When the Mn concentration was low, the action potentials slightly increased in amplitude and, except the first spike, lost the oscillations on the plateau phase. It was possible to choose such a concentration of manganese

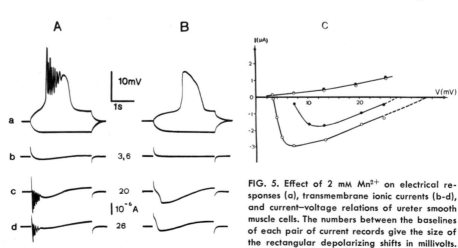

FIG. 5. Effect of 2 mM Mn²⁺ on electrical responses (a), transmembrane ionic currents (b-d), and current–voltage relations of ureter smooth muscle cells. The numbers between the baselines of each pair of current records give the size of the rectangular depolarizing shifts in millivolts. A: normal solution; B: solution with 2 mM Mn²⁺; C: transient and steady-state current–voltage relations—(●) 2mM Mn, (○) control.

(≈2 mM), that its effects on the ureter smooth muscle cells resembled those of Ca-free solution. One such experiment is illustrated in Fig. 5.

As well as Ca-free solution, Mn^{2+} abolished the oscillations on the plateau phase of the action potential and the fast component of transmembrane ionic currents, decreased the reversal potential for inward current and increased the threshold of its activation, and, finally, significantly decreased the magnitude of contractions (not illustrated). Contrary to Ca-free solution, the addition of 2 mM Mn to Ringer-Locke solution did not affect the resting potential, the membrane resistance, or the steady-state current–voltage relation.

Refractory Period

Smooth muscle cells of guinea pig ureter possess a very long relative refractory period. Complete recovery of the shape of the action potential takes

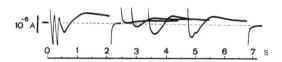

FIG. 6. Recovery of slow inward current depended on the interval between depolarizing pulses. Up to 5 to 7 sec, recovery of fast inward currents does not occur.

about 30 sec. Moreover, applying test stimulation at different time intervals, we observed first a gradual recovery of the slow potential, and then the appearance of repetitive discharge of spikes on its top. The number of these spikes depended on the interval between stimuli. In voltage-clamp conditions, if test depolarizing shifts were applied earlier than 500 to 600 msec after termination of the preceding one, then the depolarization was not accompanied by inward current (Fig. 6). Gradual increase of this interval resulted in the appearance, and then a gradual increase, of the slow inward current. The fast component of the transmembrane current began to recover only 5 to 7 sec after termination of the preceding shift of the membrane potential.

DISCUSSION

Application of the double sucrose-gap technique for voltage-clamping of smooth muscles is based on the utilization of the cable properties of smooth muscle. These properties have first been established for smooth muscle of frog stomach and cat intestine (20,21), and have then been confirmed for other kinds of smooth muscles (22–25).

Kuriyama et al. (24) showed that space constant of the ureter smooth muscle of guinea pig is 2.5 to 3.0 mm. Therefore, we thought it was possible to restrict the nodal area to 500 μm. In this case the calculated longitudinal nonuniformity of potential distribution does not exceed 2%. However, because the membrane resistance decreases during the active state, the nonuniformity increases. And so the question arose as to whether increased nonuniformity was the reason for oscillatory behavior of transmembrane currents observed in our experiments. There is no direct answer to this question, but, judging from some indirect evidence, we venture to hope that this is not so. Thus, Keynes et al. (26) also observed oscillatory transmembrane currents in barnacle muscle fibers in spite of the fact that there was demonstrably good longitudinal space clamp control. Also, barnacle muscle fibers generate action potentials, the shape of which resemble those of guinea pig ureter smooth muscle cells.

It seems to be a characteristic feature that multicomponent transmembrane currents appear, as a rule, in those excitable tissues which generate a complex action potential with plateau phase, for example, in frog atrial (27) and ventricular (28) muscle fibers.

Another possible reason for the oscillatory behavior of transmembrane currents may be a transverse nonuniformity of potential control due to the cleft resistances. The only way to exclude the effect of cleft resistance on voltage control is to use muscle strips of the smallest possible size. But in our experiments, results obtained with about 100-μm-wide muscle strips were qualitatively the same as the results obtained with pieces of ureter about 400 μm in diameter.

The results of our experiments suggest the existence of two systems with different kinetics transferring ionic currents through the membrane.

The slow inward current, which exists in Ca^{2+}-free solution and has a lower reversal potential than the fast one, is carried probably by sodium ions. It must be mentioned that the system transferring the sodium current is not indifferent to the external Ca^{2+} concentration. Exposure to Ca^{2+}-free solution for longer than 15 min not only inhibited the fast component of inward current having oscillatory behavior, but also greatly reduced the slow inward current. These late changes may be related to the decrease of Ca absorption in the membrane which, according to the hypothesis of Frankenhaeuser and Hodgkin (29), controls the resting membrane permeability to sodium and activates the regenerative process for the spike generation. The initial selective inhibition of the burst of discharge on the plateau phase and the fast inward currents in Ca-free solution may be related to the reduction of the extracellular ionized Ca^{2+} concentration, the half-time for which is less than 1 min (30–32).

Mn ions in low concentrations, competing with Ca for binding sites in the membrane, can probably partly replace Ca^{2+} for its stabilizing action, but cannot take part in carrying the current through the membrane.

Despite the fact that the inward current has a sodium component, this current is insensitive to TTX. In recent years many investigators have noted that resistance of the transmembrane currents to TTX cannot be taken as a proof of nonsodium origin (2,33,34). It is known that the blocking actions of TTX on the initial inward currents are not associated with the kinds of ions carrying this current, but with the molecular organization of the channels in the membrane (35,36). Accordingly, the resistance of ureter smooth muscle to TTX may be explained by peculiarities in the molecular organization of its membrane.

The fast component of the transmembrane ionic current, observed for a long time even in Na-free solution and inhibited by Mn ions, appeared to be carried by Ca ions. The reduction of the fast inward current in Na-free solution may be due to the significant decrease of ionic strength of the solution in which NaCl was substituted by sucrose and to the increase of the membrane resistance in those conditions.

The fact that the reversal potential for the slow inward current is lower than that for the fast one is also in accordance with the idea that, because of the low intracellular Ca concentration, the equilibrium potential for Ca^{2+} would be considerably higher than that for Na^{2+}.

The possibility of separating sodium and calcium currents suggests the existence of two different channels in the membrane. This assumption is also supported by the fact that the systems providing the slow and fast current transfer had very different times of recovery and different kinetic characteristics.

The experiments with simultaneous registration of electrical and contractile activities showed that the contractions depended on the entry of Ca^{2+} into the muscle cells during the generation of spike discharge on the plateau phase of the action potential. The inhibition of the fast transmembrane currents in Ca-free solution and in the presence of Mn^{2+} was always accompanied by a significant decrease in the magnitude of contraction of the muscle strip.

The question arose as to whether the inward Ca current can create the intracellular Ca concentration required for activating the contractile mechanism, of whether Ca^{2+} released from intracellular storage sites (such as sarcoplasmic reticulum) also takes part in this process.

Our measurements of transmembrane current enable us to estimate the intracellular Ca^{2+} concentration created by the inward calcium current. If we take the average magnitude of the first inward peak to be 2×10^{-6} A, and its duration -50 msec, then the electric charge carried by this current will be 10^{-7} C, which corresponds to 5×10^{-12} M of Ca^{2+}. The cell volume can be estimated from the tissue volume in the test compartment equal to 6×10^{-5} cm³ (400-μm diameter and 500-μm length). Taking into account the extracellular space of 35%, the cell volume will correspond to 4×10^{-5} cm³. Then the intracellular concentration created by the first peak of inward current will be equal to 5×10^{-12} mol/4×10^{-8} liter $= 1.2 \times 10^{-5}$ mol/liter. According to Goodford (37) the threshold concentration of Ca^{2+} for triggering contraction does not exceed 2.5×10^{-5} mol/liter. Therefore, two or three impulses of the fast inward current will be quite enough for triggering the contractile mechanism of the cell. In such conditions the role of sarcoplasmic reticulum appears to be insignificant. The possible small amount of Ca released from sarcoplasmic reticulum during excitation may be responsible for that negligible contraction of ureter smooth muscle which occurs in Ca^{2+}-free and in Mn^{2+} solutions. This suggestion is supported by the data of Devine et al. (38), which indicates that the tonic smooth muscles which contract even after prolonged exposure to Ca-free solution have a relatively large volume of sarcoplasmic reticulum, whereas the phasic muscles (to which ureter smooth muscle belongs) have a small amount of sarcoplasmic reticulum.

On the basis of the data presented in this paper it is tempting to speculate that in normal Krebs solution the role of the slow inward current carried by Na^{2+} is to develop depolarization to facilitate the activation of the system, transferring the fast calcium current. The inward calcium current, in turn, triggers the contraction of the muscle cell, which is the general aim of the whole process.

REFERENCES

1. Anderson, N. C. (1969): *J. Gen. Physiol.*, 54:145–165.
2. Anderson, N. C., Ramon, F., and Snyder, A. (1971): *J. Gen. Physiol.*, 58:322–339.
3. Kao, C. Y., McCullough, J. R., and Davidson, H. L. (1970): *Pharmacologist*, 12:250.
4. Kao, C. Y., McCullough, J. R., and Davidson, H. L. (1969): *Fed. Proc.*, 28:637.
5. Kumamoto, M., and Horn, L. (1970): *Microvasc. Res.*, 2:188–201.
6. Mironneau, J., and Lenfant, J. (1971): *C. R. Acad. Sci. (Paris)*, 272:436–439.
7. Mironneau, J., Lenfant, J., and Gargouil, Y. M. (1971): *C. R. Acad. Sci. (Paris)*, 273:2290–2293.
8. Mironneau, J., and Lenfant, J. (1972): *C. R. Acad. Sci. (Paris)*, 274:3269–3272.
9. Mironneau, J. (1973): *J. Physiol. (Lond.)*, 233:127–141.
10. Shuba, M. F., and Bury, V. A. (1971): *Proc. 1st Eur. Biophys. Congr., Baden*, pp. 265–268.
11. Brading, A., Bülbring, E., and Tomita, T. (1969): *J. Physiol. (Lond.)*, 200:637–654.
12. Kobayashi, M. (1965): *Am. J. Physiol.*, 208:715–719.
13. Kobayashi, M. (1969): *Am. J. Physiol.*, 216:1279–1285.
14. Kuriyama, H., and Tomita, T. (1970): *J. Gen. Physiol.*, 55:147–162.
15. Kochemasova, N. (1971): *Bull. Exp. Biol. Med.*, 9:9–12.
16. Bury, V. A., and Shuba, M. F. (1972): *Proc. 4th Int. Biophys. Congr. Moscow*, pp. 239–240.
17. Bury, V. A. (1973): *Fiziol. Zh. SSSR*, 10:1608–1613.
18. Bury, V. A., and Shuba, M. F. (1974): *Fiziol. Zh.*, 8:1288–1297.
19. Berger, W., and Barr, L. (1969): *J. Appl. Physiol.*, 26:378–382.
20. Shuba, M. F. (1961): *Biofizika*, 6:56–64.
21. Vorontzov, D. S., and Shuba, M. F. (1966): In: *Physical Electrotonus of the Nerves and Muscles.* Naukova Dumka, Kiev.
22. Tomita, T. (1966): *J. Physiol. (Lond.)*, 183:450–468.
23. Tomita, T. (1967): *J. Physiol. (Lond.)*, 189:163–176.
24. Kuriyama, H., Osa, T., and Toida, N. (1967): *J. Physiol. (Lond.)*, 191:225–238.
25. Abe, Y., and Tomita, T. (1968): *J. Physiol. (Lond.)*, 196:87–100.
26. Keynes, R. D., Rojas, E., Taylor, R. E., and Vergara, J. (1973): *J. Physiol. (Lond.)*, 229:409–455.

27. Gargouil, Y. M., Rougier, O., Coraboeuf, E., and Vassort, G. (1968): *Proc. Int. Union Physiol. Sci.,* 7:150.
28. Mascher, D., and Peper, K. (1969): *Pflügers Arch.,* 307:190–203.
29. Frankenhaeuser, B., and Hodgkin, A. L. (1957): *J. Physiol. (Lond.)* 137:218–244.
30. Schatzmann, H. J. (1961): *Pflügers Arch.,* 274:295–310.
31. Goodford, P. J. (1965): *J. Physiol. (Lond.),* 176:180–190.
32. Lammel, E., and Golenhofen, K. (1971): *Pflügers Arch.,* 329:269–282.
33. Kao, C. Y. (1972): *Fed. Proc.,* 31:1117–1123.
34. Narahashi, T. (1972): *Fed. Proc.,* 31:1124–1132.
35. Hagiwara, S., Hayashi, H., and Takahashi, K. (1969): *J. Physiol. (Lond.),* 205:115–129.
36. More, J. W., Blaustein, M. P., Anderson, N. C., and Narahashi, T., (1967): *J. Gen. Physiol.,* 50:1401–1411.
37. Goodford, P. J. (1967): *J. Physiol. (Lond.)* 192:145–157.
38. Devine, C. E., Somlyo, A. V., and Somlyo, A. P. (1973): *Philos. Trans. R. Soc., Lond. [Biol. Sci.],* 265:17–23.

Physiology of Smooth Muscle, edited by
E. Bülbring and M. F. Shuba.
Raven Press, New York © 1976.

Transmembrane Ionic Currents in Guinea Pig Myometrium During Excitation

Guy Vassort

Laboratoire de Physiologie Comparée, Université Paris XI, F 91405, Orsay

THE ELECTRICAL ACTIVITY OF GUINEA PIG MYOMETRIUM

The experiments were performed on small bundles, 70 to 150 μm in diameter, isolated from the uterus of estradiol-treated guinea pigs. These bundles were set up in a double sucrose-gap apparatus in which the test compartment is about 100 μm wide. The electrical circuits allow both current-clamp and voltage-clamp experiments (1).

The electrical activity of the smooth muscle cells is generally auto-regenerative. A typical action potential (AP) is shown in Fig. 1. It has a

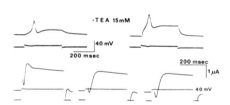

FIG. 1. Electrical activity of smooth muscle cells and the effects of TEA (15 mM). The action potential (upper row) obtained in Krebs solution is, after 10 min application of TEA, increased in amplitude and slower in its repolarization, with no undershoot. Under voltage-clamp conditions a 40-mV depolarization elicits an inward current followed by a hump (lower row, left). The two other recordings illustrate the action of TEA after 3 min (lower row, middle) and in steady state after 10 min (lower row, right). The amplitude of the inward current is enhanced while the hump (fast outward current) disappears.

rather low amplitude of 50 to 60 mV. It has typically a fast rate of repolarization, sometimes faster than the rate of depolarization, and is followed by a short undershoot. The shape of the AP varies over a wide range, not only in the initial spike component, but also in the long plateau that follows, in which the amplitude is sometimes very low, or can reach 40 mV.

Voltage-clamp experiments were carried out with the holding potential at the resting potential. Thus, in the following description, E is the absolute membrane potential and V is the displacement from the resting potential E_r. In normal solution several ionic currents could be distinguished. At least three different currents overlapped, so that the situation was very complicated.

THE EFFECTS OF TETRAETHYLAMMONIUM CHLORIDE IONS ON THE ELECTRICAL ACTIVITY

In the presence of tetraethylammonium chloride (TEA) 15 mM, large spikes with a slow repolarization phase were elicited and the undershoot was abolished (Fig. 1). The lower set of records in Fig. 1 shows the evolution of the ionic currents for a given depolarization (40 mV) during the action of TEA. Two main observations were made: the inward current increased in amplitude while the hump (fast outward current) was abolished. After less than 10 min the situation was stable and an analysis of the inward current and of the late current, slowly decreasing with time, was then possible.

CALCIUM INWARD CURRENT

The ionic nature of the early inward current has already been investigated (2–8). My results support the hypothesis that, in normal conditions, this current is carried mainly by Ca ions. It is insensitive to TTX, but is inhibited by Mn ions (5 mM) or D 600 (0.22 mM); it is unchanged in Na-deficient solution, whatever the substitute (Li, K, or sucrose), but is strongly dependent upon the external Ca concentration.

SODIUM- AND POTASSIUM-DEPENDENT CURRENT

The Ca-inward current was followed, for depolarizations larger than 30 mV, by a slow outward current for which some characteristics are depicted in Fig. 2. It reached a peak value after about 150 msec and then inactivated with a time constant of 390 msec for the 38mV depolarization imposed. The right side of the figure shows that the reversal potential of this current occurred at 18 mV depolarization (i.e., E_{rev} about −35 mV).

When the ionic composition of the bathing solution was changed, evidence was obtained that Na and K ions are the main charge carriers of this current, since E_{rev} was less negative in a K-rich medium, and more negative in a

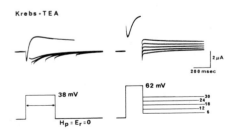

FIG. 2. Some characteristics of the Na- and K-dependent current in TEA solution. Left: Time course (time to peak and inactivation) of this current. The inactivation time constant is determined from the instantaneous inward tails when clamping back to the resting membrane potential after different pulse durations ($H_p = E_r$). Right: The reversal potential of this current is about −35 mV(V = +18 mV).

Na-poor medium. This is an observation comparable to that of the synaptic reversal potential at the neuromuscular junction. Though the Cl equilibrium potential (as determined from tracer experiments) is in this potential range, removing Cl ions did not noticeably modify this current. Actually, the Na- and K-dependent current was most clearly seen in fibers whose AP exhibited a large plateau.

EXISTENCE AND CHARACTERISTICS OF THE FAST OUTWARD CURRENT

In normal solution, the inward current was followed by a hump which increased in amplitude according to the amplitude of the depolarization

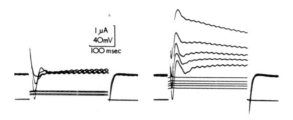

FIG. 3. Typical records of the ionic currents obtained in normal solution under voltage clamp. Notice the occurrence of a fast outward current after the calcium inward current for a 15-mV depolarization and its larger amplitude with further depolarizing steps.

(Fig. 3). Its nature is a subject of controversy. It might be of a purely artifactual origin (damped oscillation). However, as shown in Fig. 1 the transient outward current was suppressed by TEA. Further characteristics of this current are as follows.

Reversal Potential

After a 64-mV-depolarizing pulse lasting for 40 msec, a duration at which the hump is maximal, different potential steps were imposed (Fig. 4). Because of the large capacitive current which occurred with large hyperpolarizations, the inversion of this current was not clear. However, the plot of the amplitude of the tail currents vs. the amplitude of the test potential suggested a reversal potential at about -95 mV ($V = -40$ mV). Thus, the transient current responsible for the hump is probably carried by K ions. This interpretation was supported by experiments in K-rich media (KCl $\times$ 10, NaCl partly substituted). When the membrane potential was held at the original resting membrane potential before exposure to high K, applied depolarizations elicited currents which differed from those in the normal solution mainly by a smaller hump and by a smaller Na–K-dependent current. Using the same experimental procedure, an inversion of the fast tail currents was obtained, the reversal potential appearing at about -45 mV ($V = 8$ mV), suggesting a shift of 50 mV for a 10-fold increase in the external K concentration. It will be noted that in high-K solution the K conductance is larger.

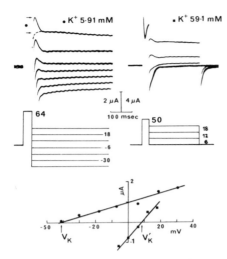

FIG. 4. Reversal potential of the transient outward current in normal K (5.91 mM) and in K-rich (59.1 mM) solutions. The direction and the amplitude of the fast tail current in these double-step experiments allowed the determination of the reversal potential of the current flowing at the end of the first step. The plots (including one other recording) of the maximal amplitude of the tail currents (corrected for the capacitive current) versus the amplitude of the test pulses show that the reversal potential is $E_K = -95$ mV ($V_K = -40$ mV) in normal K solution (●) and $E'_K = -47$ mV ($V'_K = 8$ mV) in K-rich solution (■).

Availability of the Transient Current

The occurrence of a fast outward current just after or simultaneously with the Ca-inward current is further supported by studying the availability of the inward current). Examples of the influence of a conditioning step on the inward current elicited by a given depolarization (43 mV) are illustrated in Fig. 5A. The experiment was performed in the presence of TEA. Condition-

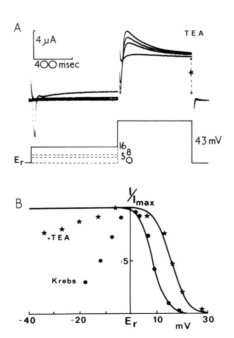

FIG. 5. Voltage dependence of the availability of the measured inward current. A: Typical recordings of ionic currents obtained for a 43-mV depolarization in TEA solution. This illustrates the influence of a conditioning polarization which modifies the availability of the inward current (and of the Na–K current). B: The amplitudes of the measured inward current are plotted versus the conditioning polarizations as the relative values of the current recorded in the absence of conditioning polarization. (●) Normal solution; (*) TEA solution. The curves are drawn according to the equation $I/I_{max} = 1/1 + \exp[(V_H - V)/K]$, where $V_H = 8.5$ mV and $K = -2.5$ in normal solution, and $V_H = 15.5$ mV and $K = -3$ in TEA solution.

ing depolarizations of 5, 8, and 16 mV decreased and then abolished the inward current (and also the Na–K-dependent current). Conditioning hyperpolarizations slightly decreased its amplitude. Similar experiments were performed in Krebs solution. The results, expressed as the relative value to the current recorded without conditioning pulse, are plotted in Fig. 5B. Conditioning depolarizations inactivated the inward current as in TEA, but the inactivation of the Ca conductance was very much greater for conditioning hyperpolarizations. This can be explained if we assume that a previous hyperpolarization, though ineffective on the calcium current itself, makes a transient outward current more available. The Ca-inward current is more or less masked by this fast outward current which occurs nearly at the same time. This is shown by comparing the current recordings in normal and in TEA solution shown in Fig. 1. The difference between the two recordings gives an idea of the time course of the transient outward current in this condition.

CONCLUSIONS

A transient current carried by potassium ions, similar to that found in the present investigation, has been described in gastropod neurons (9,10) and in crab muscle fibers (11), cells which also exhibit calcium spikes. This outward current accounts for the relatively low amplitude of the AP and its fast repolarization, while in TEA solution, when this current is inhibited, the amplitude of the AP is increased and the repolarization slowed. Moreover, in TEA, after conditioning hyperpolarization, action potentials of nearly 90 mV can be elicited.

ACKNOWLEDGMENT

This work was carried out during the tenure of a European Science Exchange Programme at the University Department of Physiology, Oxford, and has been supported by the Medical Research Council.

REFERENCES

1. Vassort, G. (1975): *J. Physiol.,* in press.
2. Anderson, N. C. (1969): *J. Gen Physiol.,* 54:145–165.
3. Kumamoto, M., and Horn, L. (1970): *Microvasc. Res.,* 2:188–201.
4. Anderson, N. C., Ramon, F., and Snyder, A. (1971): *J. Gen. Physiol.,* 58:322–339.
5. Kao, C. Y. (1971): In: *Research in Physiology,* edited by F. F. Kao, K. Koizumi, and M. Vassalle, pp. 365–372. Aulo Gaggi.
6. Mironneau, J., and Lenfant, J. (1971): *C. R. Acad Sci. Paris,* 272:436–439.
7. Shuba, M. F., and Bury, V. A. (1971): *Proc. 1st Eur. Biophys. Congr.,* pp. 265–268.
8. Vassort, G. (1973): *J. Physiol.,* 237:50–51P.
9. Connor, S. A., and Stevens, C. F. (1971): *J. Physiol.,* 213:31–53.
10. Neher, E. (1971): *J. Gen. Physiol.,* 58:36–53.
11. Mounier, Y., and Vassort, G. *J. Physiol.,* in press.

Physiology of Smooth Muscle, edited by
E. Bülbring and M. F. Shuba.
Raven Press, New York © 1976.

Voltage-Clamp Studies on Rat Portal Vein

C. Daemers-Lambert

Département de Biologie Générale, Institut E. Van Beneden, B-4000 Liège, Belgium

More voltage-clamp information about smooth muscle would lead to a better knowledge of its physiology and pharmacology. At present, few satisfying and clear results have been obtained, because smooth muscle is far from providing the theoretical conditions for an ideal voltage clamp. The uterine muscle seems to be the best tissue for making observations most easily (as described elsewhere in this volume).

A few years ago, I became interested in membrane control of the tonic contraction of some vascular smooth muscle the ATP and phosphorylcreatine metabolisms of which had been studied (1). In 1970, with Dr. Vassort, we tried to record the action potential and ionic currents from longitudinal strips of rat portal vein, using Vassort's double sucrose-gap method which has been used for cardiac muscle (2). Results were unsatisfactory: graded action potentials of scarcely 30 mV amplitude, and, under voltage-clamp, a small inward current and a large leak current.

Since this time, the purpose of my investigations has been to find experimental conditions in which an all-or-none action potential could be recorded by the double sucrose-gap as it was by microelectrode (3,4). Conditions in which a functional muscle preparation may be obtained have finally been discovered. Results on rat portal vein will be compared with those obtained on earthworm muscles, which also have calcium spikes (5,6).

EXPERIMENTAL CONDITIONS

Longitudinal strips of rat portal vein between 100 and 200 μm wide were able to contract spontaneously for a few hours in normal physiologic solution at temperatures greater than 20°C without oxygen. Dissected bundles of these dimensions were incubated for 3 to 4 hr at 18°C in a physiologic solution of the following composition (mM): NaCl 131, KCl 5.6, NaHCO$_3$ 11.9, NaH$_2$Po$_4$ 1.2, and MgCl$_2$ 1 or 2; it is known that a solution without calcium and containing 1 to 2 mM MgCl$_2$ prevents the depolarization which occurs in a Ca-free medium as shown by Bülbring and Tomita (7). In this way full relaxation was quickly obtained, but stored calcium seemed to exchange with magnesium very slowly. This treatment was necessary to avoid contracture and to have a good electrical isolation of node and KCl pools by the sucrose gaps. Another necessary condition was a relatively low temperature (18° to 20°C), which allowed a better isolation with the Vaseline seals and prevented the solutions from mixing with one another.

Following incubation in this solution, the muscle strip was placed in the double sucrose-gap apparatus and the node region perfused with a test solution, composition as follows (mM): NaCl 131; KCl 5.6; $NaHCO_3$ 11.9; NaH_2Po_4 1.2; $SrCl_2$ or $CaCl_2$ 4.0 or 4.6; and tetraethylammonium chloride (TEA) 10 or 20. All-or-none action potentials could be recorded only when TEA was present. Calcium or strontium were necessary to support the spike or inward current. The experiments were performed in normal sodium concentration, although spikes could be recorded in Na-free solution with sucrose substitution; no record could be obtained with Tris-chloride substitution.

RESULTS

Rat Portal Vein

Under voltage clamp, there were some variations in the ionic currents which could be interpreted as resulting from different levels of the resting permeability of the membrane to potassium and from different calcium contents of the preparations. In an attempt to alter these results, the Mg concentration in the incubating solution and the TEA concentration or the presence of Ca or Sr in the test solution were changed.

Strontium Spikes

Figure 1 shows a recorded action potential and ionic currents in a muscle preparation, which was first incubated with 1 mM $MgCl_2$ but without calcium, for 4 hr. Following this a test solution containing 4 mM $SrCl_2$ and 10 mM TEA was used. The current–voltage relations were plotted (potential measured as displacement from resting level) at 50, 100, 150, and 200 msec clamp duration without any correction for leakage current. The drawn curves are only indicative.

The action potential had an amplitude of 35 mV and 200 msec duration. Under voltage clamp a 10-mV depolarizing pulse elicited an early inward current which increased in amplitude to a maximum at 50 msec pulse duration, and was inactivated at 100 msec in such a way that there was no net current at this time.

Measured at 50 msec under voltage clamp at 33-mV depolarization, there was no obvious current, as shown in the current–voltage relation in Fig. 1. The reduction and reversal of the fast inward current was associated with the activation of outward current which may have become progressively earlier with increasing depolarization. The apparent late inward current, appearing at 150-msec pulse duration and at 10- and 20-mV depolarization, could be a

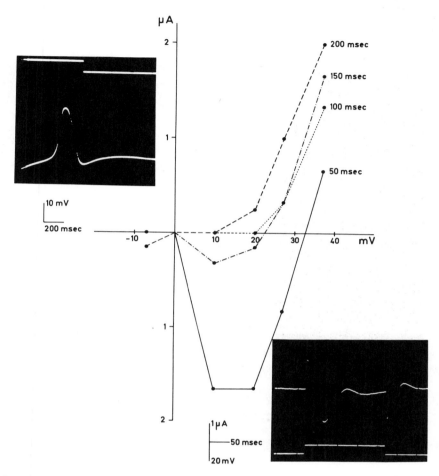

FIG. 1. Rat portal vein: action potential and membrane currents in response to a 10-mV depolarizing impulse; corresponding current–voltage relations (18° C). The incubation solution was Ca-free and contained 1 mM $MgCl_2$. The test solution contained 4 mM $SrCl_2$ and 10 mM TEA chloride.

fall due to the closure of potassium channels responsible for the resting membrane conductance, or a reopening of inward current channels despite the constant potential (see also Fig. 5).

These results were confirmed in a second type of record (Fig. 2), obtained with the same experimental conditions, but in the presence of 20 mM TEA. Capacitive current was larger and faster and the peak in inward current occurred earlier (25 msec). Again as inward current declined and then reversed with stronger depolarization, outward current appeared progressively earlier. It will be noted that reversal of inward current and the peak of the spike coincided at 30 mV depolarization. Investigations over the hyper-

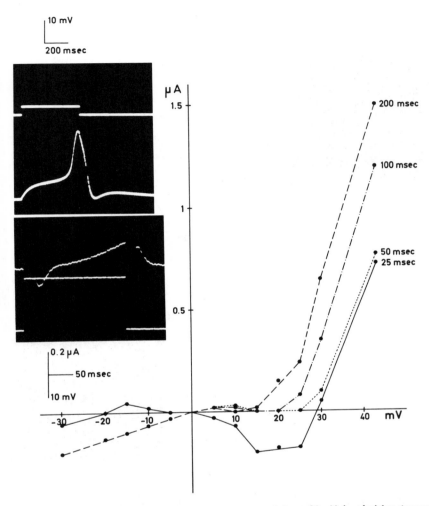

FIG. 2. Rat portal vein: action potential and membrane currents during a 20-mV depolarizing step; corresponding current–voltage relations (18° C). The incubation solution was Ca-free and contained 2 mM MgCl₂. The test solution contained 4 mM SrCl₂ and 20 mM TEA.

polarizing range showed an early outward current, which was deactivated quickly and reversed its sign at −20mV hyperpolarization.

Calcium Spikes

In the presence of 4.6 mM calcium and 20 mM TEA (Fig. 3) an all-or-none action potential could be elicited of 45-mV amplitude and 200-msec duration, with the already described features in voltage clamp. Moreover, it

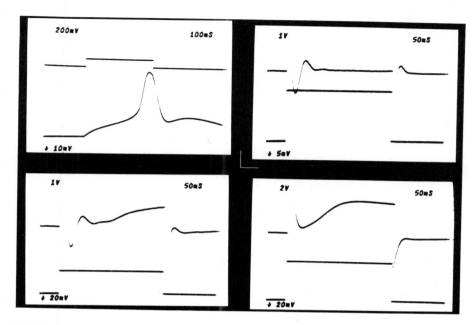

FIG. 3. Rat portal vein: action potential and membrane currents. The incubation solution was Ca-free and contained 1 mM MgCl₂. The test solution contained 4.6 mM Ca²⁺ and 20 mM TEA. The figures on the panels give the oscilloscope calibrations per division.

gave an early outward current which appeared at 59-msec pulse duration as a hump after the apparent inactivation of the inward current.

The value of the maximal amplitude of the outward current hump measured at 59 msec was the same as the value seen after 28 msec with a 47-mV imposed depolarization where inward current had reversed (Fig. 4). Therefore, it is believed that there may be only one kind of potassium channel controlling the permeability of this membrane and showing depolarizing inactivation over a limited and defined voltage range, i.e., the range of the action potential. This early outward current, which is typical of many voltage-clamp results on smooth muscles, reflects the decay of a potassium current already activated at the resting membrane potential. The long latency of the spike when an imposed current is applied, is explained by slow potassium inactivation which must take place before opening the gate further for inward ionic current.

Earthworm Muscles

The following results were obtained with earthworm muscles which were studied to obtain more information about calcium spikes under better voltage-clamp conditions.

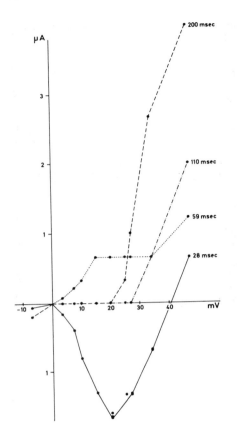

FIG. 4. Rat portal vein: current–voltage relations corresponding to the results illustrated in Fig. 3.

The retractor muscle of the bristles (setae) of earthworm (*Lumbricus terrestris*), which is a little muscle 50 to 150 μm wide and 2 or 3 mm long, can be dissected easily without injury. If the muscle is allowed to reach total relaxation, which takes a few hours in physiologic solution, an all-or-none action potential can be recorded. It is 70-mV amplitude and less than 100-msec duration (Fig. 5, top).

An inward current starts five times faster than in the best case on rat portal vein, with good resolution of capacitative currents. It is followed by delayed activation of the outward current. In some muscle, only graded responses can be recorded (Fig. 5, bottom). In this case voltage-clamp pictures also showed a hump following the peak inward current before activation of the delayed outward current.

Conditioning hyperpolarization increased both early and late outward current in such a way that the initial inward current was less. A 15-mV conditioning depolarization gave the maximum inward current at the test pulse, although a partial activation of this inward current had already occurred. A 25-mV conditioning depolarization inactivated the fast inward current of the

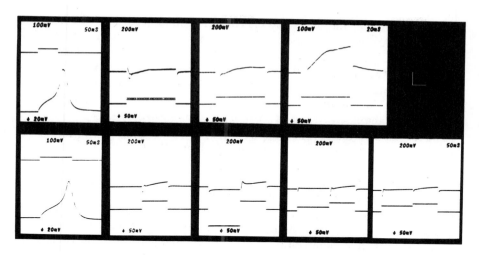

FIG. 5. Setae retractor muscle of the earthworn. Top: All-or-none action potential and associated membrane currents. Bottom: Graded action potential and associated membrane currents. Effect of a conditioning polarization of 120 msec duration. The test solution contained (mM): NaCl 140, KCl 2.5, NaHCO$_3$ 2.4, and CaCl$_2$ 1.8. The figures on the panels give the oscilloscope calibrations per division.

test pulse. Such a conditioning depolarization abolished the early outward current hump of the test pulse.

Thus, the amplitude of the inward current depended on the steady-state potassium conductance of the membrane. If the resting potassium permeability is high, potassium inactivation upon depolarization cannot take place before inactivation of the inward current occurs. In this case, no spike can be elicited.

DISCUSSION

In conclusion, the calcium spike in these smooth muscles requires a low intracellular calcium content associated with a low resting potassium permeability as was shown by Hagiwara et al. on crustacean muscles (8). This is particularly crucial when the double sucrose-gap recording technique is used; it is likely, when resting potassium conductance is high, that there is some variation in potential between the various fibers of the node.

The effect of TEA on rat portal vein fibers remains unclear. As in crustacean muscle (9) its pharmacologic action is very fast. It is possible that TEA selectively reduces the resting potassium permeability and delays the outward K current activation, but it does not seem to affect the late outward current itself. Spike prolongation is rarely seen.

The very small leak current after incubation in Mg-containing Ca-free saline for several hours is believed to be caused by a decrease of the membrane chloride conductance. This may be brought about by a loss of chloride from the interior of the fibers associated with a modification of the Donnan equilibrium. An increasing content of nonpermeant anions in the fibers may occur. Indeed, it has been shown that smooth muscle myosin has a high negative charge at neutral pH (10), due to a high content in dicarboxylic amino acids, and that Ca-free medium leads to a solubilization of the myosin filaments (11). This would increase the intracellular concentration of soluble nonpermeant anion.

The results described in this chapter are believed to explain some peculiarities of those smooth muscles. Under current clamp, the threshold of the active response requires that more current must be supplied if the resting potassium permeability is high. Threshold can be reached when depolarizing inactivation of the potassium channels responsible for the resting permeability has taken place. At this time calcium ions penetrate the channel without competing with potassium ions, giving all-or-none responses. Otherwise, graded responses are produced. Further depolarization quickly opens the gates for potassium ions and, as a consequence, seems to close the calcium channel.

Under voltage clamp, instantaneous depolarization does not allow slow potassium inactivation. This is why preparations which give all-or-none spikes can show, under voltage clamp at some low imposed depolarizations, the appearance of an early outward current ("hump") which falls later because the potassium conductance inactivates.

ACKNOWLEDGMENT

This work and symposium participation have been supported by the Fonds National de la Recherche Scientifique de Belgique.

REFERENCES

1. Daemers-Lambert, C. (1969): *Angiologica* (Symp. Int. Biochim. Paroi Vasc., Fribourg, 1968), Part II, pp. 1–17 (185–201).
2. Rougier, O., Vassort, G., and Stämpfli, R. (1968): *Pflügers Arch.*, 301:91–108.
3. Golenhofen, K., and Loh, D., v. (1970): *Pflügers Arch.*, 319:82–100.
4. Ito, Y., and Kuriyama, H. (1971): *J. Physiol. (Lond.)*, 214:427–441.
5. Hidaka, T., Ito, Y., and Kuriyama, H. (1969): *J. Exp. Biol.*, 50:387–403.
6. Ito, Y., Kuriyama, H., and Tashiro, N. (1970): *J. Exp. Biol.*, 52:79–94.
7. Bülbring, E., and Tomita, T. (1970): *J. Physiol. (Lond.)* 210:217–232.
8. Hagiwara, S., Hayashi, H., and Takashi, K. (1969): *J. Physiol. (Lond.)*, 205:115–129.
9. Hagiwara, S., Chichibu, S., and Naka, K. I. (1964): *J. Gen. Physiol.*, 48:163–179.
10. Hamoir, G. (1969): *Angiologica*, 6:190–226.
11. Fay, F. S., and Cooke, P. H. (1973): *J. Cell Biol.*, 56:399–411.

Physiology of Smooth Muscle, edited by
E. Bülbring and M. F. Shuba.
Raven Press, New York © 1976.

Spontaneous Activity and Functional Classification of Mammalian Smooth Muscle

K. Golenhofen

Department of Physiology, University of Marburg/Lahn, West Germany

DIFFERENTIATION IN EVOLUTION

The evolution of motor processes can be taken as a natural guideline for a functional classification of smooth muscle. The simplest type of motility is the protoplasmic movement which occurs in protozoa in connection with the development of special contractile proteins (molecular specialization, step 1 in Table 1) (1). Step 2 is characterized by the development of organelles such

TABLE 1. *Classification of motility*

Step	Differentiation	Motor element
1	Protoplasmic movement	Molecular specialization
2	Myoid: cilia and flagella	Cell organelle
3	Full myogenic automaticity	Single muscle cell, group of similar cells
4	Differentiated myogenic automaticity	Many muscle cells with different specialization
5	Nervously controlled movement	Muscle and nerve cells

as cilia and flagella which are specialized for motility. They can also act after they have been isolated from the cell (2). The occurrence of whole cells specialized for movement signifies step 3: full myogenic automaticity. The single muscle cell, or a group of similar cells, is in this case the "motor element," which means the minimal prerequisite for a full motor act, including generation of excitation. Examples are the many types of intestinal smooth muscle and some types of vascular smooth muscle. Differentiation between muscle cells within a muscular tissue characterizes the step "differentiated myogenic activity" (step 4). Examples are heart muscle for striated muscle, and the pyeloureteral system for smooth muscle. At the last step muscle cells have become specialized for contraction only, as generation of excitation and control is fully delegated to the nervous system. Such specialization, which is typical for skeletal muscle, can also be found in the smooth muscle system (vas deferens, some blood vessels, etc.).

One dimension for a functional classification is evident in evolution: the degree of automaticity which becomes smaller over a range extending from full myogenic automaticity via differentiated myogenic automaticity to full nervous control, with many intermediate forms, of course.

QUALITY OF AUTOMATICITY

The second dimension is the quality of automaticity, which I would like to describe in more detail, as we have investigated this in comparative studies with many types of smooth muscle (steps 3 and 4).

The fastest rhythmic process of smooth muscle is the spike potential, an electrical impulse of 10- to 50-msec duration at half amplitude. The intrinsic frequency of the spike is visible in the action potential of guinea pig ureter muscle, as shown in an intracellular recording in Fig. 1. A series of spikes

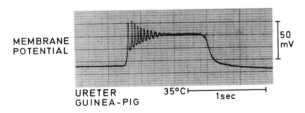

FIG. 1. Action potential of guinea pig ureter, intracellular recording of the membrane potential (3).

with a frequency of 10 to 30/sec occurs as a damped oscillation on top of a plateau depolarization. The next slower rhythm is shown in Fig. 2, an intracellular recording from guinea pig taenia coli. There are sinusoidal oscillations with a preferred period duration of about 1 sec, which are therefore called "second-rhythm" (SR). The intrinsic frequency could be demonstrated by a resonance effect during rhythmic stretch (4,5). Spikes appear to be triggered by the SR oscillations. The basis for these two fast events is called the "basic membrane potential." During long-term recording under normal conditions the basic membrane potential of taenia coli shows slow fluctuations with a period duration around one minute, called "minute-rhythm" (MR) (6). This modulates the spike discharges and consequently the tension development (Fig. 3A). An even slower type of rhythmical activity is well pronounced in guinea pig uterus, as shown in Fig. 3B, and it is also present in the fundus of guinea pig stomach (7). This slow rhythm can be called "hour-rhythm" (HR). In addition to these rhythms we find other types of rhythmic activity of more organ-specific character, such as stomach peristalsis, segmentation rhythm of the small bowel, ureteral peristalsis etc. They lie with their frequency between SR and MR, and we summarize this group under the term "basic organ specific rhythms" (BOR). Figure 4 shows a very pro-

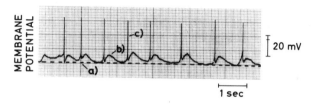

FIG. 2. Intracellular recording of membrane potential in guinea pig taenia coli. (a) Basic membrane potential; (b) second-rhythm; (c) spike.

nounced rhythm of this type, the peristaltic activity in the antrum of guinea pig stomach. Large fluctuations of the basic membrane potential, 30 to 40 mV in size, characterize this BOR. All described rhythms are myogenic in origin; they can be measured in isolated preparations and also when nervous activity is suppressed, for example by tetrodotoxin.

A spectrum of all rhythms is shown in Fig. 5, with a diagrammatic illustration of some of their frequency characteristics. A multimodal histogram of period duration with harmonic relations between the peak values is typical for the MR. The frequencies of BOR and SR are species-specific (for MR probably not), and values from the guinea pig have been taken for the figure. A shift to the right occurs with larger species; for man the period durations have to be multiplied by about 2. It is characteristic for BOR that the frequency of a single rhythm is strictly fixed, e.g., stomach peristalsis, but a rather great variation exists in the whole group. There is some evidence that in the guinea pig the specially indicated value of 6/min is a sort of fundamental value for the whole group (8). Spike intervals vary over a wide

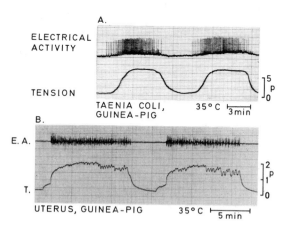

FIG. 3. (A) Normal spontaneous activity of an isolated guinea pig taenia coli. Electrical activity measured intracellularly (6). (B) Spontaneous electrical and mechanical activity of an isolated strip from guinea pig uterus. Pronounced hour-rhythm (HR), with minute-rhythm (MR) superimposed. Electrical activity recorded extracellularly. (Golenhofen and Neuser, *unpublished*.)

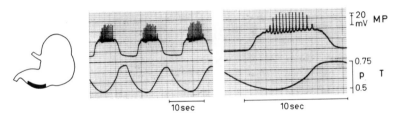

FIG. 4. Spontaneous electrical and mechanical activity of an antrum preparation from guinea pig stomach, localization indicated in the diagram. Membrane potential measured intracellularly. (7)

range; their frequency is usually determined by the SR or other processes. The intrinsic frequency of the spike process is marked in addition.

This five-type classification is a minimal necessity for an adequate description of mammalian smooth muscle activity. The five main types can be clearly separated from each other, partly directly by the differences in their appearance in electrical and mechanical recordings, partly by differences in their sensitivity to temperature (9), in their dependence on sodium and calcium (10), by selective effects of various drugs (11), and other functional characteristics. Further subdivision is easily possible, particularly in the group BOR.

TONIC ACTIVITY

The classification of phasic activity described above is, with slight modifications, identical with that which we proposed in 1970 (7,12,13). The only major alteration is that I have now added the "tonic activity" which means an active tension development remaining smooth and constant over many minutes, under constant conditions, without visible rhythmic fluctuations in

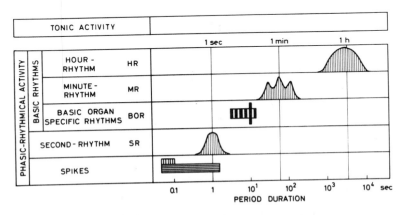

FIG. 5. Classification of myogenic automaticity of mammalian smooth muscle. Frequency characteristics of the different rhythms are indicated diagrammatically, after values obtained from the guinea pig. Values of BOR and SR are shifted to the right in larger species.

the mechanical recording. This tonic type is not a sixth type in addition to the five types of phasic activity, but rather another type in contrast to the whole group of phasic rhythmic events, and it is justified to describe the differentiation between phasic and tonic types of smooth muscle as a third dimension in a functional classification.

A clear demarcation of smooth muscle tone has only recently become possible when we observed that a group of substances, described as calcium antagonists, can selectively suppress phasic activity leaving a special tonic

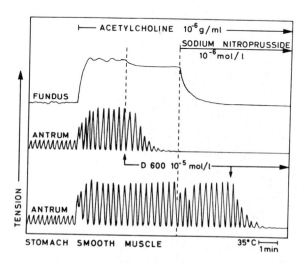

FIG. 6. Diagrammatic illustration of selective blockade of phasic and tonic activity in stomach muscle. Mechanical activity of isolated strips from the fundus and from the antrum, stimulated with acetylcholine. D 600 inhibits selectively the phasic activity of the antrum and the weak phasic component in the fundus; sodium nitroprusside inhibits selectively the tonic activation of the fundus. [After results from Boev, Golenhofen, and Lukanow (14).]

activation mechanism uninfluenced, as shown in Fig. 6 for stomach smooth muscle. The phasic activity of the antrum preparation, but not the tonic activation of the fundus strip, is suppressed by D 600, and the latter can be selectively suppressed by sodium nitroprusside (see Boev et al., and Golenhofen, this symposium; also Refs. 14,15).

It can be concluded that different activation mechanisms exist for phasic and tonic processes; the tonic mechanism operates usually without spikes or other electrical rhythms. The functional principle is the same as described by Kuffler and Vaughan Williams for frog skeletal muscle (16–18). The only difference is that the phasic–tonic differentiation is connected with a cellular specialization in frog skeletal muscle and with a subcellular specialization in mammalian smooth muscle.

The first studies using the new tool of selective phasic blockade have shown that the tonic activation is much more important than previously expected. Even in clearly "spike-generating types" (19) of smooth muscle, as portal vein or uterus, strong tonic mechanisms exist but are normally masked by additional phasic activity and spike discharges (20,21).

DISCUSSION

The entire "symphony of smooth muscle," to modify Code's expression (22), can be described as composed of the six elements listed in Fig. 5. A complex smooth muscle organ such as the stomach plays with the full orchestra, all types are present, with a systematic topical differentiation of preferred localizations, just as the instruments in an orchestra.

A detailed discussion of other classifications is not possible at this juncture (22–24). All rhythmic events slower than the spike can be found in the literature under the term "slow wave"—each group of workers has its own slow waves today. Many other terms exist, particularly for the group BOR, for example: basic electrical rhythm (BER) (25); electrical control activity (ECA) (26); gastric control wave (27); pacesetter potential (PP) (28); multispike complexes (MSC) (29,30).

Prosser (22) has recently presented a classification which is in some respects quite similar to ours. One difference is that his system does not contain a special type which is comparable to the SR. His "slow waves" correspond well to our group BOR. He called the MR "ultra-slow rhythms" and the HR "hourly bursts." He also mentioned tonic contractions as a special type but without a clear demarcation against the other types of activity.

Further important dimensions for smooth muscle classification are intensity and quality of external control, both nervous and humoral, but it is beyond the scope of this paper to describe these aspects in detail.

CONCLUSION

An adequate classification of mammalian smooth muscle is only possible using a multidimensional system. Three of the main dimensions have been described in some detail. Variations in the different dimensions are more or less independent of each other, in this way generating the characteristic diversity in the smooth muscle system.

REFERENCES

1. Taylor, D. L., Condeelis, J. S., Moore, P. L., and Allen, R. D. (1973): *J. Cell Biol.*, 59:378–394.
2. Sleigh, M. A. (1974): *Cilia and Flagella.* Academic Press, London and New York.
3. Golenhofen, K., and Hannappel, J. (1974): *Experientia*, 30:33.
4. Golenhofen, K. (1965): *Pflügers Arch. Ges. Physiol.*, 284:327–346.
5. Hannappel, J., and Golenhofen, K. (1969): *Pflügers Arch.*, 307:R84–R85.
6. Golenhofen, K., and v. Loh, D. (1970): *Pflügers Arch.*, 314:312–328.

7. Golenhofen, K., v. Loh, D., and Milenov, K. (1970): *Pflügers Arch.*, 315:336–356.
8. Hannappel, J., and Golenhofen, K. (1974): *Pflügers Arch.*, 348:65–76.
9. v. Loh, D., and Golenhofen, K. (1970): *Pflügers Arch.*, 318:35–50.
10. Golenhofen, K. (1970): Slow rhythms in smooth muscle. In: *Smooth Muscle*, edited by E. Bülbring, A. F. Brading, A. W. Jones, and T. Tomita, pp. 316–342. Arnold, London.
11. Golenhofen, K., and Lammel, E. (1972): *Pflügers Arch.*, 331:233–243.
12. Golenhofen, K. (1971): Intrinsic rhythms of the gastrointestinal tract. In: *Gastrointestinal Motility*, edited by L. Demling and R. Ottenjann, pp. 71–81. Thieme–Academic Press, Stuttgart and New York.
13. Golenhofen, K., and v. Loh, D. (1970): *Pflügers Arch.*, 319:82–100.
14. Boev, K., Golenhofen, K., and Lukanow, J. (1973): *Pflügers Arch.*, 343:R56.
15. Golenhofen, K. (1973): *Pflügers Arch.*, 343:R57.
16. Kuffler, S. W. (1953): *Arch. Exp. Pathol. Pharmakol.*, 220:116–135.
17. Kuffler, S. W., and Vaughan Williams, E. M. (1953): *J. Physiol. (Lond.)*, 121: 289–317.
18. Kuffler, S. W., and Vaughan Williams, E. M. (1953): *J. Physiol. (Lond.)*, 121: 318–340.
19. Somlyo, A. V., Vinall, P., and Somlyo, A. P. (1969): *Microvasc. Res.*, 1:354–373.
20. Golenhofen, K., Hermstein, N., and Lammel, E. (1973): *Microvasc. Res.*, 5:73–80.
21. Golenhofen, K., and Neuser, G. (1974): *Pflügers Arch.*, 347:R15.
22. Prosser, C. L. (1974): *Proc. 4th Int. Symp. Gastrointestinal Motility*, pp. 21–37. Mitchell Press, Vancouver.
23. Bozler, E. (1948): *Experientia*, 4:213–218.
24. Burnstock, G. (1970): Structure of smooth muscle and its innervation. In: *Smooth Muscle*, edited by E. Bülbring et al., pp. 1–69. Arnold, London.
25. Bass, P. (1968): In vivo electrical activity of the small bowel. Handbook of Physiology, Sect. 6, Vol. 4, 2051–2074, Amer. Physiol. Soc., Washington, D.C.
26. Sarna, S. K., and Daniel, E. E. (1973): *Am. J. Physiol.*, 225:125–131.
27. Sarna, S. K., and Daniel, E. E. (1974): *Am. J. Physiol.*, 226:749–755.
28. Code, C. F., and Szurszewski, J. H. (1970): *J. Physiol. (Lond.)*, 207:281–289.
29. Holman, M. E. (1969): *Ergebn. Physiol.*, 61:137–177.
30. Holman, M. E., Kasby, C. B., Suthers, M. B., and Wilson, J. A. F. (1968): *J. Physiol. (Lond.)*, 196:111–132.

Physiology of Smooth Muscle, edited by
E. Bülbring and M. F. Shuba.
Raven Press, New York © 1976.

Types of Slow Rhythmic Activity in Gastrointestinal Muscles

C. Ladd Prosser, W. A. Weems, and J. A. Connor

Department of Physiology and Biophysics, University of Illinois, Urbana, Illinois 61801

The term "slow waves" was first used and continues to be used as a purely descriptive name for rhythmic electrical oscillations in gastrointestinal musculature. The synonymous terms "pacesetter potentials" and "control potentials" imply functional meaning. In terms of cellular mechanisms, it now appears that several kinds of slow potentials can be identified in intestinal smooth muscles. Each is rhythmic and endogenous, i.e., myogenic in nature (1). This chapter lists four identifiable types of slow potentials and briefly summarizes evidence concerning the ionic mechanisms of two of them.

MINUTE RHYTHMS

Rhythmic electrical waves of 1 min or more in duration have been described by Golenhofen (2) and have been termed "minute rhythms." We have observed similar minute-long electrical waves in the intestine of a lizard (*Tiliqua*) and a fish (*Opsanus*), and in the stomach of a toad (*Bufo*) and a salamander (*Necturus*). Sometimes faster 5- to 7-sec waves are superimposed on the minute ones (3). The very slow waves are remarkably unaffected by ionic changes that alter the faster ones, and we have no knowledge of the ionic mechanisms causing them.

PREPOTENTIALS

Many gastrointestinal muscles show spontaneous spikes; the rising phase of these appears to involve an increase in calcium conductance. In taenia coli of guinea pig, in isolated circular intestinal muscle of cat, and in uterine muscle of rat, some cells show prepotentials at the base of spikes. These prepotentials resemble pacemaker potentials of certain cardiac tissues and of some spontaneously active neurons. Longitudinal intestinal muscle may show prepotentials at the base of those spikes that occur at the peak of the 5-sec slow waves. Frequently, as in taenia, the prepotentials do not reach the threshold for spiking and under these conditions have been called "slow waves" (4). However, the falling phase of a spike eliminates the prepotential, whereas the 5-sec slow waves are unaffected by this part of the spike. The prepotentials can be blocked by the β-action of catecholamines (5), and in guinea pig taenia have been suggested to be caused by rhythmic changes in g_{Na} (4). In cat intestine in a low-calcium medium the prepotentials are abolished along with spikes (6).

SLOW WAVES

The most extensively studied "slow waves" are of 5- to 7-sec duration and occur in longitudinal intestinal muscle of cat, rat, rabbit, and dog. Waves resembling these occur in the circular layer of colon and in some diagonal fibers of stomach, and also in the longitudinal layer of intestine of some primitive orders of mammals (marsupial mice), but not in guinea pigs (3). It appears that well-defined slow waves occur only in those species in which the ratio of thickness of the longitudinal to circular layers is at least 1:2 or 1:3; in those species where the longitudinal layer is very much thinner relative to the circular layer, slow waves of this type have not been observed (3).

It was shown some years ago (9) that conduction down a segment of the intestine is interrupted if either the ring of circular muscle fibers or the strands connecting the two muscle layers are transected. In strips of longitudinal muscle, devoid of any circular fibers, slow waves can be recorded in local regions, but there is no conducted wave in a regenerative sense. There appear to be multiple pacemaker regions at a given time, and the waves spread electrotonically (passively) into neighboring cells. In the intact intestine there is some sort of cycling between the two muscle layers which permits longitudinal conduction.

It was suggested some years ago (6,7) that the slow waves in longitudinal muscle of cat intestine may represent a rhythmic active extrusion of sodium ions from the muscle fibers. The evidence was as follows: high sensitivity to Na–K pump blocking agents such as ouabain and K-free medium, active extrusion of radioactive sodium at the beginning of the repolarization phase, reduction in amplitude after prolonged exposure to Na-free medium, high sensitivity to metabolic poisons and to oxygen concentration, high temperature coefficient, and increased amplitude when intracellular Na^+ concentration was raised iontophoretically. Recently we have shown (8) that the slow waves are maximal in amplitude at K_o concentrations at which the pump contribution to the resting potential is maximal.

Definitive evidence for the rhythmic electrogenic Na-pump mechanism has recently been presented (10,11) on the basis of data obtained by means of a double sucrose gap with cat duodenal and jejunal longitudinal muscle strips.

It was possible to voltage-clamp by means of the sucrose gap and the reliability of the clamping was established according to several criteria (11). The method employed very thin strips of longitudinal muscle (< 100 μm wide), a node much smaller (100 to 200 μm) than the space constant (1,000 to 1,500 μm). Except for attempts to determine whether membrane resistance changed during a slow wave, measurements under clamp were generally made in steady state or slowly changing potential (5 mV/sec), small currents were used and only events of slow time course were clamped. When such strict conditions were applied, a nodal region of a strip of longitudinal intestinal muscle could be voltage-clamped with some confidence, and when the membrane potential was held at the trough potential, current pulses appeared with frequency and waveform similar to the voltage waves. In no instance of voltage clamping of an active preparation did current pulses fail to appear (10).

Examples of rhythmic current pulses during clamping of the slow waves are shown in Fig. 1. In Fig. 1A a single slow wave is shown, then five current pulses, while the voltage was clamped at the trough of the slow wave. Figure 1B represents the current and voltage of a slowly rising slow-wave pattern to

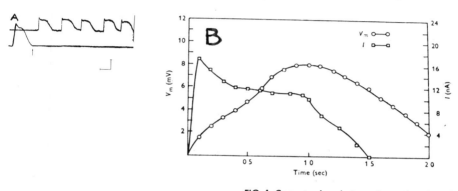

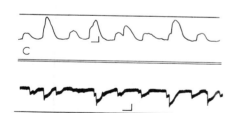

FIG. 1. Current pulses during voltage clamping of rhythmic slow potentials. A: Lower record single voltage curve, upper record series of inwardly directed current transients during voltage clamp. Calibration: 2.5 sec, 5 mV, 10 nA. B: Plot of membrane current at left and slow-wave voltage at right for a single wave. C: Upper trace voltage curves, lower trace current pulses while voltage was clamped immediately following upper trace recording. This preparation showed two pacemaker loci within the node and current trace shows contribution of each. Calibration: 2 sec, 4 mV, 10 nA. [(A) and (B) from ref. 11.]

show the correspondence. Figure 1C is from a preparation in which there were two pacemaker loci within the node; hence, some of the recorded slow waves were double as were the corresponding current pulses. In each example the inwardly directed currents rise faster than the slow potential because of the charging of the membrane capacitance.

Furthermore, the current pulses were reduced or abolished by the same agents and concentrations (ouabain, zero K) as abolish the slow waves. Both ouabain and zero K cause some depolarization, but if the membrane is repolarized after the slow waves have been abolished by these treatments, the slow waves do not reappear, thereby showing that the blocking is not caused by a shift in membrane potential (11).

In agreement with prior results (12), it was found that no changes in membrane resistance could be demonstrated during slow waves by direct measurement. This is equivocal because for pulsital current stimuli, the long membrane time constant (200 to 400 msec) prevents pulses from reaching true steady states during an oscillating membrane potential. Under voltage clamp, faster measurements could be taken, but the noise level is higher. Figure 2 shows a recording in which small voltage steps (lower trace) were imposed beginning just after the onset of the spontaneous current transient (shown here as a quick downward deflection). Membrane resistance as measured by $\Delta V/\Delta I$ does not change, to the limits of reading accuracy, from its resting value. Experiments in which $[Na]_o$ was reduced to approximately 10% of its normal value showed no change in the amplitude or rate of rise of the slow wave over a time period much longer than the extracellular space washout for these small bundles. A sizeable effect would be expected if a small conductance change to this ion accounted for slow wave depolarization.

FIG. 2. Record of current corresponding to a single slow wave under voltage clamp (upper trace). Depolarizing pulses of 4-mV amplitude were applied at different times in the current wave (lower trace). *Calibration:* 250 msec, 1 nA, 2 mV.

The voltage waveform of the slow wave could be reconstructed from the current waveform under voltage clamp and the values of membrane resistance and capacitance measured at the slow-wave trough, both constants. The voltage waveform was computed using the standard convolution integral. Figure 3 shows the results of a particular simulation with $R_m = 1 \times 10^6$ Ω and $C_m = 3.1 \times 10^{-7}$ F. The good agreement between measured and simulated waveforms for this and other preparations makes it highly unlikely that the voltage-clamp currents are artifactual or that there are conductance changes of significance generating the slow waves. In order to account for the data in another way, one would have to suppose that the sucrose-gap test node consisted of at least two functional parts, an active, voltage-generating portion isolated from the rest of the node and the voltage end-pool by a relatively high resistance. Such an active region could be very insensitive to imposed voltage control and continue to generate spontaneous activity which would appear as current deflections in the clamp current record. Such a situation, however, would require, if the active region were the sole generator, that its waveform would have an amplitude much greater than any slow wave ever observed by any recording technique since one would demand a considerable voltage drop across the isolating resistance; or, if both regions normally were generators, that the voltage-clamp currents should be entirely too small to simulate the voltage waveform. Other explanations have been explored, but to make a satisfactory account of the data in totality requires assumptions about the tissue and experimental configuration which are unwarranted and a consistency of these factors which it is unreasonable to expect from preparation to preparation.

Both the slow waves and current pulses showed similar voltage dependences. There is no tendency to approach an equilibrium potential for any ion. The frequency increases and amplitude decreases with depolarization up to threshold for spiking, and conversely the frequency decreases and amplitude slightly increases on hyperpolarization by 10 to 30 mV; greater hyperpolarization eliminates slow waves. Thus there is a voltage range within which

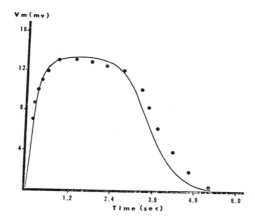

FIG. 3. Comparison of slow wave computed from a model of an electrogenic pump in parallel with a fixed resistance and capacitance (*continuous curve*) and measured slow wave (*dots*) (11).

the pump shows rhythmic activity. Voltage dependences have been noted for other Na–K pumps (13).

Much evidence has been presented (15,8) that a significant contribution to the resting potential is made by the Na–K pump. At optimal concentration of K_0 this resting pump potential is of the same order as the slow-wave amplitude recorded with intracellular microelectrodes (8).

Finally, current–voltage curves of longitudinal smooth muscle are linear over a wide range, with rectification as the depolarization approaches a spiking threshold (11). Pacemaker systems which are based on voltage- and time-dependent conductance changes (e.g., heart, molluskan neurons, crustacean axons) show regions of anomalous rectification. We have recently found that rhythmically active circular muscle shows anomalous rectification by the same recording technique.

It has been suggested (14) from microelectrode recordings that there may be two components of the 5-sec waves, fast and slow, separated by a notch in some recordings, and that the fast component may result from an increase in g_{Na}, the slow one from an increase in g_{Cl}. An alternative explanation of the notch is that an intracellular electrode records not only the potential changes in the penetrated cell, but also a conducted wave in neighboring cells which are coupled by low-resistance connections. The evidence presented for changes in g_{Na} is indirect and our results fail to give any evidence for the quantitative relation between Na_0 and amplitude that would be expected on a g_{Na} hypothesis. It is claimed (14) that the late component of the slow wave seen in those preparations with notches is reduced in duration when Cl^- is replaced by propionate. Our unpublished data (C. Connor) show that in the sucrose gap, propionate leads to a deterioration of the preparation, irreversible drop in resistance, and loss of slow waves and spikes. Replacement of Cl^- by SO_4^{2-} causes initial depolarization with recovery to a slightly depolarized level. No systematic change in resistance occurs after the plateau potential is reached and no change in slow waves is seen.

It is concluded that the most rigorous evidence shows that the basic driving function for slow waves in cat longitudinal muscle of the small intestine is a rhythmic electrogenic Na–K pump. This does not imply that the rhythmicity resides in the pump per se. It is much more likely that the pacemaker process is metabolic and slow waves may be determined by the rate at which ATP is made available to the pump. It is postulated that between the periodic slow waves the electrogenic system is operating at basal level, that at a given time the Na–K transport rate drops, either because the total transport decreases or the coupling ratio of inward to outward flux approaches unity. This decrease in net outward flux leaves an unbalanced inward leakage flux which depolarizes the membrane and creates the rising phase of the slow wave. Under voltage clamp such a decrease in outward flux appears as an inward current. Following the decrease in transport rate, the pump level starts to return to the basal level and the membrane voltage, which lags behind the current because of membrane capacitance, reaches a peak and then declines.

Furthermore, it is postulated that spread of slow waves to neighboring cells and into the circular layer is by way of low-resistance connections. We have been unable to obtain convincing evidence that the slow waves are conducted in a regenerative sense. However, in a passive RC network a voltage of whatever origin spreads according to the resistive coupling between cells.

DEPOLARIZATION-INDUCED WAVES IN MAMMALS

Intestinal muscle of guinea pig shows slow rhythms with multiple spikes, and these slow waves are refractory to ouabain. These waves are blocked by atropine and by nerve blockers, such as TTX, which have no effect on the slow waves of cat (16). When quiescent strips of guinea pig longitudinal muscle are treated with carbachol (or acetylcholine) rhythmic slow waves are initiated (17,18). Acetylcholine is known to be liberated normally from

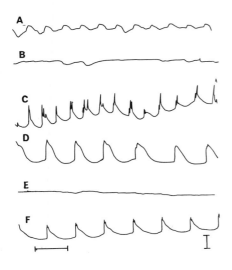

FIG. 4. Pressure electrode recordings from intact cat duodenum. (A) Normal Krebs solution; (B) ouabain at 10^{-6} w/v; (C) ACh 10^{-7} w/v plus ouabain; (D) K-free Krebs plus ACh and ouabain; (E) Na-free Krebs (Li replacement) plus ACh and ouabain; (F) normal Krebs. *Calibration:* 10 sec, 0.2 mV (3).

plexus neurons of the guinea pig ileum. Slow waves of guinea pig are much more sensitive to Na levels than are those of cat. We have found (3) that when the slow waves of cat are abolished by ouabain, addition of acetylcholine can initiate waves, usually of higher than normal frequency. These ACh-induced waves are insensitive to reduced levels of Ca_o that eliminate spikes, but they are more readily blocked by reduced levels of Na_o than are the 5-7-sec waves (Fig. 4). Hence, it appears that the cat muscle is capable of the ACh-induced type of wave when the pump-type waves are blocked and that in guinea pig the ACh-induced waves may be the normal ones. In quiescent strips of cat muscle in the sucrose gap, depolarization could induce pacemaker-like waves which clearly differed in form from the 5-sec ouabain-sensitive waves (11).

DEPOLARIZATION-INDUCED WAVES IN LOWER VERTEBRATES

The distribution of the two types of rhythmic electrical waves in different vertebrates is now being investigated. It was shown by van Harn (19) that in frog stomach, spontaneous waves are like prolonged spikes and are Ca-sensitive. We now report new observations on muscle of the stomach and spiral intestine of a skate, *Raja clavata,* and of a toadfish, *Opsanus tau.* When first mounted, electrical and mechanical rhythms were observed (Fig. 5). Intracellular recordings showed large spike-like waves of 10- to 12-sec duration, 5-sec rise time and 5 to 7 sec repolarization time. It was difficult to maintain microelectrodes inside a cell for longer than a half-minute. Pressure electrode recordings showed similar depolarizing waves, somewhat more prolonged and of lower amplitude than the microelectrode recordings. The waves often showed two levels of amplitude, some of low voltage and others 5 to 10

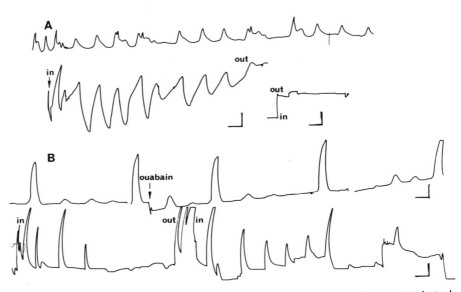

FIG. 5. Rhythmic electrical activity recorded from pyloric end of skate stomach by pressure electrode (upper records) simultaneously with intracellular microelectrode (lower records). "In" and "out" indicate penetration and emergence of microelectrode. A: Spontaneous rhythmic activity, calibration 20 sec, 0.5 mV for pressure electrode, 20 mV for microelectrode. B: Two levels of spontaneous activity recorded by pressure electrode; variable spontaneous spikes recorded by microelectrode and related in part to varying penetration. Ouabain at 2×10^{-5} M applied at arrow. Calibration: 20 sec, 0.2 mV for pressure, 10 mV for microelectrode. Preparations bathed in a skate Ringer's solution prepared according to analysis of skate blood (21) concentrations (mM): NaCl 254; KCl 8; $CaCl_2$ 6; $MgCl_2$ 2.5; urea 320, buffered at pH 7.5 with $NaHCO_3$ and NaH_2PO_4.

times greater (Fig. 5B). The low-level activity consisted of oscillatory electrical waves of 12 to 14 sec, interrupted at intervals of 25 to 30 sec by large spikes of 10- to 12-sec duration each. The most common activity was spikes of 8- to 12-sec duration each, at intervals of 25 to 35 sec. All electrical activity, including the low-level waves, was reflected in contractions.

In many preparations the rhythmic activity declined after a variable number of minutes. It could then be initiated by one of several methods. Stretch in either axis was often effective; changing the position of a pressure electrode or reapplying pressure often initiated activity (Fig. 6). Occasionally application of cold Ringer's solution started activity. Also, addition of KCl to raise the concentration from 8 to 18–28 mM caused one or more of the "slow spikes" and contractions. The most consistent rhythms were started by addition of acetylcholine to give a concentration of 5×10^{-6} to 1×10^{-5} M (Figs. 5 and 6). These procedures have in common the bringing about of depolarization, an effect which is more prolonged with the use of acetylcholine than with the other methods. The 10-sec spikes initiated by these treatments could occur singly or in trains with approximately 30-sec intervals between spikes. Contractions were regularly recorded at 30-sec intervals. The basic rhythm

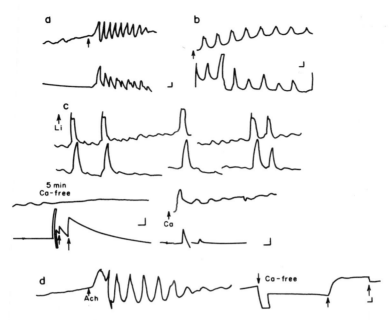

FIG. 6. Simultaneous electrical recordings (*upper records*, using pressure electrode) and mechanical recordings (*lower records*, using Grass force transducer) of skate gastrointestinal muscle. a, b: Effects of ACh (10^{-5} M) applied at times indicated by arrows. Calibration 1 mV, 10 sec. a: Pyloric stomach. b: Cardiac stomach. c: Stomach preparation, bathed in Ringer's solution in which LiCl replaced NaCl 15 min prior to record, spontaneous large spikes and waves after 5 min in Ca-free Ringer's, contractions lost and electrical waves reduced, stretches (*arrows*) failed to elicit responses, when Ca-Ringer was reintroduced (*arrow* Ca) rhythmic electrical and mechanical activity returned. Calibration 20 sec, 0.5 mV. d: Electrical response to ACh 10^{-5}M (*arrow*), Ca-free Ringer then applied (*2nd arrow*) and activity ceased, ACh (*3rd arrow*) and applied pressure (*4th arrow*) elicited no responses. *Calibration: 2.5 mV, 10 sec.*

of the skate lower stomach was, therefore, at a period of about 30 sec, although occasional oscillations at 12 to 14 sec were observed.

The spontaneous rhythmicity was not reduced or blocked by treatment with ouabain (10^{-5} M) for periods up to 40 to 60 min (Fig. 7). Similarly, it was not reduced by atropine or by xylocaine (Fig. 7). Replacement of NaCl by LiCl in the Ringer's solution had no effect in more than 20 min (Fig. 6). However, replacement of normal Ringer's solution by Ca-free solution rapidly stopped all contractions and reduced the rhythmic electrical activity and after several washes eliminated all activity; the effect of zero Ca was readily reversible (Fig. 6). The spontaneous waves were also blocked by cobalt at 20 mM.

It is concluded that depolarizing agents lead to rhythmic electrical activity in skate gastrointestinal muscle consisting of prolonged spikes, and that these are probably caused by increased conductance of calcium. Under normal physiologic conditions mechanical stretch may be the initiating agent. The retractor of the spiral intestine of dogfish is readily stimulated by slight stretch to contract and to maintain tension while continuing to give spikes (20).

Recordings were also made from the digestive tract of the teleost toadfish, *Opsanus*. Rhythmic activity was recorded from stomach, small intestine, and colon, but in none of these did activity last as long as in the skate. Prolonged spikes, 8 to 12 sec in duration, sensitive to Ca_0 were recorded from preparations under slight tension; and spiking was enhanced by acetylcholine. Two preparations showed sinusoidal low-level waves of a 14-sec period. Also two preparations showed very long (90 sec) waves with 8- to 10-sec spikes superposed on them. This latter pattern resembles the minute-long waves in toad and *Necturus* but was not observed in the skate. It is concluded that the predominant electrical activity in toadfish, as in skate and frog stomach, consists

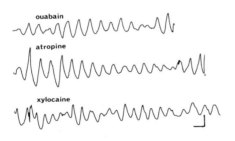

FIG. 7. Continued rhythmic electrical activity recorded by pressure electrodes in different preparations in presence of agents: ouabain 2.5×10^{-5} M for 40 min before recording, atropine 2×10^{-4} w/v for 10 min, xylocaine 2×10^{-4} w/v added before this recording. *Calibration:* 0.1 mV, 20 sec.

of very prolonged spikes (8 to 10 sec), that there may be two types of rhythm in toadfish intestine—some at 13 to 15 sec and others at 90- to 100-sec intervals.

It is postulated from the preceding evidence that the rhythm of the skate gastric and intestinal muscle may result from voltage- and time-dependent conductance changes. This rhythmicity may resemble that of heart, except that calcium, rather than sodium, may be the critical ion for conductance change. The depolarization-induced waves in guinea pig intestine and in cat after blocking the Na–K pump may also be similar in ionic mechanism to cardiac rhythms.

CONCLUSIONS

The meaning of slow rhythmic potential waves in gastrointestinal muscles is as follows: For normal digestive propulsion there must be (a) rhythmic contractions, and (b) synchrony of thousands of muscle fibers in a ring. Different animals adapt various preexisting mechanisms for the intestinal rhythms. Some mammals adapt the Na–K pump, which normally contributes to the resting potential by making it oscillate. Others, such as guinea pig, may use continuously liberated acetylcholine to trigger a rhythm, which is possibly caused by time-dependent Na-conductance changes. Many, perhaps most, vertebrates use any of several methods of depolarization, commonly stretch, to initiate rhythmic prolonged "spikes," possibly caused by Ca-conductance change. Experiments are now in progress to test these hypotheses for the depolarization-induced waves. In all cases there appears to be electrical coupling between cells, although the precise mechanisms of spread and synchronization remain uncertain.

ACKNOWLEDGMENTS

Supported by National Institutes of Health grant AM 12768. The experiments with skate and toadfish were conducted at Marine Biological Laboratory, Woods Hole, Massachusetts.

REFERENCES

1. Prosser, C. L. (1974), *Ann. Rev. Physiol.,* 36:503–535.
2. Golenhofen, K., von Loh, D., and Milenov, K. (1970), *Pflügers Arch.,* 315: 336–356.
3. Prosser, C. L. (1974): In: *Proc. 4th Int. Symp. Gastrointestinal Motility,* edited by E. Daniel, pp. 21–37. Mitchell Press, Vancouver.
4. Kuriyama, H., and Tomita, T. (1970), *J. Gen. Physiol.,* 55:147–162.
5. Bülbring, E., and Tomita, T. (1969), *Proc. R. Soc. Lond. B.,* 172:103–119.
6. Liu, J., Prosser, C. L., and Job, D. D. (1969), *Am. J. Physiol.,* 217:1542–1547.
7. Job, D. D. (1969), *Am. J. Physiol.,* 217:1534–1541.
8. Connor, C., and Prosser, C. L. (1974), *Am. J. Physiol.,* 226:1212–1218.
9. Kobayashi, M., Nagai, T., and Prosser, C. L. (1966), *Am. J. Physiol.,* 211:1281–1291.
10. Weems, W. (1973) *Voltage clamp studies on slow waves in cat longitudinal muscle* (Ph.D. Thesis). Univ. of Illinois, Urbana.
11. Connor, J. A., Prosser, C. L., and Weems, W. (1974), *J. Physiol. (Lond.),* 240: 671–702.
12. Kobayashi, M., Prosser, C. L., and Nagai, T. (1967), *Am. J. Physiol.,* 213:275–286.
13. Kostyuk, P. G., Krishal, O., and Pikoplichko, V. I. (1973), *J. Physiol. (Lond.),* 226:373–392.
14. El Sharkawy, T. Y., and Daniel, E. E. (1974): In: *Proc. 4th Int. Symp. Gastrointestinal Motility,* edited by E. Daniel, pp. 39–51. Mitchell Press, Vancouver.
15. Casteels, R., Droogmans, G., and Lund, G. F. (1973), *Philos. Trans. R. Soc. Lond. B.,* 265:47–56.
16. Kuriyama, H., Osa, T., and Toida, N. (1967), *J. Physiol. (Lond.),* 191:259–270.
17. Bolton, T. B. (1972), *J. Physiol. (Lond.),* 220:647–671.
18. Bolton, T. B. (1972), *Pflügers Arch.,* 335:85–96.
19. Van Harn, G. (1968), *Am. J. Physiol.,* 215:1351–1358.
20. Parnas, I., Prosser, C. L., and Rice, R. (1974), *Am. J. Physiol.,* 226:977–981.
21. Hartman, F. A., Lewis, L. A., Brownell, K. A., Shelden, F. F., and Walther, R. F. (1941), *Physiol. Zool.,* 14:476–486.

Physiology of Smooth Muscle, edited by
E. Bülbring and M. F. Shuba.
Raven Press, New York © 1976.

Intestinal Slow-Wave Propagation Velocity as a Function of Longitudinal Muscle Impedance

Alexander Bortoff

Upstate Medical Center, State University of New York, Syracuse, New York 13210

The relationship between intestinal slow waves and intestinal contractions is well established (1–3). During slow-wave depolarization spike potentials may be elicited which in turn trigger the contractile process. Thus, contractions are initiated during the period of slow-wave depolarization and their magnitude is a function of the number of spike potentials occurring during that period. It follows that propagated slow waves will result in propagated, or peristaltic, contractions; indeed, this has been shown to be the case in both dog and cat (4,5).

In the small intestine peristaltic contractions occur most frequently in the duodenum and upper jejunum. In this region of the intestine slow-wave frequency is quite constant, but slow-wave propagation velocity decreases aborally. Thus, in one study on anesthetized dogs the slow-wave propagation velocity was found to decrease from about 14 cm/sec in the duodenum to less than 2 cm/sec in the jejunum (4). Similar results have been obtained from the cat, although the relative decrease in propagation velocity is not as marked as it is in the dog. In a recent study on cats, for example, the propagation velocity decreased from 3.8 cm/sec in the duodenum to 2.6 cm/sec in the mid-jejunum (6). The purpose of the present study was to determine what factors might be responsible for the decrease in propagation velocity along the proximal small intestine.

On the assumption that slow-wave propagation occurs by means of local circuit currents three primary factors could determine the velocity at which propagation occurs: (a) the rate of depolarization, (b) the duration of the refractory period, and (c) the resistance to the flow of local circuit current. Although a careful study of the relative rates of slow-wave depolarization in duodenum and jejunum was not made, visual inspection of monophasic records obtained from the two areas of the cat intestine revealed some variability in each area, but no marked difference between the two areas. Therefore, attention was focused on possible differences in refractory period and resistance.

STUDIES OF SLOW-WAVE FREQUENCY VERSUS PROPAGATION VELOCITY

The first series of experiments was designed to study the relationship between slow-wave frequency and propagation velocity in the duodenum and jejunum. The general characteristics of this relationship are shown by each of the three curves shown in Fig. 1A, in which the reciprocal of propagation velocity in seconds per centimeter is plotted against the reciprocal of frequency, or inter-slow-wave interval, in seconds. The upper limiting frequency occurs at zero velocity, and is determined by the duration of the functional refractory period, i.e., the duration of the functional refractory period is the reciprocal of the upper limiting frequency. The upper limiting velocity occurs at zero frequency, and is largely determined by the resistance to flow of local circuit current, since the tissue is not refractory at this time. At intermediate

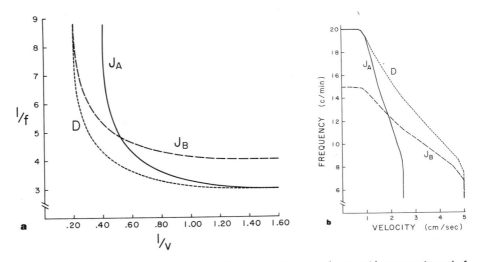

FIG. 1. a: Hypothetical plots of the reciprocal of slow-wave frequency (in seconds) versus reciprocal of propagation velocity (in cm/sec) for duodenum (D) and jejunum (J_A and J_B). The jejunal curves represent hypothetical conditions where decreased propagation velocity is due only to increased resistance to flow of excitatory current (J_A) or only to increased duration of functional refractory period (J_B). b: Same data as in a, but plotted as slow-wave frequency versus velocity. Note that in the mid-frequency range the slopes of curves D and J_A converge at higher frequencies, whereas the slopes of curves D and J_B *diverge* at higher frequencies.

slow-wave frequencies propagation velocity is influenced by both resistance and relative refractoriness.

If the difference in propagation velocity in duodenum and jejunum were entirely caused by differences in resistance to local circuit current flow, the condition could be described by curves D (duodenum) and J_A (jejunum) (Fig. 1). The upper limiting frequencies at zero velocity would be essentially the same because of similar functional refractory periods. At zero frequency, because of the greater resistance to the flow of local circuit current in jejunum, its slow-wave propagation velocity would be lower than in the duodenum. The condition in which the lower propagation velocity in jejunum results from a relatively longer functional refractory period is described by the curves

D and J_B. In this case, the upper limiting frequency would be lower in the jejunum, but the propagation velocity at zero frequency would equal that in the duodenum.

Figure 1b shows these same data plotted in a more conventional manner, as slow-wave frequency in cycles per minute against propagation velocity in centimeters per second. Again, in the case of differences in resistance to local circuit current, the velocities at zero frequency differ, but the upper limiting frequency is the same. The reverse is true in the case where the only difference is in length of refractory period. Consequently, the slopes of the curves

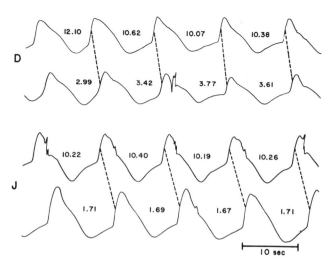

FIG. 2. Spontaneous slow waves simultaneously recorded from duodenum (D) and jejunum (J) of anesthetized cat with suction electrodes. Amplifier time constant, 0.8 sec. The figures between slow waves in the top of each pair of tracings represent instantaneous slow-wave frequency, and the figures between slow waves in the bottom of each pair represent corresponding slow-wave propagation velocity. The dashed lines represent propagation from the oral (top) to the aboral (bottom) electrodes.

in the mid-frequency range differ so that the duodenal and jejunal curves converge at higher frequencies in the case of resistance differences, and they diverge in the case of differences in refractory period.

In order to determine which of these alternatives is the correct one, slow waves were simultaneously recorded from the duodenum and proximal jejunum of five cats anesthetized with sodium pentobarbital. Recordings were obtained with three flexible suction electrodes on the mid-duodenum and three suction electrodes on the jejunum. Tracings were amplified and displayed on an 8-channel polygraph, using a time constant of 0.8 sec and a paper speed of 10 mm/sec. Although the relatively short time constant produced some distortion in slow-wave configuration, it facilitated the measuring procedure by eliminating drift and sharpening the upsweep of the recorded potentials.

Spontaneous activity was recorded for 4 to 6 hr, after which the electrodes were removed and the distances between the hematomized spots left by the electrodes were measured with a caliper. Sections of the tracings were selected for measurement in which slow waves simultaneously recorded from both regions of the intestine were of uniform amplitude but in which the instantaneous frequency showed some variability. Such recordings were obtained near, but not on, regions of waxing and waning, and provided a sufficient range of frequencies and propagation velocities. Examples of such tracings are shown in Fig. 2. In addition, measurements were made at random from sections of tracings in which both amplitude and frequency were relatively uniform. Two kinds of measurement were made. First, the time between similar points on successive slow waves at a single locus was determined (the inter-slow-wave interval). When divided into 60, this gives the instantaneous slow-wave frequency in cycles per minute. (These frequencies are indicated by the numbers between the slow waves in the first and third tracings of Fig. 2.) Second, the time difference between similar points on the second of the two slow waves and the next slow wave occurring at the adjacent aboral electrode was determined. This time difference divided into the distance between the electrodes was taken as the propagation velocity in centimeters per second and is

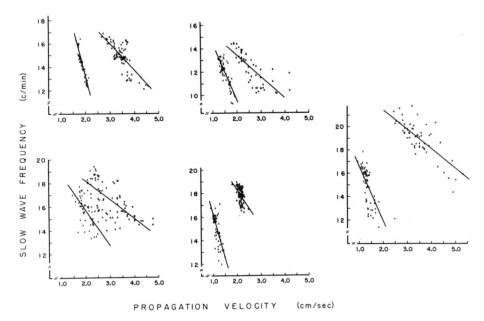

FIG. 3. Actual plots of slow-wave frequency vs propagation velocity from each of five cats. Note that for any frequency propagation velocity is higher in the duodenum. Regression lines converge at higher frequencies (cf. Fig. 1B, J_A and D) indicating the lower propagation velocity in jejunum is due to greater resistance to flow of excitatory current. (X) Jejunum; (O) duodenum.

indicated by the numbers between the slow waves in the second and fourth tracings of Fig. 2. Between 50 and 75 of these paired values were obtained from duodenum and from jejunum in in each of five cats and plotted as shown in Fig. 3. The regression lines were determined by a programmed computer. The data from each animal indicate (a) that for a given slow-wave frequency the propagation velocity is higher in the duodenum than in the jejunum, and (b) that the regression lines converge at higher frequencies. Thus, these data are consistent with the condition previously described in which the lower slow-wave propagation velocity in the jejunum is caused by greater resistance to local circuit current flow, as compared to duodenum. It is also consistent with earlier data showing the same upper limiting slow-wave frequencies in strips of longitudinal muscle from cat duodenum and jejunum (7).

IMPEDANCE STUDIES

If the decrease in slow-wave propagation velocity is indeed due to an increased resistance to flow of excitatory current, it should be possible to demonstrate a difference of tissue impedance between duodenum and jejunum. We attempted to do this, using a modification of a method described by Tomita (8). Strips of longitudinal muscle about 2 mm wide and 2 to 3 cm long were carefully dissected from segments of cat duodenum and upper jejunum beginning about 10 cm distal to the ligament of Treitz. The strips were tied at both ends with 6–0 surgical silk suture and threaded through a specially constructed Lucite chamber, 1 mm in diameter and 30 mm in length. Platinum-black electrodes, 10 mm apart, are permanently mounted in the chamber in such a way that each electrode makes contact with the tissue. One end of the tissue was fixed and the other was attached to a force transducer so that tension could be monitored throughout the experiment. The vertically oriented chamber was perfused through the bottom with either Tyrode solution or 50% Tyrode: 50% isotonic sucrose solution at a rate of 2 to 3 ml/min. The platinum electrodes were connected to the unknown terminals of an impedance bridge and an external generator was used to deliver alternating current to the bridge at frequencies between 50 and 10,000 Hz (Fig. 4).

In determining tissue impedance, first the specific resistances of Tyrode and sucrose-Tyrode solutions were determined, then the resistance of each solution was measured between the platinum electrodes as it flowed through the chamber. A "cell constant" could then be determined, by means of which impedances measured in the chamber could be expressed as impedance per unit volume. Next, with the tissue in the chamber, impedances were measured at several frequencies between 50 and 10,000 Hz, first while being perfused with Tyrode solution, and after equilibration, while being perfused with sucrose-Tyrode. The two solutions, having different specific resistances, made

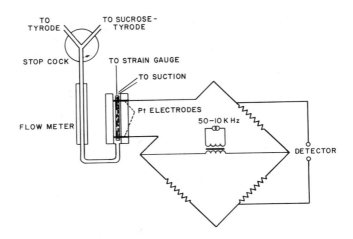

FIG. 4. Diagram of experimental setup for tissue impedance measurements. See text for details.

it possible to determine the relative volumes occupied by the solution and the tissue between the electrodes. The specific resistance of the tissue could then be calculated.

In all, tissue impedance was determined for 19 duodenal strips and 20 jejunal strips from five cats. Analysis of variance indicated no significant differ-

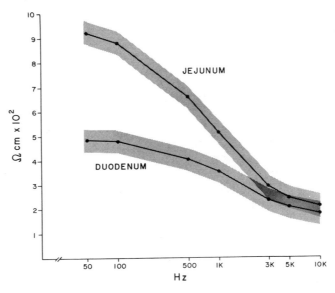

FIG. 5. Plot of mean specific impedance values at indicated current frequencies for longitudinal muscle from cat duodenum and proximal jejunum. Stippled areas indicate 95% confidence limits determined by analysis of variance. Differences in mean values are highly significant below 3,000 Hz, insignificant at 5,000 Hz and above.

ence between animals. Mean specific impedance values at each frequency for duodenum and jejunum are shown in Fig. 5. Confidence limits of 95% are indicated by the stippled areas. Analysis of variance indicates highly significant differences between duodenum and jejunum at frequencies below 3,000 Hz and no difference at frequencies of 5,000 Hz and above.

The shapes of these curves are similar to those obtained by Tomita (8) for the guinea pig taenia coli. Both curves approach plateaus below 100 Hz and above 5,000 Hz. One of the simplest equivalent circuits which could account for the shape of these curves is shown in the inset of Fig. 6. The impedance curves are again plotted as solid lines. R_i represents the myoplasmic resistance, R_n the junctional resistance, and C_n the junctional capacitance. At low frequencies the capacitive reactance is very high, so that the measured impedance represents the sum of the two resistances R_i and R_n. At high frequencies, the capacitive reactance approaches zero, so that the measured impedance represents only R_i, the myoplasmic resistance. As indicated previously, the specific impedances of duodenum and jejunum above 5,000 Hz are not significantly different from one another. Hence, we conclude that the specific myoplasmic resistance of duodenum and jejunum are the same. On the other hand, the impedances measured at low frequencies are very different for the two tissues. We attribute this to differences in the junctional resistance, R_n, which, according to the model, is represented by the difference between the low- and high-frequency impedances.

Using the impedance values obtained at 50 and 10,000 Hz, we calculate a specific myoplasmic resistance of 181 and 210 $\Omega \cdot$ cm, and a junctional resistance of 304 and 710 $\Omega \cdot$ cm for duodenum and jejunum, respectively. The myoplasmic values are very close to the 190 $\Omega \cdot$ cm calculated by Tomita for guinea pig taenia coli. Although the junctional resistance is somewhat higher than the 180 $\Omega \cdot$ cm calculated for the taenia, the total specific longitudinal resistance of 485 $\Omega \cdot$ cm for duodenum compares favorably with 470 $\Omega \cdot$ cm obtained by Weidmann (9) and 463 $\Omega \cdot$ cm obtained by Kleber (10) for mammalian ventricle, another tissue in which the cells are tightly coupled electrically.

What these data indicate is that the increased resistance to excitatory current flow in the jejunum is localized in the junctions between the smooth muscle cells. The myoplasmic resistance is the same in the two regions of the intestine as are, supposedly, the extracellular and membrane resistances. An increase in junctional resistance implies that there is a reduction in junctional surface area representing low resistance pathways. This could occur either by a reduction in total number of junctions, or a reduction in the average junctional area, or both. In any case, one would expect an increase in junctional resistance to be accompanied by a decrease in junctional capacitance. It is possible to calculate the capacitive reactance, and hence the capacitance, in a circuit such as this if R_i, R_n, and the total impedance are known at any particular frequency. Using the values obtained for R_i and R_n and the total impedance values obtained at 500, 1,000, 3,000, and 5,000 Hz, capacitive reactance was calculated for both duodenum and jejunum at each of these intermediate frequencies. From these values, an average specific capacitance of 5.9×10^{-7} F/cm was calculated for duodenum and 3.4×10^{-7} F/cm for jejunum. The values are somewhat less than the 1 to 3×10^{-6} F/cm calculated by Tomita (8) for taenia coli, but this is to be expected, considering the larger junctional resistance values of the cat duodenum and jejunum.

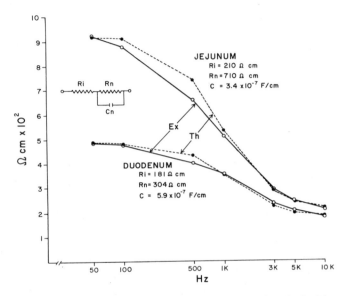

FIG. 6. (○) Replotted impedance data from Fig. 5. *Inset:* probable equivalent circuit of tissue. Myoplasmic resistance (R_i); junctional resistance (R_n); junctional capacitance (C_n). Values for jejunum and duodenum were obtained as described in text. (●) Theoretical curves calculated from values indicated. Difference in impedance between duodenum and jejunum is primarily caused by greater junctional resistance of jejunum.

Using the resistance and capacitance values in Fig. 6, the impedance values can be calculated for the equivalent circuit for each of the frequencies between 50 and 10,000 Hz. These values are indicated by the closed circles connected by the dashed lines in Fig. 6. Only one point (500 Hz, jejunum) lies outside the 95% confidence limits shown in Fig. 5.

CONCLUSIONS

The results of these experiments indicate that the decrease in slow-wave propagation velocity which occurs from duodenum to jejunum may be attributed to a relatively larger resistance to the flow of local circuit current in the jejunum, rather than to a relatively longer refractory period. This increased resistance apparently occurs at the junctions that provide low-resistance pathways between smooth muscle cells, i.e., the cells in the jejunum are not as tightly coupled electrically as are those in the duodenum.

ACKNOWLEDGMENTS

This work was supported in part by Public Health Service Research Grant AM-06958 from the National Institute of Arthritis, Metabolism and Digestive Diseases, and in part, by the Division of Research Facilities and Resources of the National Institutes of Health through Grant RR-00353. Statistical analysis by Dr. Henry Slotnick and technical assistance by Robert Lansing are gratefully acknowledged.

REFERENCES

1. Bortoff, A. (1972), *Ann. Rev. Physiol.*, 34:261–290.
2. Daniel, E. E. (1969), *Ann. Rev. Physiol.*, 31:203–226.
3. Daniel, E. E., and Chapman, K. M. (1963), *Am. J. Dig. Dis.*, 8:54–102.
4. Armstrong, H. I. O., Milton, G. W., and Smith, A. W. M. (1956), *J. Physiol. (Lond.)*, 131:147–153.
5. Bortoff, A., and Sacco, J. (1974), In: *Proc. 4th Int. Symp. Gastrointestinal Motility*, edited by E. E. Daniel, pp. 53–60. Mitchell Press, Vancouver.
6. Weisbrodt, N. W., and Christensen, J. (1972), *Gastroenterology*, 63:1004–1010.
7. Specht, P. C., and Bortoff, A. (1972), *Am. J. Dig. Dis.*, 17:311–316.
8. Tomita, T. (1969), *J. Physiol. (Lond.)*, 201:145–159.
9. Weidmann, S. (1970), *J. Physiol. (Lond.)*, 210:1041–1054.
10. Kleber, A. G. (1973), *Pflügers Arch.*, 345:195–205.

Physiology of Smooth Muscle, edited by
E. Bülbring and M. F. Shuba.
Raven Press, New York © 1976.

The Electrophysiological Characteristics of the Smooth Muscles of the Ureter

S. A. Bakuntz and V. Ts. Vantsian

Orbeli Institute of Physiology, Academy of Sciences of Armenian SSR, Yerevan, 375028, USSR

Smooth muscles differ from other excitable tissues in the great variability of functional properties. This is especially noticeable when one compares smooth muscles of different visceral organs and systems; of the same organs in different animal species; or the change of functional properties produced by changes in the physicochemical environment.

There are also difficulties in exploring excitatory and contractile processes of smooth muscles because of the intricate organization within the intraorgan structures and the well-known difficulties in using the microelectrode technique in this field.

Taking into account the above-mentioned factors, it was important to find a suitable tissue in which functional properties, especially excitatory and contractile processes, could be studied. The most convenient tissue for an investigation of automatic activity and conduction of excitatory and contractile processes was found to be the upper urinary tract, where automatic processes as well as rhythmic activity are well regulated. It is known that the time course of excitatory and contractile waves in ureter has a strictly linear relationship and brings about rhythmic peristaltic waves with some decrement during propagation.

The rhythm of ureter activity is so regular that Bozler (12) in his well-known paper compared it with the rhythmicity of cardiac muscle. This is in good agreement with our opinion. We also find great similarities in the principal organization of the origin of the rhythm in these two organs, although some differences exist in their hydrodynamic activity. The most characteristic common feature is the existence of a local pacemaker, generating rhythmic slow-wave deflections which give rise to action potentials, the latter being propagated to the other parts of organ.

The presence of slow waves in smooth muscles has been established for nearly all visceral organs and systems (7,8,13–18,33,34,36–38,42–44,46,49, 51).

Kuriyama and Tomita (34) suggested that slow potentials and action potentials were generated by different parts of the smooth muscle cell membrane. Kuriyama (33), classifying the different types of slow-wave activity in visceral organs, divided them into three main groups.

The first research work on this problem was done by Orbeli and Brücke (41) in 1910. Bozler (9–11) systematically investigated the excitatory processes in the ureter, including the generation of rhythm in the pacemaker part. The results of later investigations (2,3,11,20,21,25,26,39,47,48,50, 52,53) established the main properties of this organ: excitatory and contractile processes, excitability, refractoriness, responses to different physiologically active drugs, nervous control, and urodynamics.

A series of experiments performed by various microelectrode techniques (6,22,24,27–32,35) determined the functional properties of single cells and their ionic basis.

Our own investigations (4,5) of ureter physiology were mainly concerned with comparative physiology. Using dogs, cats, rabbits, rats, and guinea

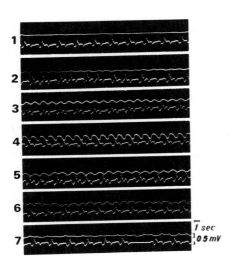

FIG. 1. Patterns of spontaneously generated slow potentials, recorded successively from points situated in the pelvis, at the pelviureteral junction and along the ureter of the cat (1–7, top). The action potentials recorded from the ureter (1–7, bottom).

pigs we studied the generation of rhythm in this organ, the localization of the pacemaker, the interrelationship between slow waves and action potentials, and some aspects of urodynamics. Different registration techniques of ureter biopotentials *in situ* were employed: intraluminal, extraureteral, and intramuscular recording from the pacemaker zone by the transrenal access or in nephrectomized conditions. We found that the most proximal zone where rhythmic excitatory waves with proper frequency for a given animal species could be generated is the pelviureteral junction. In more proximal parts of the renal pelvis, no rhythmic bioelectrical processes synchronous with the rhythm of ureter activity were ever observed in any animal species.

Investigating the topography of the pacemaker zone by recording biopotentials from many points of the renal pelvis and the initial part of the ureter, we found that slow waves with the highest amplitude were observed

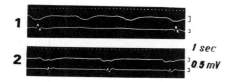

FIG. 2. Spontaneous activity from ureters of rabbit (1) and guinea pig (2). Top: Slow potentials from the pelviureteral junction. Bottom: Action potentials of the ureter. Time marker: 1 sec.

in the region of the pelviureteral junction, and that from here the rhythmicity of distal parts of the ureter was controlled (Fig. 1).

In more proximal regions than this, the amplitude of slow waves diminished with abrupt decrement in all animal species. This is probably the result of electrotonic spread from the generation zone to the adjacent smooth muscle cells. Comparison of the parameters of slow waves in different kinds of animals shows that the cause for the comparatively rare spontaneous contractions of rabbit and guinea pig ureters was the slow rhythm of their pacemakers, whereas in the other three species (dog, cat, and rat) the high frequency of motor activity of the ureter was due to the high frequency of their pacemaker activity.

The duration of single slow waves of ureter of rabbits and guinea pigs was long, and there were long diastolic pauses between them (Fig. 2). In contrast, the slow-wave deflections of ureter pacemakers in dogs, cats, and rats were approximately sinusoidal (Fig. 3).

The automaticity of the pacemaker, and its independence from more proximal parts in the upper urinary tract, was also shown in a series of experiments done under conditions of nephrectomy. Removal of the kidney, so that only a small part of the pelvis remained, never disturbed spontaneous slow-wave activity of the pacemaker and the generation of action potentials in the ureter. These results are in some disagreement with the cystoid theory of organization of urodynamics by the upper ureteral tract (19,40) and also with the conclusion of Sleator and Butcher (50) that the zone of ureteral rhythm generation is situated in the upper part of the urinary tract, more proximal than the pelviureteral junction. But they are in good agreement with the manometrical observations of Kiil (26) who put forward the hypothesis about the important role of the ureteral cone in urodynamics.

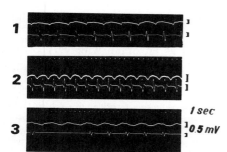

FIG. 3. Spontaneous activity from ureters of dog (1), cat (2) and rat (3). Top: Slow potentials from the pelviureteral junction. Bottom: Action potentials of the ureter. Time marker: 1 sec.

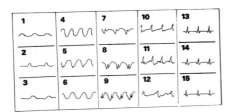

FIG. 4. Various patterns of the slow potentials (1–6), relation between slow waves and action potentials (7–12), and action potentials (13–15) of the dog ureter. Time marker: 1 sec.

During the advancement of an intraluminal electrode toward the lower part of the ureter, the amplitude of the slow waves diminished until they disappeared with simultaneous enhancement of the spike component (Fig. 4). Measurements of the propagation velocity of action potentials in different parts of the ureter showed that in the lower part of the ureter propagation is somewhat slower than in other parts, and it is mainly in this region that the great diminution of action potentials generated in the pacemaker zone takes place, so that they are not able to reach the bladder end of the organ.

During the simultaneous recording of electroureterogram, contraction of ureteral wall, urinary flow, respiratory rhythm, and heart activity, some interrelationships were observed: e.g., strict rhythmicity of the ureter, correlation between action potentials and contractile waves of this organ, and differences between the speed of propagation of excitation in different parts of ureter (Fig. 5).

From the point of view of comparative physiology, it should be mentioned that there are also certain species differences in the configuration of action potentials. For example, the duration of the action potentials in rabbits and guinea pigs is considerably longer, and some additional low-amplitude oscillatory deflections can be observed in the main deflection. In contrast, the action potentials of ureters of dogs, cats, and rats are fast and consist of the main deflection only. In investigations performed by intracellular microelectrodes on single smooth muscle cells of guinea pig ureter (22,23,35) additional oscillations on the main wave of the action potential were also observed. These data permit us to suggest the existence of species differences.

The frequency of ureteral action potentials of rabbits and guinea pigs is considerably lower than that of dogs, cats, and rats. Nevertheless, with each peristaltic wave in rabbits and guinea pigs, more urine is transported to the urinary bladder.

In conclusion, the ureter must be considered as an organ with very com-

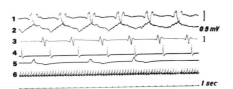

FIG. 5. Simultaneous recording of electrical (1, 3) and mechanical (2) activity of the ureter, urinary flow (4), respiratory movements (5), and electrocardiogram (6) from the dog. Time marker: 1 sec.

plicated functional organization and a characteristic pacemaker automaticity, which has the function of emptying the renal pelvis and the transport of urine into the urinary bladder.

REFERENCES

1. Agar, W. T. (1948): *Aust. J. Exp. Biol. Med. Sci.*, 26:253–257.
2. Bäcklund, L. (1963): *Acta Physiol. Scand.*, 59, Suppl. 212:1–86.
3. Baker. R., and Huffer, J. (1953): *J. Urol.*, 70:874–883.
4. Bakuntz, S. A. (1962): *Int. Congr. Physiol. Sci., 22nd, Leiden*, p. 284.
5. Bakuntz, S. A. (1970): *The Questions of the Physiology of Ureters*. Nauka Press, Leningrad.
6. Bennett, M. R., Burnstock, G., Holman, M. E., and Walker, J. W. (1962): *J. Physiol. (Lond.)*, 161, 47–48.
7. Bogach, P. G. (1968): *Int. Congr. Physiol. Sci., 24th, Washington*, pp. 99–100.
8. Bortoff, A. (1961): *Am. J. Physiol.*, 201:203–208.
9. Bozler, E. (1938): *Am. J. Physiol.*, 124:502–510.
10. Bozler, E. (1942): *Am. J. Physiol.*, 136:543–552.
11. Bozler, E. (1947): *J. Physiol. (Lond.)*, 149:229–231.
12. Bozler, E. (1948): *Experientia*, 4:213–218.
13. Bülbring, E. (1955): *J. Physiol. (Lond.)*, 128:200–221.
14. Bülbring, E. (1962): *Physiol. Rev.*, 42, Suppl. 5:160–178.
15. Burnstock, G., and Prosser, C. L. (1960): *Am. J. Physiol.*, 199:553–559.
16. Burnstock, G., Holman, M. E., and Prosser, C. L. (1963): *Physiol. Rev.*, 43:482–527.
17. Daniel, E. E. (1965): *Gastroenterology*, 49:403–412.
18. Daniel, E. E., Sehdev, H., and Robinson, K. (1962): *Physiol. Rev.*, 42, Suppl. 5:228–260.
19. Fuchs, F. (1931): *Z. Urol. Chir.*, 33:1–144.
20. Hukuhara, T., Nanba, R., and Fukuda, H. (1964): *Jap. J. Physiol.*, 14:197–209.
21. Ichikawa, S., and Ikeda, O. (1960): *Jap. J. Physiol.*, 10:1–12.
22. Irisawa, H., and Kobayashi, M. (1962): *Proc. Jap. Acad.*, 38:171–175.
23. Irisawa, H., and Kobayashi M. (1963): *Jap. J. Physiol.*, 13:421–430.
24. Irisawa, H., Kobayashi, M., and Irisawa, A. (1963): *J. Physiol. Soc. Jap.*, 25, 52–53.
25. Kawasaki, M. (1963): *Bull. Yamaguchi Med. Sch.*, 10:39–48.
26. Kiil, F. (1957): *The Function of the Ureter and Renal Pelvis*. Saunders, Philadelphia, Pa.
27. Kobayashi, M. (1964): *Tohoku J. Exp. Med.*, 83:220–224.
28. Kobayashi, M. (1965): *Am. J. Physiol.*, 208:715–719.
29. Kobayashi, M., and Irisawa H. (1964): *Am. J. Physiol.*, 206:205–210.
30. Kochemasova, N. G. (1971): *Bull. Exp. Biol. Med.*, 9:9–13.
31. Kochemasova, N. G., and Shuba, M. F. (1972): *Fiziol. Zh.*, 58:426–433.
32. Kochemasova, N. G., and Shuba, M. F. (1972): *Fiziol. Zh.*, 58:1287–1294.
33. Kuriyama, H. (1970): In: *Smooth Muscle*, edited by E. Bülbring, A. F. Brading, A. W. Jones, and T. Tomita, pp. 366–395. Edward Arnold, London.
34. Kuriyama, H., and Tomita, T. (1965): *J. Physiol. (Lond.)*, 178:270–289.
35. Kuriyama, H., Osa, T., and Toida, N. (1967): *J. Physiol. (Lond.)*, 191:225–238.
36. Marshall, J. M. (1959): *Am. J. Physiol.*, 197:935–942.
37. Marshall, J. M. (1962): *Physiol. Rev.*, 42, Suppl. 5:213–227.
38. Milton, G. W., and Smith, A. W. M. (1956): *J. Physiol. (Lond.)*, 132:100–113.
39. Murnaghan, G. F. (1961): *Br. J. Urol.*, 33:251–260.
40. Narath, P. A. (1951): *Renal Pelvis and Ureter*. Grune and Stratton, New York.
41. Orbeli, L. A., and Brüke, E. Th. (1910): *Pflügers Arch.*, 133:341–364.
42. Orlov, R. S. (1963): *Fiziol. Zh.*, 49:115–121.
43. Orlov, R. S. (1967): *The Physiology of Smooth Muscle*. Medicine Press, Moscow.

44. Papazova, M. (1970): *An Electrophysiological Study of the Motor Activity of the Stomach.* Publishing House of the Bulgarian Academy of Sciences, Sofia.
45. Prosser, C. L. (1954): *Am. J. Physiol.,* 179:663–663.
46. Prosser, C. L. (1962): *Physiol. Rev.,* 42, Suppl. 5:193–212.
47. Prosser, C. L., Smith, C. E., and Multon, C. E. (1955): *Am. J. Physiol.,* 181:651–660.
48. Shiratori, T., and Kinoshita, H. (1961): *Tohoku J. Exp. Med.,* 73:103–117.
49. Shuba, M. F. (1965): *Biofizika,* 10:67–76.
50. Sleator, W., Jr., and Butcher, H. R. (1955): *Am. J. Physiol.,* 180:261–276.
51. Steedman, W. M. (1966): *J. Physiol. (Lond.),* 186:382–400.
52. Weinberg, S. R. (1964): *J. Urol.,* 91:482–487.
53. Weinberg, S. R., and Hoffman, R. E. (1961): *J. Appl. Physiol.,* 16:933–934.

Physiology of Smooth Muscle, edited by
E. Bülbring and M. F. Shuba.
Raven Press, New York © 1976.

The Role of the Intrinsic Nervous System in the Correlation Between the Spike Activities of the Stomach and Duodenum

E. Atanassova

Bulgarian Academy of Sciences, Institute of Physiology, Sofia, Bulgaria

The correlation between the motor and the electrical activities of the stomach and duodenum is a problem of particular interest. Several authors have reported a myogenic transmission of slow waves from the stomach to the initial part of the duodenum (7,8,13). Allen et al. (1) found a correlation between the duodenal spike activity and the basic electrical rhythm of the stomach. Since coordination was observed only after feeding, the authors believe that feeding is an unlocking mechanism.

We have established that the correlation is one between the spike activities of the stomach and duodenum (2). Experiments were made on dogs in which bipolar ball-shaped silver electrodes were chronically implanted subserously (19) on the wall of the stomach and duodenum. During periods of relative rest, the bioelectrical activity of the muscle wall of the stomach of fasted dogs is characterized by slow potential changes of 4, 5, to 6 cycles per minute, whereas slow waves of 17 to 19 cycles per minute are recorded from the duodenal wall. Thus, there is no coordination between the slow activities of the gastric and the duodenal walls. However, during periods of hunger peristalsis, when groups of spike potentials follow slow potentials of the gastric wall, there are bursts of spike potentials on some slow duodenal waves which follow every gastric slow potential with spikes (Fig. 1A). Hence, there is a correlation between the spike activities. A similar correlation exists after mechanical stimulation of the stomach wall by inflating a balloon introduced through a fistula (Fig. 1B), or after feeding (Fig. 1C). It was concluded that coordination of the bioelectrical activities of the gastric and duodenal walls existed in all instances of a clearly manifested spike activity in the gastric wall, even when the stomach contained no food that, by propulsion in the duodenum, could stimulate the duodenal walls.

The question arises: What is the manner in which this correlation is effected?

We studied the bioelectrical activity of the stomach and duodenal walls before and after thoracic bilateral vagotomy and in cases of drug-induced sympathectomy. In both cases, the correlation between the spike activities of the stomach and the duodenum remained unchanged. Thus the extrinsic

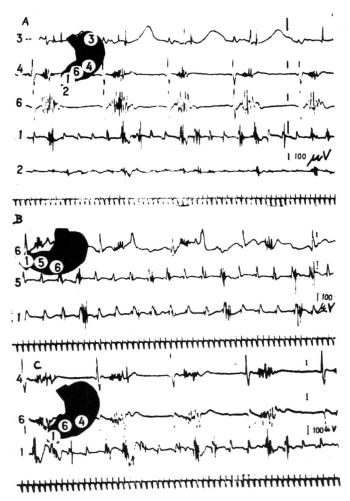

FIG. 1. Coordination of the spike activity of the stomach and duodenum in dogs with intact gastroduodenal junction (A) during what is known as hunger peristalsis; (B) after the distension of a balloon in the stomach; (C) 16 min after feeding with meat. The schemes in the upper left corners show the position of the electrodes. Time in seconds.

nervous system cannot play a decisive part in the correlation between the activities of the stomach and duodenum.

In order to prove that the intactness of gastric and duodenal walls was necessary for the manifestation of the above-mentioned correlation, we made a section at the level of the sphincter pylori in dogs. Cutting through the gastroduodenal junction abolished the coordination of gastric and duodenal activities. There were periods in which groups of spike potentials appeared with the slow potentials of the stomach, while either only slow

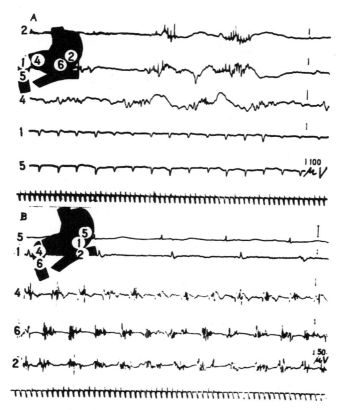

FIG. 2. Dissociation of the spike activity between the stomach and duodenum after cutting at the level of sphincter pylori. (A) Spike activity from the gastric wall alone; (B) spike activity only from the duodenal wall, while the stomach is in a state of relative rest. Schemes in upper left corners show position of electrodes. Time in seconds.

waves were led off from the duodenal wall (Fig. 2A), or groups of spike potentials occurred with the duodenal slow waves only (Fig. 2B). Another indication of this dissociation was the considerably increased spike activity of the duodenal wall compared with that of the duodenum of dogs with intact gastroduodenal junction. This fact was corroborated by the work of Bedi and Code (5).

Acute experiments on dogs were performed to establish the pathway for the coordination of the spike activity (Fig. 3). Direct current was applied to stimulate the muscle wall of the stomach. Spike activity in response to stimulation was recorded only from the region through which the current had been passed (Fig. 3B). Obviously, myogenic conductance plays no decisive role in effecting the correlation between the spike activities of stomach and duodenum.

In another group of animals we cannulated one of the small arteriae

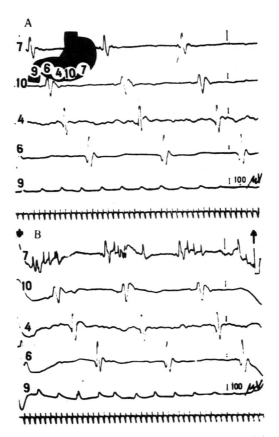

FIG. 3. Evoked spike activity of the stomach by stimulation with direct current. (A) Background; (B) during stimulation. The arrows show the moments of turning on and switching off of the current. Scheme in upper left corner shows position of electrodes. Time in seconds.

gastricae breves and the arteria gastroepiploica (Fig. 4). Infusing carbachol into the a. gastrica brevis, we produced a focus of excitation in the stomach wall. The pulses originating from the focus evoked bursts of spike potentials with the slow potential changes of the stomach and also the duodenum (Fig. 4B). Then we blocked the intramural ganglion cells of both the stomach and duodenum with hexamethonium introduced into the a. gastroepiploica. Ten minutes later carbachol injected into the a. gastrica brevis did not evoke spike activity either from the stomach, or from the duodenum (Fig. 4C). It was of interest to know whether the smooth muscle cells could still generate spike potentials. Therefore, after blocking the intrinsic ganglia with hexamethonium, we injected carbachol into the a. gastroepiploica, thereby stimulating the postganglion terminals: both stomach and duodenum responded immediately by intensification of their activities (Fig. 4D).

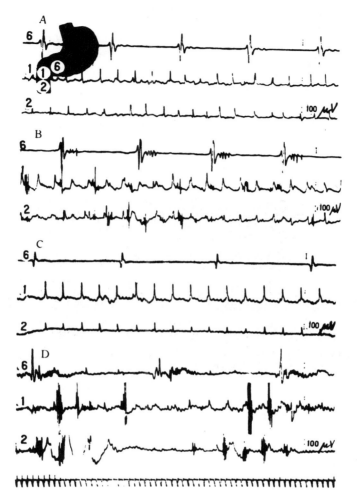

FIG. 4. The influence of blocking the intramural nervous system on the correlation between the spike activities of the stomach and duodenum. (A) Background; (B) 3 min after the administration of carbachol in arteria gastrica brevis; (C) 10 min after the injection of hexamethonium in arteria gastroepiploica; 2 min after the administration of carbachol in a. gastrica brevis; (D) 1 min after injecting carbachol in a. gastroepiploica. Scheme in upper left corner shows position of electrodes. Time in seconds.

Consequently, the intrinsic nervous system was considered to play a basic role in bringing about the correlation between the spike activities of the stomach and duodenum (3). To elucidate this mechanism further we used the microelectrode technique.

Microelectrode investigations were made on the gastroduodenal area of cats with Professor Gonella (Marseille, France) in his laboratory. A 2-mm-wide muscle strip containing 2 mm of stomach, the gastroduodenal junction, and 5 to 6 mm of proximal duodenum was placed in a two-section chamber

(20). The gastric part was fixed in the stimulation chamber with a platinum needle, which was also the stimulating electrode. Transmural stimulation was effected by single or repeated square-wave pulses of 0.2 to 0.5-msec duration. The electrical activities of single duodenal smooth muscle cells were detected by glass microelectrodes with a resistance of 30 to 50 MΩ. Atropine sulfate 10^{-6} g/ml, guanethidine 10^{-6} g/ml, and phentolamine 10^{-7} g/ml were added to the nutrient medium during the experiments.

The pattern of spontaneous electrical activity recorded from single smooth muscle cells of the proximal duodenum resembled that reported in the literature (6,14,17).

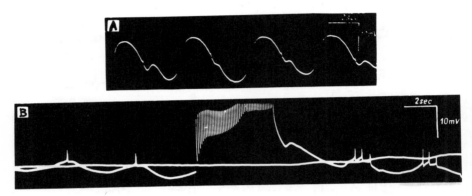

FIG. 5. (A) Excitatory junction potentials evoked during the repolarization phase of the duodenal slow waves. (B) Depolarization and increase in the number of spike potentials superimposed on the duodenal slow waves in response to repeated stimulation with frequency of 10 Hz.

Single short-duration pulses applied to the gastric part of the strip produced depolarization, i.e., the excitatory junction potential (EJP) (Fig. 5A). Its amplitude depended on the degree of membrane polarization at the time of stimulus presentation. When the stimulus fell into the phase of more complete repolarization the amplitude of the junction potential was higher. In some cases the depolarization provoked by single stimuli was so large that it exceeded the critical depolarization level, and bursts of spike potentials preceded a contraction. Sometimes a larger depolarization led to the appearance of spike potentials on several slow waves which preceded contraction waves of higher amplitude. The EJP had an amplitude of 2.5 to 15 mV with a latency of 100 to 150 msec. Junction potentials, but with a smaller amplitude, were also obtained when the recording microelectrode was moved 4 to 6 mm away from the stimulating electrode. Repeated electrical stimulation with a frequency of 10 Hz resulted in a long-lasting depolarization and an increase in electrical activity; the number of spike potentials and the amplitude of the contraction waves were increased (Fig. 5B), or spike potentials were initiated where not previously observed.

The addition of atropine sulfate to the nutrient medium altered the response pattern. Single transmural stimuli caused a hyperpolarization, i.e., the inhibitory junction potential (IJP), which preceded relaxation (Fig. 6A). The latency of these potentials was about 150 to 160 msec and their amplitude 5 to 10 mV.

Repeated electrical stimulation produced a long-lasting hyperpolarization (Fig. 6B). Immediately after stopping the stimulation a recovery process began, as a result of which the membrane became depolarized beyond the initial level. This was evidenced by a high-amplitude spike (rebound excitation) and a subsequent increase in slow-wave amplitude. During the period of hyperpolarization, there occurred a long-term relaxation, after

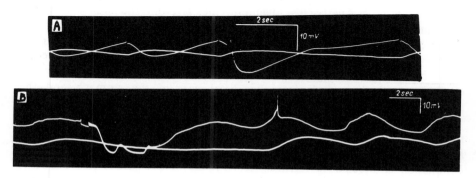

FIG. 6. (A) Inhibitory junction potential in response to single transmural stimulation after atropine sulfate. It precedes relaxation. (B) Hyperpolarization in response to repeated stimulation after atropine sulfate. Recovery and rebound excitation are seen after the stimulation. Relaxation is associated with the hyperpolarization and contractile waves follow rebound excitation.

which the rebound excitation preceded a high-amplitude contraction followed by several higher amplitude waves.

Inhibitory junction potentials were recorded from smooth muscle cells also without blocking the cholinergic system.

The hyperpolarizations in response to single or repeated transmural stimulation were not abolished by α- and β-adrenergic blocking agents (phentolamine and propranolol) added to the nutrient medium. However, neither IJP nor EJP was recorded after blockade of nervous conduction with procaine.

In order to show that the responses obtained resulted from conduction along the intrinsic nervous elements in the gastric and duodenal wall and were not coming from the outside, we performed thoracic vagotomy in some animals. Twenty-five days later, we made experiments on isolated preparations from the vagotomized cats. In response to transmural stimulation, well-defined junction potentials were obtained which were not affected by sympatholytic drugs or by α- and β-antagonists.

The junction potentials, excitatory and inhibitory, observed during our experiments, were similar in pattern to the junction potentials recorded from single smooth muscle cells of guinea pig duodenum, jejunum, ileum, colon, and rectum (18); cat and rabbit duodenum and jejunum (6); and guinea pig taenia coli (10,11) and stomach (4). Because these junction potentials were obtained from the duodenal smooth muscle cells in response to transmural stimulation of the stomach by short-lasting impulses (9,12), they proved the conduction of the impulses from the stomach to the duodenum across the gastroduodenal junction through nerve fibers. The existence of such fibers was established by Horton (16). Using the method of reconstruction with thousands of transections across the distal part of the stomach, sphincter pylori, and the proximal part of the duodenum, Horton showed the continuity of Auerbach's myenteric plexus from the stomach to the duodenum across the gastroduodenal junction. However, in addition to the intrinsic nervous system, a common nerve supply to the stomach and duodenum is also provided from the outside. Hollinshead (15) described a small bundle of the vagus, branching from its anterior trunk below the diaphragm, which spreads into the distal part of the stomach and the proximal part of the duodenum. The degenerative processes developing after thoracic vagotomy abolish the effect of these preganglionic fibers; thus the junction potentials which we obtained in response to transmural stimulation of preparations from vagotomized animals suggest that the nerve conduction is effected by intrinsic nerve fibers in the gastric and duodenal wall. Conduction along those nerves might be the basis for coordination between the spike activities in stomach and duodenum.

The slow potentials which are transmitted myogenically increase rhythmically the depolarization of duodenal smooth muscle cells (7,8,13). We assume that, upon increasing the gastric activity, bursts of impulses are conducted along nervous pathways and augment the depolarization of duodenal cells to exceed the critical firing level and generate spike potentials. Hence the correlation between the spike activities of stomach and duodenum is achieved through the intrinsic nervous system.

ACKNOWLEDGMENT

I wish to thank Professor Gonella from the Institute of Neurophysiology and Psychophysiology of C.N.R.S., Marseille, France, for his guidance and criticism in the microelectrode investigations.

REFERENCES

1. Allen, C. L., Poole, E. W., and Code, C. F. (1962): *Fed. Proc.,* 21:261.
2. Atanassova, E. (1970): *Bull. Inst. Physiol.,* 13:211.
3. Atanassova, E. (1970): *Bull. Inst. Physiol.,* 13:229.

4. Atanassova, E., Vladimirova, I. A., and Shuba, M. F. (1972): *Neirofiziologiia,* 4:216.
5. Bedi, B. S., and Code, C. F. (1972): *Am. J. Physiol.,* 222:1295.
6. Bortoff, A. (1961): *Am. J. Physiol.,* 201:203.
7. Bortoff, A., and Weg, N. (1965): *Am. J. Physiol.,* 208:531.
8. Bortoff, A., and Davis, R. S. (1968): *Am. J. Physiol.,* 215:889.
9. Bülbring, E., and Tomita, T. (1966): *J. Physiol. (Lond.),* 185:24P.
10. Bülbring, E., and Tomita, T. (1967): *J. Physiol. (Lond.),* 189:299.
11. Burnstock, G., and Holman, M. E. (1961): *J. Physiol. (Lond.),* 155:115.
12. Day, M. D., and Warren, P. R. (1967): *J. Pharm. Pharmacol.,* 19:408.
13. Fudjii, Y. (1971): *Am. J. Physiol.,* 221:413.
14. Gonella, J. (1965): *C. R. Acad. Sci. Paris,* 260:5362.
15. Hollinshead, W. H. (1956): *Anatomy for Surgeons,* New York.
16. Horton, B. (1931): *Arch. Surg.,* 22:437.
17. Kobayasi, M., Prosser, C. L., and Nagai, T. (1967): *Am. J. Physiol.,* 213:275.
18. Kuriyama, H., Osa, T., and Toida, N. (1967): *J. Physiol. (Lond.),* 191:257.
19. Papasova, M., and Milenov, K. (1965): *Bull. Inst. Physiol.,* 9:17.
20. Vladimirova, I. A., and Shuba, M. F. (1970): *Neirofiziologiia,* 2:544.

Physiology of Smooth Muscle, edited by
E. Bülbring and M. F. Shuba.
Raven Press, New York © 1976.

Spectral Analysis of Spontaneous Activity in Smooth Muscles

E. Başar and C. Eroğlu

Institute of Biophysics, Hacettepe University, Ankara, Turkey

The purpose of our studies was to perform a component analysis of spontaneous and evoked contractions of smooth muscles of the portal vein and the taenia coli of guinea pig.

Relevant studies on the mechanical rhythmicity of smooth muscles already exist in the literature. We may point out the studies of Bülbring (1), Bülbring and Kuriyama (2), Golenhofen and Loh (3), Funaki and Bohr (4), Johannson and Ljung (5), and Holman et al., (6), who have carefully studied the mechanical rhythmicity in the portal vein and the taenia coli. However, we did not encounter any study on the mechanical activity that was carried out using the time series analysis method. Time series analysis methods and some other systems theory tools with ideal filtering, which we have applied in this study, enable the investigator to determine the exact spectral composition of different rhythms when the oscillatory system has a multiperiodical behavior. The determination of amplitude frequency characteristics of smooth muscle contractions upon quick stretch should give new information on the quantification of evoked contractile oscillations.

The smooth muscle preparations used in our experiments consisted of isolated strips of guinea pig portal vein and taenia coli with approximate lengths of 2 cm. The experimental setup and methods of preparation are described elsewhere (7,8).

SPONTANEOUS CONTRACTION PATTERNS OF THE PORTAL VEIN AND THE TAENIA COLI

Portal Vein

Figure 1A illustrates five typical records of spontaneous fluctuations of the tension developed in the portal vein. They are typical in the sense that the activities recorded from 20 different portal vein preparations during a total of 250 hr of observation could be identified most of the time as one of the types shown in Fig. 1A. It should also be noted that during a long recording period of about 8 hr almost all the activity patterns recorded from each individual preparation fit one of these forms. Pattern I of Fig. 1A has a frequency of about 0.1 Hz. Pattern II has a frequency of about 0.01 Hz,

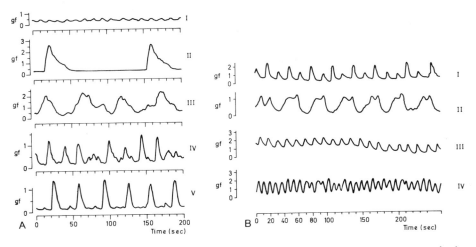

FIG. 1. A (I–V): Typical records (time histories) of spontaneous tension changes of guinea pig portal vein. Transfer spring constant used was 3 g/cm. Tension scale shows directly the force created in the portal vein strip. B (I–IV): Typical records (time histories) of spontaneous tension changes of guinea pig taenia coli. Transfer spring constant used was 3 g/cm.

and pattern III presents a contraction period with a frequency of 0.02 Hz. Pattern II (Fig. 1A) has a simple periodic form without wavelets (or superimposed waves), whereas pattern III has a more complicated shape. Patterns IV and V are complex waveforms and in these cases it is difficult to define a unique contraction frequency. Pattern V was observed in 28% of our experiments, pattern IV in 23%, pattern III in 19%, and pattern II in only 12%; the simple higher frequency pattern I was recorded in 18% of our data.

Another remarkable observation was the following: Any increase in the passive tension (stretch) of the preparations resulted in oscillations of shorter periodicities. Dominant periodicities of about 1 to 2 min were usually observed later in the experiment. This important point is discussed later.

Taenia Coli

Figure 1B illustrates four typical records of spontaneous fluctuations of the tension developed in taenia coli. Visual inspection of these records indicated also that the rhythmicities in these curves are of a complex nature. As in portal vein contractions, the rhythmicity in taenia coli shows multiperiodicity. We used taenia coli preparations from 20 different guinea pigs, and examined recordings of about 300 hr of observations. The activities recorded could always be identified as one of the activity types shown in Fig. 1B.

SPECTRAL ANALYSIS OF SMOOTH MUSCLE CONTRACTIONS

Portal Vein

Plots of power spectral density versus frequency (S_{xx} vs f) are presented in Fig. 2. These plots are called power spectra. For the methods of power spectra evaluation, the reader is referred to the literature (7,8). Curves I–V of Fig. 2 were obtained using time histories I–V of Fig. 1A, respectively. In power spectrum I of Fig. 2 only one maximum around 0.1 Hz is seen. The dominant maximum is about 0.01 Hz for curve II, and about 0.02 Hz for curve III. Power spectrum V contains three maxima centered at frequencies of about 0.02, 0.06, and 0.1 Hz. Power spectrum IV depicts two maxima at frequencies of 0.05 and 0.1 Hz, but no maximum around 0.01 to 0.03 Hz (100 power spectra were evaluated).

According to the percentages of patterns given in the previous section and combining the information contained in the power spectra, the following results were obtained: (a) The 0.01 to 0.03 Hz component (which corresponded to the minute rhythm) existed in 88% of all the curves; (b) the

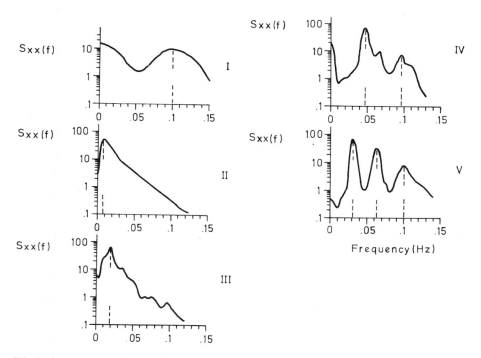

FIG. 2. Power spectra of tension changes of portal vein which are obtained using the time histories of Fig. 1A. Curves I–V correspond to the patterns I–V of Fig. 1A, respectively. Along the abscissa is the frequency in Hz, along the ordinate is the power spectral density $S_{xx}(f)$ in logarithmic relative units.

0.05 to 0.08 Hz component existed in 60% of all our recordings; (c) the faster 0.09 to 0.2 Hz component was observed in 47% of all the recordings.

Taenia Coli

Power spectra of the spontaneous mechanical activity of taenia coli are illustrated in Fig. 3. Curves I, II, III, and IV were obtained using the time histories I, II, III, and IV of Fig. 1B, respectively. The dominant peak is about 0.02 Hz for curve II, 0.07 Hz for curve III, and 0.13 Hz for curve IV. For some of the patterns which are not presented in this study, a maximum was noted at about 0.01 Hz.

The results show that the rhythmic contractions of the portal vein and of the taenia coli are found mainly in three frequency ranges: 0.01 to 0.03 Hz, 0.05 to 0.08 Hz, and 0.09 to 0.2 Hz. These contractions appear either as single rhythmicities, or in the form of complex contraction patterns. Time histories of Fig. 1A(IV, V) and of Fig. 1B(I, II) show complex contraction patterns. Accordingly, the corresponding power spectra contain several maxima: the spectrum of Fig. 2(IV) depicts two maxima and the spectrum of Fig. 2(V) depicts three maxima. In other words, smaller superimposed waves in the dominant periodical element of the contraction pattern give rise to various maxima in the power spectrum. Therefore, the power spectrum of Fig. 2(IV) indicates the existence of two, and the power spectrum of Fig. 2(V) indicates the existence of three distinct components

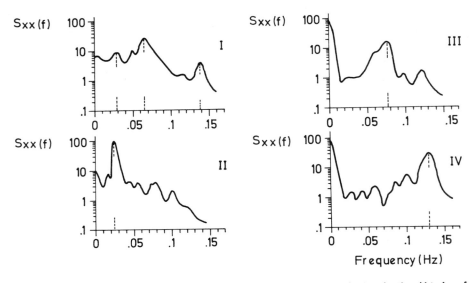

FIG. 3. Power spectra of tension changes of taenia coli which are obtained using the time histories of Fig. 1B. Curves I–IV correspond to the patterns I–IV of Fig. 1B, respectively. Along the abscissa is the frequency in Hz, along the ordinate is the power spectral density $S_{xx}(f)$ in relative units.

in the complex contraction patterns. A detailed example for the super-imposed waves on the dominating component is given in Ref. 8.

The methods encountered in the smooth muscle literature enable one to determine the approximate period of only the dominating component or the dominating rhythm. But, by applying this mathematical tool (evaluation of the power spectra) all the discrete periodicities contained in a complex pattern of spontaneous contractions can be detected and interpreted.

Our discussion may even lead one to interpret various forms of multi-periodical contraction as different combinations of these components with varying relative magnitudes and phases during different stages of smooth muscle contractions and under different experimental conditions. This consideration will be further explained in the next section.

CONTRACTILE COMPONENTS OF SMOOTH MUSCLE IN TIME AND FREQUENCY DOMAINS—STAGES OF CONTRACTION OF SMOOTH MUSCLE

The previous sections revealed that the shape of time histories of contraction patterns alone cannot allow any exact statement on the multiperiodicity of the mechanical spontaneous activity. The power spectra, however, showed that in most of the contraction patterns various periodicities with different weights are present. For example, visual inspection of the pattern of Fig. 1B,III does not reveal the existence of a rhythm of 0.08 Hz, which we can distinguish only in the power spectrum (Fig. 3,III). Similarly, visual inspection of patterns of Fig. 1B,I does not allow us to distinguish the three periodicities which are seen in the power spectrum of Fig. 3,I. Therefore, we can assume that different contraction patterns are due to the fact that the equilibrium of strengths of different rhythmicity components perpetually changes. As a result we observe different contraction patterns, although usually all the components are present with different weights. According to this interpretation we introduce the concept of *different stages of spontaneous activity of smooth muscle.*

In the previous section we demonstrated the existence of three different (rhythmic) components of portal vein and taenia coli. These rhythmic components were centered in three similar frequency ranges in portal vein and taenia coli. Using an ideal theoretical filtering method, which will be explained below, we can select or reject components of contraction patterns presented in Fig. 1A,B. Let us mention the example of Fig. 1B,I,II.

The ideal mathematical filtering consists of the following steps:

1. The frequency band limits of theoretical filters are chosen according to the frequency and bandwidth of the power spectra $S_{xx}(f)$.

2. The weighting function, $g_F(t)$ of the ideal filter characteristic $G_F(j\omega)$ is computed with the inverse Fourier transform

$$g_F(t) = \frac{1}{2\pi} \int\limits_{-\infty}^{+\infty} \{|G_F(j\omega)|\, e^{-j\omega\tau}\} e^{j\omega t}\, d\omega$$

3. The experimentally obtained contraction patterns response, $c(t)$, is filtered using the convolution integral:

$$c_F(t) = \int\limits_{-\infty}^{+\infty} g_F(\tau) c(t-\tau)\, d\tau$$

where $c_F(t)$ is the filtered pattern.

For detailed information, see our recent reports (9).

Application of theoretical filters to two of the patterns of Fig. 1B gave the following results:

1. By application of a low pass filter with a cutoff frequency of 0.04 Hz to the time histories of Fig. 1B,I we reject all activities higher than 0.04 Hz.

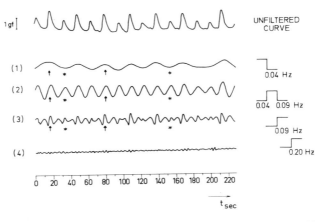

FIG. 4. Theoretical filtering of the curve of Fig. 1B(I) with different ideal filters. The band limits of the filters (filter characteristics) applied are schematically shown on the right side of the filtered curves.

After rejection of frequency components higher than 0.04 Hz, we observe the oscillatory behavior of curve 1 in Fig. 4. These results confirm the existence of a component below 0.04 Hz, although such a component could not be detected by visual inspection of the unfiltered pattern. Application of a passband filter between 0.04 and 0.09 Hz selects another oscillatory component (curve 2 in Fig. 4). A smaller component between 0.09 and 0.2 Hz must also exist, since by application of a high pass filter which rejects activities slower than 0.09 Hz a smaller oscillatory behavior is seen (curve 3 in Fig. 4), whereas no activity is left by application of a high pass filter of 0.2 Hz (curve 4 in Fig. 4).

2. Similar theoretical filters were applied to the contraction pattern (II)

of Fig. 1B. The largest component was the component selected by application of a low pass filter with a cutoff frequency of 0.05 Hz (curve 1 in Fig. 5). The oscillatory component between 0.05 and 0.09 Hz (curve 2 in Fig. 5) and the component larger than 0.09 Hz were small in comparison to the slow component in curve 1.

The theoretical filtering gave us the same results as those obtained by the study of the power spectra. If we compare the power spectra of Fig. 3 and the filtered curves of Figs. 4 and 5, we see that in both results the largest periodical component is the activity around 0.02 to 0.03 Hz. The analysis, which we extended to all curves obtained, gave further support to the power spectra. We mention here only two examples.

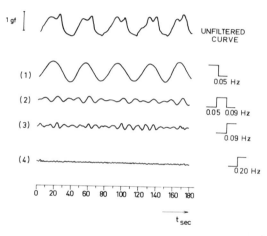

FIG. 5. Theoretical filtering of the curve of Fig. 1B(II) with different ideal filters. The band limits of the filters (filter characteristics) applied are schematically shown on the right side of the filtered curves.

The isolated components (obtained with theoretical filtering) of both spontaneous activities shown in Figs. 4 and 5 have similar oscillatory time courses, although no resemblance of the unfiltered contraction patterns could be detected by visual inspection. This is because of the fact that the corresponding oscillatory components (shown in Figs. 4 and 5) have different amplitudes (weights) and different relative phases during different stages of taenia coli contractions and different experimental conditions. (If the components are in opposite phase they cancel each other.) In other words, subsystems or mechanisms, which are responsible for different contraction components, change their activities or magnitudes during different contraction stages. Therefore dissimilar contraction patterns are observed, although the periodicities of the components are basically similar. This is another explanation (or support) for the concept of the *equilibrium of strengths of different rhythmic components* stated in the previous section.

It is also possible to assume "contractile reverberation circuits." These reverberatory contractions might be due to at least three different contraction components. The filtered components of Figs. 4 and 5 reveal a very important point: The interactions between the contraction components seem to be very strong. One can easily see the following correlation: When the tension maxima in components 1 and 2 of Fig. 4 are in phase, component 3 (the component with activity higher than 0.09 Hz) has the largest magnitude (see the arrows in Fig. 4). When components 1 and 2 are in counterphase (asterisks in Fig. 4), component 3 has a small magnitude. This fact can be explained as follows: When some smooth muscle filaments do contract, they may act as mechanical stimulators for other contractile filaments. This stimulation effect can be compensated when different contraction components are acting in counterphase.

This point emphasizes the necessity for systems theoretical component analysis of the rhythms, especially for the difficult task of classification of the smooth muscle rhythms. The superimposition and the mutual influence of the rhythms merit important consideration for the functional classification of the rhythms. Therefore, as an important remark, we want to emphasize that the application of systems theoretical component analysis methods adds an important point of view to the analysis of smooth muscle rhythmicity. Using these methods, the experimenter can define the compound contraction patterns in terms of their frequency components. Accordingly, in the classification of rhythmicities, compound patterns should be considered with their spectral compositions and not with the frequency of the dominating contraction rhythm.

TIME AND FREQUENCY CHARACTERISTICS OF THE TENSION INDUCED BY PASSIVE STRETCH IN SMOOTH MUSCLES

We have performed experiments in order to determine time and frequency characteristics of the tension developed in smooth muscle to passive stretch. Step functions (stretch stimulation in the form of a step function) are applied to portal vein and taenia coli preparations. After measuring the step responses (responses to sudden stretch) we evaluated the amplitude frequency characteristics of the tension induced by using a Laplace transform of the following form:

$$|G(j\omega)| = \int_0^\infty e^{-j\omega t}\, d[c(t)]$$

where $|G(j\omega)|$ is the amplitude frequency characteristic, and $c(t)$ is time response to sudden stretch (step response).

Upon sudden passive stretch, the portal vein responds with a quick increase of tension. The tension reaches maximal values between 2 and 4 sec after quick stretch; it then decreases and reaches a steady-state value after some small fluctuations. (When the spring constant of the force transducer is low, the fluctuations have larger magnitudes.) The response of taenia coli to sudden stretch is basically similar. Our results are in good agreement with earlier measurements of Burnstock and Prosser (10). Figure 6 shows two typical amplitude frequency characteristics of the tension developed in the portal vein. The abscissa is the frequency in logarithmic scale; the

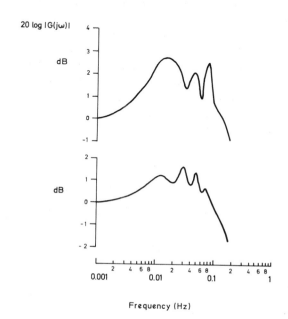

Frequency (Hz)

FIG. 6. Typical amplitude frequency characteristics of the tension induced by passive stretch in the portal vein. Along the abscissa is the frequency in logarithmic scale. Along the ordinate is the amplitude of the tension in relative units and decibels (20 log = 1 dB).

ordinate is the amplitude (tension amplitude) in relative units and decibels (20 log = 1 dB).

The amplitude characteristics depict maxima centered usually at frequencies ~0.01 to 0.03 Hz, ~0.05 to 0.06 Hz, or ~0.1 Hz. In other words, the amplitude characteristics show stretch-sensitive components in the same frequency bands seen in the spontaneous mechanical activity (cf. Figs. 2, 3, and 6). However, it is interesting to note that the high-frequency components (or amplitude maxima) in the vicinity of 0.1 Hz have relatively larger magnitude in comparison to the components of power spectra obtained using the spontaneous mechanical activity. If one assumes that the amplitude maxima in Fig. 6 are due to myogenic resonance phenomena, it can be concluded that the resonance in the higher frequency range of 0.1 Hz is larger than the resonance in the low-frequency range of the minute rhythm. This means that forced oscillations have greater selectivities in higher frequencies. In the first section we mentioned that any increase in the passive tension of the preparations resulted in oscillations of shorter periodicities. The amplitude frequency characteristics serve to obtain a more objective quantification of this observation. In simpler words, the evaluating of frequency characteristics support the following assumption: Passive stretch of smooth muscles augments the frequency of smooth muscle oscillations.

We will not explain here in detail the different uses of the amplitude frequency characteristics. However, we have to remember that vascular smooth muscles are subjected to a form of stretch when pressure is increased in the circulatory system: an increase of pressure in the peripheral circulatory system corresponds to a passive increase of tension in the arterial wall. The vascular smooth muscle amplitude characteristics fit well together with the frequency characteristics of peripheral vasoconstriction in organs such as the rat kidney and the coronary system (11,12).

REFERENCES

1. Bülbring, E. (1955): Correlation between membrane potential, spike discharge and tension in smooth muscle. *J. Physiol. (Lond.)*, 128:200–221.
2. Bülbring, E., and Kuriyama, H. (1973): The action of catecholamines on guinea-pig taenia coli. *Philos. Trans. R. Soc. Lond. [Biol. Sci.]* 265:115–121.
3. Golenhofen, K., and Loh, D. V. (1970): Intracellulare Potentialmessungen zur normalen Spontanaktivität der isolierten Portal Vene des Meerschweinchens. *Pflügers Arch.*, 318:82–100.
4. Funaki, S., and Bohr, D. (1964): Electrical and mechanical activity of isolated vascular smooth muscle of the rat. *Nature*, 203:192–194.
5. Johansson, B., and Ljung, B. L. (1967): Sympathetic control of rhythmically active vascular smooth muscle as studied by a nerve–muscle preparation of portal vein. *Acta Physiol. Scand.*, 70:299–311.
6. Holman, M. E., Kasby, C. B., Suthers, M. B., and Wilson, J. A. F. (1968): Some properties of the smooth muscle of rabbit portal vein. *J. Physiol. (Lond.)*, 196: 111–132.
7. Başar, E., Eroğlu, C., and Ungan, P. (1974): Time series analysis of guinea-pig taenia coli spontaneous activity. *Pflügers Arch.*, 347:19–25.
8. Başar, E., Eroğlu, C., and Ungan, P. (1974): An analysis of portal vein spontaneous contractions. *Pflügers Arch.*, 352:135–143.
9. Başar, E., and Ungan, P. (1973): A component analysis and principles derived for the understanding of evoked potentials of the brain: Studies in the hippocampus. *Kybernetik*, 12:133–140.
10. Burnstock, G., and Prosser, C. L. (1960): Responses to quick stretch; relation of stretch to conduction. *Am. J. Physiol.*, 198:121–125.
11. Başar, E. (1974): Systems theory of coronary and renal hemodynamics. In: *Regulation and Control in Physiological Systems*, edited by A. S. Iberall and A. C. Guyton, pp. 394–396.
12. Başar, E., Ruedas, G., Schwarzkopf, H. J., and Weiss, Ch. (1968): Untersuchungen des zeitlichen Verhaltens druckabhängiger Änderungen des Strömungswiderstandes im Coronargefäßsystem des Rattenherzens. *Pflügers Arch.*, 304:189–202.

Physiology of Smooth Muscle, edited by
E. Bülbring and M. F. Shuba.
Raven Press, New York © 1976.

Investigation of Contractile and Electrical Activity of Smooth Muscle of Lymphatic Vessels

R. S. Orlov, R. P. Borisova, and E. S. Mundriko

Department of Physiology, Medical Institute of Sanitation and Hygiene, Leningrad, USSR

It has long been suggested that lymphatic vessels contract. However, no special investigations of smooth muscle functions in lymphatic vessels have been reported. Findings recently published deal with contractile activity of mesenteric lymphatic vessels (2,3). In our experiments, we have studied contractile and electrical activity of smooth muscles in the main lymphatic vessels of rats, such as the upper, middle, and lower third of the thoracic lymphatic duct; the cistern; and the efferent vessels of the mesenteric lymphatic nodules. Contractions were recorded by mechanotransducer, the membrane potential, with the double sucrose-gap technique (1).

It was found that the main lymphatic vessels have spontaneous contractile rhythmic activity and respond to electrical stimulation and catecholamines by contracture-like contractions. Spontaneous contractions are regularly registered in the cistern and the cranial part of the thoracic lymphatic duct, but they are usually not observed in its middle and lower third. At the temperature of 37°C the frequency of spontaneous rhythmic contractions average 12 contractions per minute. The appearance of spontaneous activity depends not only on the regions of the lymphatic vessel examined, but also on a number of other factors: the composition of the medium, the initial degree of extension, and the temperature. For example, at the beginning of an experiment, spontaneous rhythmic activity was sometimes not observed in regions in which it usually occurs. However, under the influence of some factors (ions, mediators) rhythmic contractions appeared later and persisted during the remainder of the experiment. Furthermore, spontaneous rhythmic activity could be induced by combined application of Ca^{2+} and K^+ ions and by acetylcholine even in regions (lower and middle thirds of the thoracic lymphatic duct) which usually do not exhibit it.

Epinephrine and norepinephrine in concentration 1×10^{-6} M had two effects: they caused a contracture and they increased the frequency of the spontaneous contractions. As a rule, the positive chronotropic effect developed on the top of the contracture. It was accompanied by a reduction of the amplitude of the separate spontaneous contractions (Fig. 1A). Phentolamine (1×10^{-6} M) caused an increase in the frequency of spontaneous rhythmic contractions (positive chronotropic effect) followed by a progres-

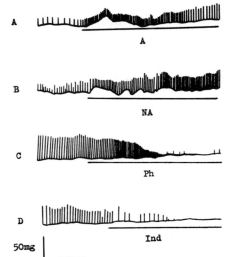

FIG. 1. Effects of epinephrine (A, adrenaline) and norepinephrine (NA, noradrenaline) and adrenergic blocking agents on spontaneous activity. (A) E (epinephrine 1×10^{-6} M); (B) NE (norepinephrine 1×10^{-6} M); (C) Ph (phentolamine 1×10^{-6} M); (D) Ind [Inderal [propanolol] 1×10^{-6} M].

sive reduction of their amplitude (Fig. 1C) until they were almost abolished. Inderal [propanolol] (1×10^{-6} M) from the very beginning reduced both the frequency and the amplitude of spontaneous contractions (Fig. 1D).

Acetylcholine in concentration of 1×10^{-6} M also caused a contracture and had a positive chronotropic effect (Fig. 2A). It increased the amplitude of the spontaneous rhythmic contractions. In some quiescent preparations

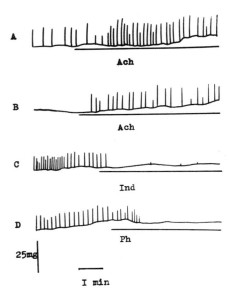

FIG. 2. Effect of acetylcholine on spontaneous activity (A, B) and blockade of the acetylcholine induced activity by adrenergic blocking agents (C, D). (A, B) ACh (acetylcholine 1×10^{-6} M); (C) Ind (Inderal [propanolol] 1×10^{-6} M); (D) Ph (phentolamine 1×10^{-6} M).

acetylcholine initiated rhythmic spontaneous activity (Fig. 2B). Such activity, induced by acetylcholine, was inhibited by adrenergic blocking agents (Fig. 2C,D).

Spontaneous rhythmic activity of smooth muscle cells of lymphatic vessels and its response to norepinephrine depended on the level of Ca^{2+} ions in the medium (Fig. 3A). In Ca^{2+}-free solution the rhythmic spontaneous activity ceased (Fig. 3B) and the addition of Ca^{2+} restored it (Fig. 3C). The extracellular calcium concentration influenced the responses of the lymphatic vessels to catecholamines as well. For instance, neither epinephrine

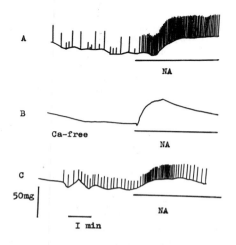

FIG. 3. Action of norepinephrine on spontaneous activity in normal and Ca^{2+}-free solution. (A) Spontaneous contraction in normal solution: NE (norepinephrine 1×10^{-6} M); (B) after 10 min in Ca^{2+}-free solution: NE (norepinephrine 1×10^{-6} M); (C) after addition of $CaCl_2$ in Ca^{2+}-free solution: NE (norepinephrine 1×10^{-6} M).

nor norepinephrine restored the rhythmical spontaneous activity in Ca^{2+}-free solution (Fig. 3B), but they still produced a contracture.

Electrical stimulation evoked tonic contractile responses of lymphatic vessels. At a frequency of 15 imp./sec the contraction increased gradually during stimulation and receded slowly after the stimulation was stopped. We used the term "tonic contraction" because it differed essentially from the faster spontaneous ones. As the frequency of stimulation was increased the rate of rise and the amplitude of the tonic contraction were increased (Fig. 4A). A similar effect was observed when the other parameters of the stimulation (intensity and duration of the electrical pulses) were increased. Tonic contractions persisted in the presence of adrenergic blocking agents, they were unaffected by addition of manganese ions (Fig. 4C,D) and the removal of Ca^{2+} (Fig. 4E).

The membrane potential of smooth muscle cells in different portions of the thoracic lymphatic ducts was 25 mV [measured by the sucrose gap method (1)]; in the smooth muscle cells of the mesenteric lymphatic vessels it was up to 35 to 37 mV. Action potentials were associated with the individual contractions (Fig. 5A,B).

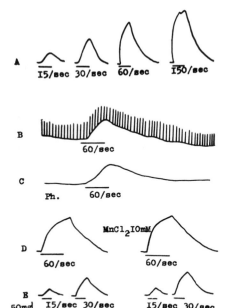

FIG. 4. Contractions triggered by electrical stimulation. (A) Effect of increasing frequency of stimulation from 15 to 150 pps; (B) effect of electrical stimulation in normal solution; (C) in solution containing phentolamine (1×10^{-6} M); (D) effect of electrical stimulation before (left) and after (right) treatment with $MnCl_2$ (10 mM); (E) effect of electrical stimulation in normal (left) and Ca^{2+}-free (right) solution.

In spontaneously active preparations of the thoracic lymphatic duct a single electrical pulse produced a contraction, the parameters of which did not differ from those of spontaneous ones. The duration and the amplitude of the evoked contractions did not change when the duration and intensity of the stimulating pulses were increased (Fig. 5C,D). However, if the

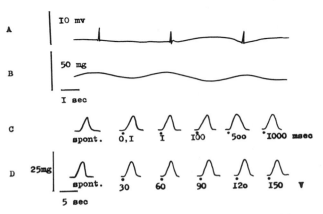

FIG. 5. Simultaneous recording of electrical (A) and mechanical (B) activity. Influence of duration (C) and amplitude (D) of stimulating pulses on the magnitude of the contractions.

interval between stimulating pulses was decreased, then the amplitude of the induced contractions became smaller (Fig. 6).

The experiments that have been described demonstrate the spontaneous rhythmic contractile activity in the upper third of the thoracic lymphatic duct, as well as in the cistern and in the lymphatic nodules of the efferent vessels. The regularity of the rhythm of spontaneous contractions may be due either to simultaneous activity of the pacemaker cells in several lymphangions, or to transmission of impulses from one lymphangion to another. The constant rhythmic activity in those parts of the lymphatic system which open into the venous system or into sacs where the lymphatic fluid coming from the viscera is stored indicates that the accumulation of lymph leads to

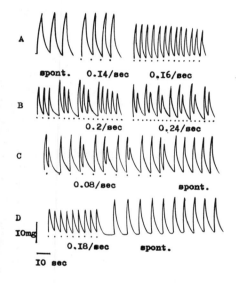

FIG. 6. Effect of interval between stimulating pulses on the magnitude of the contractions. (A) From left to right: spontaneous contractions, contractions evoked by electrical stimulation 0.14 and 0.16/sec; (B) contractions evoked by electrical stimulation 0.2 and 0.24 per sec; (C) spontaneous and evoked contractions, 0.08 per sec; (D) from left to right: evoked contractions (0.18 per sec) and spontaneous contractions.

smooth muscle distension, and this may be the most important factor underlying spontaneous rhythmic activity. Mislin's experiments (3), which show the dependence of the spontaneous contractions on the pressure inside the vessels of the mesenteric lymphatic system, may be considered as supporting evidence.

The study of the action of catecholamines and adrenergic blocking agents brought to light the existence of adrenergic receptors in the smooth muscle cells of lymphatic vessels. The dual response to catecholamines (increased frequency of spontaneous contractions and contracture) shows that the catecholamines have two different actions on the lymphatic smooth muscle. The increased frequency of spontaneous contractions may result from an action on the pacemaker cells, whereas the contracture may result from a direct action of epinephrine and norepinephrine on the contractile mechanism of smooth muscle cells. The experiments with acetylcholine showed an influence on the spontaneous rhythmical activity similar to that of cate-cholamines. Inderal [propanolol] and phentolamine block the effect of catecholamines as well as acetylcholine. This may be caused by their direct depressant action rather than an indication that cholinergic effects on lym-phatic vessels may be produced by the excitation of adrenoreceptive struc-tures in spontaneously active cells.

The disappearance of spontaneous activity in Ca^{2+}-free solution may be explained by a requirement of calcium ions for spontaneous depolarization in pacemaker cells, or by block of electromechanical coupling. However, the development of epinephrine and norepinephrine contractures in Ca^{2+}-free solution, and their contractile responses to direct electric stimulation after application of manganese ions and adrenergic blocking agents show that lymphatic vessel smooth muscle cells must contain a store of intracellular calcium.

The resting membrane potential of smooth muscle cells in lymphatic vessels was similar to that of smooth muscle in blood vessels, but considera-bly lower than that of skeletal muscle. The registration of spontaneous action potential synchronously with contractions reveals the possibility that they are triggered by action potentials.

The discovery of spontaneous rhythmical contractions in lymphatic vessels provides a physiological basis for the idea of lymph hearts. If the contraction of the muscular wall of a main lymphatic vessel is an important factor in pushing lymph through the venous system, then an individual lymphangion is an elementary physiological unit. We have not yet succeeded in carrying out experiments with an individual lymphangion, but our results comparing several lymphangions in "large and small segments" show that small segments respond to single electric stimuli with contractions which do not differ much from spontaneous rhythmical ones in all their parameters. A single electric shock apparently stimulates pacemaker cells and excitation spreads along the functional units of the lymphatic vessel—the lymphangion smooth muscle cells.

REFERENCES

1. Berger, W., and Barr, L. (1969): *J. Appl. Physiol.,* 26:378–389.
2. Mawhinney, H. J., and Roddie, J. C. (1973): *J. Physiol. (Lond.),* 229:339–348.
3. Mislin, H. (1971). *Angiologics,* 8:207–211.

Physiology of Smooth Muscle, edited by
E. Bülbring and M. F. Shuba.
Raven Press, New York © 1976.

The Effect of Changes in Osmolarity and Oxygen Tension on Excitability and Conduction of Excitation in Vascular Smooth Muscles

M. I. Gurevich, S. A. Bershtein, and I. R. Evdokimov

Although there are numerous studies, our understanding of the mechanisms of vascular tone is still essentially deficient. Recognition of the conception of a basal (1) and a peripheral (2) tone of blood vessels has led to the view that, parallel with reflex effects, local mechanisms play an essential role in providing adequate blood supply for the function of organs and tissues. The occurrence of "specific" vasodilator substrates responsible for local changes in blood flow indicates the importance of factors closely connected with metabolism. The present investigation pays special attention to two of them, i.e., deficit in oxygen and hyperosmolarity. There is much evidence for the vasodilator effect of these factors (e.g., 3–9), but the mechanisms by which they change the vascular tone is obscure. We attempted to obtain information by studying the relation between electrical and contractile responses of vascular smooth muscles evoked by the factors mentioned.

METHODS

Isolated preparations of rat portal vein and aorta were placed into a Plexiglass chamber and bathed with normal Krebs solution. The solution was aired by a mixture of O_2 95% and CO_2 5% at 35° to 37°C.

The electrical parameters of the vascular smooth muscular cells were studied by the sucrose-gap method (10,11) and by intracellular recording (12,13) using glass microelectrodes filled with 3 M KCl, with a resistance of 30 to 40 M (14). The contractile activity of the muscles was recorded with a mechanotransducer.

For a decrease in oxygen tension the Krebs solution was saturated under pressure by a mixture of O_2 4% and N_2 96%, which under normal atmospheric pressure caused a drop in P_{O_2} to 30 mmHg.

To increase osmolarity of the Krebs solution 1.5 and 2 times, dry saccharose 146 and 292 mmol/liter was added.

For some experiments, cultures of smooth muscle cells of blood vessels were grown on glass with collagen coating (15), and experiments were carried out after 6 to 7 days cultivation.

Electron-microscopic studies were conducted on the B-513 A electron microscope (Tesla). The preparations were fixed by 1% solution of osmic acid and poured into Epon S12. The sections were contrasted with lead hydroxide by the Reynolds method (16).

RESULTS

In normally oxygenated Krebs solution the electrical activity of rat portal vein represents slow waves of depolarization, the amplitude of which does

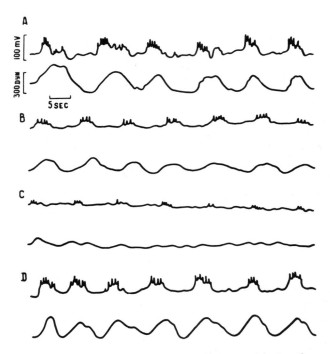

FIG. 1. The electrical (*upper curve*) and contractile (*lower curve*) activity of the smooth muscular cells of the portal vein. (A) In normally oxygenated Krebs solution; (B) 5 sec after beginning exposure to hypoxic (PO$_2$ up to 30 mmHg) solution; (C) after 2 min in hypoxic (PO$_2$ up to 30 mmHg) solution; (D) 5 min after re-admission of oxygenated solution.

not exceed 4 to 6 mV. On top of the wave a burst of 6 to 8 spike potentials is generated with an amplitude of 3 to 4 mV. The associated contractile activity consists of rhythmic phasic contractions with a duration up to 3 sec and a force of 300 to 400 dyn. The frequency of the phasic contractions reaches 8 to 12/min, and the intervals between them do not exceed 1 to 2 sec (Fig. 1A).

Usually the rhythmic contractions developed synchronously with the changes in the membrane potential, the slow waves of depolarization with the bursts of spikes preceding the phasic contractions. In most cases a direct dependence was observed between the force and duration of contractions and the amplitude and duration of the slow waves as well as the number of spikes within a burst and their amplitude.

The response to a decrease of Po_2 to 30 mmHg became noticeable 3 to 10 sec after the beginning of exposure to the hypoxic solution. It manifested itself in a decrease in the amplitude of the slow changes in the membrane potential, in a reduction of the number of spikes in the bursts and of their amplitude to 1.7 ± 0.1 mV (Fig. 1B). The force of the phasic contractions dropped on the average from 333.0 ± 7 up to 170.0 ± 3.4 dyn. Within

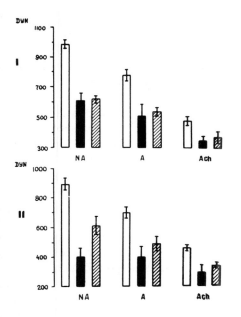

FIG. 2. The contractile responses of the aorta (I) and portal vein (II) to norepinephrine (NA, noradrenaline), epinephrine (A, adrenaline), and acetylcholine (ACh). White columns (conc. 10^{-6} g/ml) in oxygenated Krebs solution. *Black columns* (concn. 10^{-6} g/ml) in hypoxic (PO_2 up to 30 mmHg) solution. *Shaded columns* (conc. 10^{-7} g/ml) in oxygenated Krebs solution.

1 to 2 min a pronounced and stable weakening of the electrical activity and almost complete cessation of the contractile activity was observed (Fig. 1C). The amplitude of the slow waves was, on average, reduced four times; the number of spike potentials in a burst was reduced 2.5 times and their amplitude was decreased to 1 mV or less. The force of contractions usually fell below 100 dyn. Desynchronization of the electrical and contractile activities was observed, and 5 to 7 min after readmission of oxygenated solution a gradual recovery of electrical and contractile activity up to a level similar to the initial one (Fig. 1D) was observed.

The contractile reactions of the aorta and portal vein evoked by norepinephrine, epinephrine, and acetylcholine applied for 30 sec were reduced by hypoxic solution (Fig. 2). Exposure of the aorta for 1.5 to 2 min to Krebs solution, in which Po_2 did not exceed 30 mmHg, reduced the response to norepinephrine 10^{-6} g/ml 1.5 times. The decrease of the contractile response of the portal vein under the same conditions was even greater. A similar reduction of responses by aorta and portal vein was observed in response to epinephrine and acetylcholine. The contractions evoked in hypoxic Krebs solution by concentrations of 10^{-6} g/ml were of the same order or less than those evoked by these substances in concentrations of 10^{-7} g/ml in oxygenated Krebs solution (Fig. 2).

When the osmolarity of the Krebs solution was increased 1.5 and 2 times, the amplitude of portal vein contraction began to decrease noticeably within 8 to 12 sec and the rhythm was disturbed (Fig. 3). An increase in osmolarity by 50% caused a sharp decrease in the force of contractions to 15 to 20 dyn, i.e., to approximately 10% or less. An increase in osmolarity by 100% resulted in complete cessation of the rhythmic phasic contractions of the portal vein. The effect of both increases in osmolarity reached its maximum 30 to 40 sec after the beginning of exposure. An increase in the duration of exposure from 1 to 3 to 5 min had no effect on the pronounced character of the response or the time course of its development.

Hyperosmolar Krebs solution had a strong effect on the responses of the portal vein to epinephrine and acetylcholine (Fig. 4). The threshold concentration of norepinephrine was 10^{-7} and of acetylcholine 10^{-6} g/ml in normal conditions, but in hyperosmolar Krebs solutions they evoked no contraction. When the concentration was increased to 10^{-5} g/ml a small contraction ensued.

The effects of increased osmolarity on the electrical properties of the smooth muscle cells of the portal vein were studied by intracellular microelectrodes. Under natural conditions, rhythmic electrical activity was found in a great number of cells (Fig. 5A,C). In most cases, it represented slow changes in membrane potential with an amplitude of depolarization up to 10 mV and a duration of 3.0 ± 3.5 sec. On top of the slow waves action potentials were recorded as individuals or in groups (from 1 to 4) with an amplitude of 15 to 17 mV and duration up to 500 msec. The membrane potential of the electrically active cells usually did not exceed 20 to 25 mV and the resistance of their membrane was within the limits of 3 to 5 MΩ.

In some cells, we failed to record distinct rhythmic changes. These electrically inactive cells possessed a comparatively high membrane potential of 30 to 50 mV. The resistance of their membrane was 50 to 80 MΩ.

With an increase in osmolarity of the bathing solution the electrical characteristics of the smooth muscle cells of portal vein changed essentially. Depolarization of the membrane by ~8 to 10 mV was observed during the first 1 to 3 sec after admitting hyperosmolar Krebs solution in both inactive and electrically active cells. Similar data have been reported by Kuriyama et al. (17). In some of the active cells, after 15 to 20 sec in hyperosmolar solution, electrical activity disappeared (Fig. 5B,D). Usually phasic contractions continued for 20 sec or more. Other cells continued to be electri-

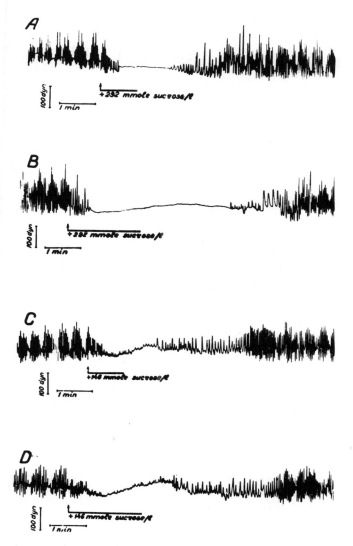

FIG. 3. Changes in the contractile activity of the portal vein smooth muscle with an increase in osmolarity of the Krebs solution twice (A, B) and 1.5 times (C, D). The arrow indicates the beginning of the effect and the bar its duration.

cally active during 50 to 60 sec after cessation of the phasic contractions.

A few experiments were carried out on vascular smooth muscle cells grown in tissue culture. In all, 69 smooth muscle cells were studied by intra-cellular microelectrodes. The mean membrane potential of these cells was 17 to 20 mV and the input impedance of the membrane did not exceed 37 to 40 MΩ. Some of these cells possessed rhythmic electrical activity (Fig.

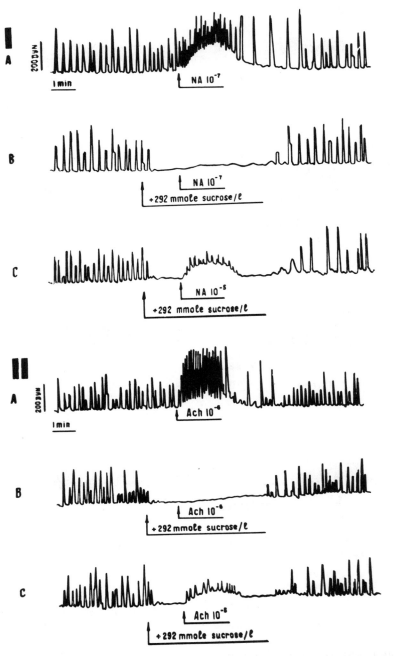

FIG. 4. Changes in contractile activity of portal vein produced by norepinephrine 10^{-7} and 10^{-5} g/ml (I), and acetylcholine 10^{-6} and 10^{-5} g/ml (II), during exposure to isosmolar (A) and hyperosmolar (B, C) solutions. The arrow indicates the beginning of the exposure and the bar its duration.

A   B

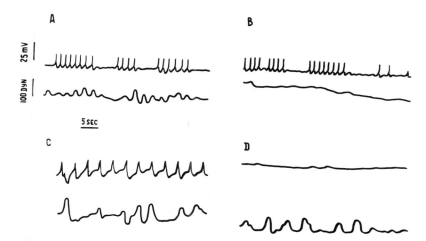

FIG. 5. The spontaneous electrical (*upper curve*) and contractile (*lower curve*) activity of the portal vein in isosmolar Krebs solution (A, C) and the change in twice the osmolarity (B, D).

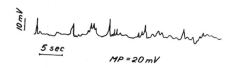

5 sec

MP = 20 mV

FIG. 6. The spontaneous electric activity of a portal vein smooth muscle cell grown in tissue culture (6th to 7th day of cultivation).

6). In other cells it was possible to evoke electrical activity by hyperpolarizing electrical stimuli.

When the cells were exposed to hyperosmolar Krebs solution their volume decreased noticeably. Under these conditions the membrane potential decreased to an average of 13 mV and the input impedance to 25 MΩ. Only 90 to 100 sec after admitting hyperosmolar Krebs solution a gradual decrease and cessation of electrical activity was observed.

An attempt was made to see whether an increase in osmolarity had an effect on the functional interrelation between cells, or on the electrical and contractile properties of individual cells. Electron micrographs of the smooth muscle layer of the vascular wall showed a rather large number of close intercellular contacts, similar to those described by Dewey and Barr (18) as "nexuses."

Electron micrographs of cultured cells from portal vein showed on day 6 or 7 some pairs of cells with contacts which resembled "nexuses."

DISCUSSION

It could be supposed that severe oxygen deficit and disturbance of oxidative metabolism might lead to exhaustion of the energy resources and reduction of contractile activity. However, many studies (e.g., 19–22) have shown that considerably less energy is necessary for the support of a definite degree of tension in smooth muscles than for the development of additional contraction. Moreover, the production of high-energy phosphates may be supported for a long time by anaerobic glycolysis. Thus, a deficiency in the energy resources is, evidently, not the main reason for the vasodilator effect of oxygen deficit in blood and tissues.

The effect of incompletely oxidized metabolites may be excluded in our experiments since we prepared thin vascular bands, free of any extravascular structures, and ensured rapid change of perfusate within fractions of a second (rate of perfusion 25 ml/min, volume of chamber less than 0.1 ml).

The contractile activity of smooth muscle cells is generally believed to depend on the level of membrane excitation. Besides, one of the main conditions for synchronous rhythmic myogenic activity is the spread of excitation throughout the population of the smooth muscle fibers; and a sufficiently high excitability of the cells is an important premise for the synchronization.

A distinct desynchronization of the electrical and contractile activity was observed in most cases with the drop of oxygen tension in the bathing solution. Moreover, the responses of both aorta and portal vein to norepinephrine, epinephrine, and acetylcholine were much reduced in hypoxic Krebs solution.

The evidence presented is consistent with our view that oxygen deficit in the medium surrounding vascular smooth muscles decreases their excitability.

Hyperosmolarity causes a rapid decrease and almost complete abolition of phasic contractions of vascular smooth muscles. In some cells this coincides with cessation of electrical activity, whereas other cells continue to be electrically active. The former may be "driven" or "passive" cells which are excited electrotonically, and the latter may be "active," according to the classification suggested by Burnstock (23).

Since muscle tension develops only if a considerable number of cells participate, the spread and the synchronization of excitation is an important premise for vascular tone. On the basis of the data presented, the vasodilator effect of hyperosmolarity may be considered as a result of disturbance in conduction of excitation.

It is generally assumed that conduction of excitation in a population of smooth muscle cells is due to the presence of close intercellular contacts with low resistance, the so-called nexuses (18,24,25). Some evidence exists (26,27) that, even after a comparatively short exposure to solutions of high osmolarity, conduction by nexuses is disturbed and is restored after normalization (28).

The electron microscopic study showed the presence of a great number of close intercellular contacts of "nexus" type in the portal vein, and similar contacts in cultured tissue. Thus, the data presented suggest that hyperosmolarity limits the spread of excitation in the smooth muscle layer of the vascular wall by disturbing the intercellular contacts or "nexuses" with low resistance. This may be the basis of the decrease in spontaneous contractile activity and also of the decrease of the responses of the portal vein to norepinephrine and acetylcholine in hyperosmotic solution.

CONCLUSION

The results of these experiments make it possible to conclude that oxygen deficit causes a decrease in excitability of the smooth muscle cells, whereas an increase in osmolarity disturbs the conduction of excitation by interrupting low-resistance pathways between cells that provide the functional interrelation.

REFERENCES

1. Folkow, B. (1949): *Acta Physiol. Scand.*, 17:289.
2. Konradi, P. (1944): *Bull. Exp. Biol. Med.*, 17:41.
3. Guyton, A. (1963): *Cardiac Output and Its Regulation.* W. B. Saunders, Philadelphia and London.
4. Smith, D. J., and Vane, J. R. (1966): *J. Physiol. (Lond.)*, 186:204.
5. Gurevich, M. I., and Bershtein, S. A. (1967). *Bull. Exp. Biol. Med.*, 63:31.
6. Gurevich, M. I., and Bershtein, S. A. (1970): In: *11th Congr. I. P. Pavlov Physiol. Soc.*, v 1, *Abstr. Rept. Symp.*, pp. 254–259. Nauka, Leningrad.
7. Detar, R., and Bohr, D. F. (1968): *Am. J. Physiol.*, 214:241.
8. Mellander, S., and Johansson, B. (1968): *Pharmacol. Rev.*, 20:117.
9. Mellander, S. (1970): *Annu. Rev. Physiol.*, 32:313.
10. Stampfli, R. (1954): *Experientia*, 10:508.
11. Shuba, M. Ph. (1963): In: *Electrophysiology of Nervous System*, pp. 143–146, Rostov-on-Don University, Rostov-on-Don.
12. Ling, G., and Gerard, R. (1949): *J. Cell. Physiol.*, 34:383.
13. Kostyuk, P. G. (1955), *Rept. Acad. Sci. USSR*, 105:858.
14. Evdokimov, I. R. (1971): *Physiol. J. Acad. Sci. SSR*, 17:487.
15. Gurevich, M. I. et al. (1974). *Physiol. J. Acad. Sci. Ukr. SSR*, 20:182.
16. Reynolds, E. S. (1963): *J. Cell Biol.*, 17:202.
17. Kuriyama, H. et al. (1971): *J. Physiol. (Lond.)*, 217:179.
18. Dewey, M. M., and Barr, L. (1964): *J. Cell Biol.*, 23:553.
19. Lundholm, L., and Mohme-Lundholm, E. (1962): *Acta Physiol. Scand.*, 55:45.
20. Lundholm, L., and Mohme-Lundholm, E. (1965): *Acta Physiol. Scand.*, 64:275.
21. Beviz, A. et al. (1965): *Acta Physiol. Scand.*, 65:268.
22. Daemers-Lambert, C., and Roland, J. (1967): *Angiologica*, 4:69.
23. Burnstock, G. (1968): In: *24th Int. Congr. Physiol. Sci.*, v 6, *Abstr. Symp. Invit. Lect.*, pp. 7–8, Washington, D.C.
24. Bennett, M. R., and Rogers, D. C. (1967): *J. Cell. Biol.*, 33:573.
25. Cliff, W. J. (1967): *Lab. Invest.*, 17:599.
26. Barr, L. M., Dewey, M., and Berger, W. (1965): *J. Gen. Physiol.*, 48:797.
27. Barr, L. M., Dewey, M., and Evans, H. (1965): *Fed. Proc.*, 24:142.
28. Nagai, T., and Prosser, C. L. (1963): *Am. J. Physiol.*, 204:215.

Physiology of Smooth Muscle, edited b
E. Bülbring and M. F. Shuba.
Raven Press, New York © 1976.

The Role of Calcium in Excitation–Contraction Coupling

Harry Grundfest

Laboratory of Neurophysiology, Department of Neurology, College of Physicians and Surgeons,
Columbia University, New York, New York 10032

More than 50 years ago, Wallace Fenn (1923) reestablished a view championed half a century earlier by Heidenhain and later by Fick (cf. Needham, 1971) that the skeletal muscle fiber is a chemical machine in which the workload regulates energy consumption and heat output. However, it is only in the past two decades that physiologists and biochemists have begun to appreciate and analyze the complex and often subtle interactions that permit the muscle fiber to behave as a self-regulatory system. A particularly encouraging result is the present tendency to correlate studies on skeletal, smooth, and cardiac muscle, rather than to compartmentalize them into separate disciplines.

Nevertheless, some older ideas tend to persist after they have become obsolete. One example is the view still widely held that excitation–contraction coupling (ECC) is effected by a direct action of depolarization on the membrane of the sarcoplasmic reticulum (SR). We have already presented a considerable amount of evidence against that view (Reuben, Brandt, Garcia, and Grundfest, 1967) in developing data that support the channeled current hypothesis (Girardier, Reuben, Brandt, and Grundfest, 1963). Two examples of that evidence are relevant here.

Procaine blocks caffeine-induced release of Ca from its store in the SR in intact muscle fibers (Chiarandini, Reuben, Girardier, Katz, and Grundfest, 1970) as well as in extracted SR fractions (Weber and Herz, 1968). Nevertheless, tensions are developed in procaine-treated crayfish muscle fibers by depolarizing stimuli which produce graded electrogenesis or spikes by depolarizing Ca activation (Fig. 16, Reuben et al., 1967; Suarez-Kurtz, Reuben, Brandt, and Grundfest, 1972). The tensions are smaller than would have been the case if release of Ca from the SR had not been blocked by the procaine. Nevertheless, they are appreciable and must result from the influx of exogenous Ca which is involved in the electrogenesis (Reuben et al., 1967).

Spike electrogenesis can also be induced when the muscle fibers are exposed to caffeine, an agent that depletes the Ca store of the SR (Chiarandini, Reuben, Brandt, and Grundfest, 1970a). Nevertheless, the spikes elicit tensions (cf. Fig. 12 in Chiarandini et al., 1970a) which must result from the entry of Ca that evokes the electrogenesis. Direct evidence correlating the influx of Ca with tension has recently been obtained with barnacle muscle fibers (Atwater, Rojas, and Vergara, 1974).

Tension may also be produced by inward current which hyperpolarize the membrane to a very considerable degree (Fig. 14, Reuben et al., 1967), and this finding has been extended recently by Uchitel and Garcia (1974) in studies on crab muscle fibers (Fig. 1). Intracellularly applied inward currents induce hyperpolarization which diminishes while the current is still applied. The decrease, which persists thereafter, denotes an increase in conductance (decrease in resistance) and this change is associated with contraction of the muscle fiber. The conductance increase is the result of hyperpolarizing Ca activation and this cation then carries some of the inward current. The Ca activation persists after the hyperpolarizing current is ended. Now, the inside-positive Ca battery contributes a small after-depolarization to the membrane potential and there is an influx of Ca under the drive of the electrochemical gradient for this cation. Although it is small, the depolarization is nevertheless accompanied by considerable tension. Depolarization by an intracellularly applied current does not induce as large a tension as does the small after-depolarization which is associated with influx of Ca.

Many chapters in this volume attest to a general consensus that contraction of smooth and cardiac muscle is associated with entry of exogenous Ca, but a similar consensus on the role of this Ca in ECC is as yet lacking. As for skeletal muscle, many workers still question any role for exogenous Ca other than to maintain the electrically excitable characteristics of the mem-

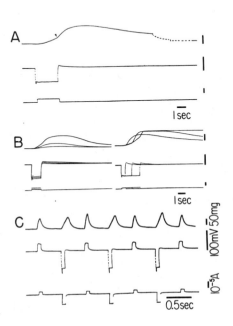

FIG. 1. Tensions induced in crab muscle fibers by intracellularly applied hyperpolarization. *Upper trace,* tension; *middle trace,* membrane potential; *lower trace,* current monitor, upward deflection for inward current A: Current applied for 2.5 sec. The initial large hyperpolarization decreased somewhat, indicating that the membrane conductance had increased. This results from hyperpolarizing Ca activation, which persists after the current is ended, giving rise to a small depolarizing electrogenesis (after depolarization) which is associated with entry of Ca down its electrochemical gradient. A slow rise in tension occurs during the hyperpolarization when Ca activation permits this cation to carry some of the inward current. The after-depolarization, though small, induces a larger tension (*arrow*). B: The induced tensions are graded with the applied current (*left*). For a constant current the amplitude of the initial tension rises as the duration of the influx of Ca is increased (*right*). C: The tensions induced by the influx of Ca during the after depolarizations are about as large as those induced by much larger depolarizations evoked by outward current. (Modified from Uchitel and Garcia, 1974.)

brane [cf. Sandow (1965) for review]. Bianchi and Bolton (1967) did suggest that the efflux of Ca stored in the SR is enhanced by entry of exogenous Ca and this view has recently been adopted by Endo, Tanaka, and Ogawa (1970) and by Ford and Podolsky (1970, 1972). We had also favored this hypothesis earlier (Reuben, Katz, and Berman, 1969; Reuben, Brandt, Katz, and Grundfest, 1970) for explaining data obtained on crayfish skeletal muscle. However, subsequent studies, particularly on skinned fiber preparations, have led us to suggest another mechanism (Reuben and Brandt, 1972; Reuben, Brandt, and Grundfest, 1974) that exogenous Ca is first rapidly taken up by the SR causing the latter to become "loaded," a condition which inhibits further uptake (Weber, 1971), and more Ca becomes available for interaction with troponin.

In the light of this new evidence we have reinterpreted the data obtained with iontophoretic microinjection of Ca with an intracellular microelectrode (Brandt and Grundfest, 1968; Reuben et al., 1969, 1970, 1974). The hypothesis that Ca loading precedes and initiates release of Ca explains an inconsistency in the alternative view, which would have predicted a regenerative release of endogenous Ca by the exogenous Ca. Delivery of Ca from the tip of a microelectrode (Fig. 2) results in a gradient in the distribution of Ca and the observed "contraction sphere" (Brandt and Grundfest, 1968; Reuben et al., 1969; April, Brandt, Reuben, and Grundfest, 1968; Forssmann, Brandt, Reuben, and Grundfest, 1974; Reuben et al., 1974) is not unexpected. We did not, however, expect the rather bizarre morphology of the contraction sphere where the sarcomeres are uniformly shortened from a resting length of about 8 μm to about 3 μm, although the thick filaments are normally about 6 μm long (cf. Forssmann et al., 1974 for further description). Rather surprising, too, was the finding that the transition zone from fully contracted to fully relaxed sarcomeres involves only two or three sarcomeres on the periphery of the sphere or a distance of about 20 μm. Once a steady-state volume and tension are attained they are maintained as long as the current is applied. The relation between tension and volume and the injected Ca is linear (Fig. 3), the steady state being related to a Ca concentration of about 0.3 mM.

These findings indicate to us that the perimeter of the contraction sphere is limited by the presence of a system that sequesters the excess injected Ca as rapidly as it is delivered from the microelectrode. Other experiments (unpublished; cf. Reuben et al., 1974) had demonstrated that crayfish muscle fibers can take up Ca to at least a concentration of 80 mM/kg wet wt. Much of this Ca is sequestered by the SR and the relative electron density of the cation in EMs outlines the SR (Fig. 4).

A monotonic rise in tension and volume is observed when the injections are brief. When Ca is injected for a longer time and at a low rate the tension and volume undergo oscillatory changes (Fig. 5A) and the oscillation is reduced to a single large overshoot, when the injection rate is high (Fig. 5B).

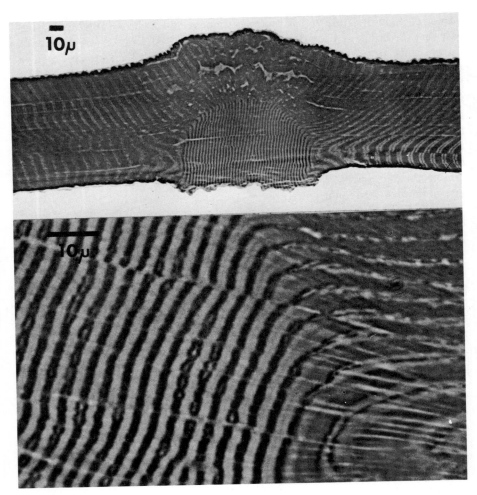

FIG. 2. Top: Contractile sphere in a crayfish muscle fiber photographed with phase contrast optics from a semithin longitudinal section. The fiber was fixed by perfusion of 0.2% glutaraldehyde while Ca was being injected. The sarcolemma was torn away on removing the electrode. The contracted sarcomeres spread radially from the site of the microelectrode. The shortened sarcomeres appear as alternating dense and light bands. Note the very small transitional zone of partially contracted sarcomeres and the stretch on the sarcomeres outside the contraction sphere. Bottom: The right edge of the contraction sphere photographed at higher magnification. Note the apparent continuity between the dark bands of the shortened sarcomeres and the light bands of the relaxed sarcomeres. (Further description in original paper by Forssmann et al., 1974.)

Both the oscillations and the overshoot are eliminated in the presence of procaine, which, as already noted, reversibly blocks the release of Ca from the SR. The plateau level of the volume and tension are not affected by procaine (Fig. 6) indicating that the delivery of Ca from the microelectrode and the uptake by the SR have attained a steady state. The oscillations and overshoot, however, must involve a variable activity of the SR. As already noted, we at first attributed these effects to a regenerative release of Ca from the SR which was induced by the entry of Ca (Reuben et al., 1969, 1970). However, a regenerative process might be expected to propagate and we have obtained propagation under certain conditions (Reuben et al., 1974). In the absence of these conditions, however, we had demonstrated (Brandt and Grundfest, 1968) that Ca injected from two microelectrodes spaced about 50 μm apart produced isolated contractile spheres which were independent of one another and coalesced only when the delivery of Ca was sufficiently large so that their perimeters joined. Thus, the interactions that might be expected of regenerative processes are absent.

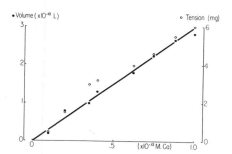

FIG. 3. Volume of the contractile sphere (*left ordinate*) and the tension it developed (*right ordinate*) as functions of the Ca delivered by 1 sec pulses to a muscle fiber bathed in potassium propionate saline. (From Reuben et al., 1974.)

Evidence that has been obtained on skinned fibers with SR intact or with the SR made nonfunctional, reversibly by caffeine, or irreversibly by Brij treatment (Reuben and Brandt, 1972; Orentlicher, Reuben, and Brandt, 1972; Orentlicher, Reuben, Grundfest, and Brandt, 1974; and *to be published*) has now led us to propose a different mechanism for a time-dependent variation in SR activity upon injection of exogenous Ca (Reuben et al., 1974). Two examples of those data are shown in Fig. 7.

The SR was initially intact in the fiber of Fig. 7A. The bathing medium was 200 mM K propionate containing 5 mM ATP and 1 mM Mg. No EGTA was used, so that when Ca was added it was buffered only by ATP. Nearly maximal tension developed almost immediately on adding 0.1 mM Ca (P$_{CA}$ = 5.3), but it was soon followed by almost complete relaxation. Thereafter the tension rose again and reached a plateau after a series of decreasing oscillations. In the presence of caffeine, which prevents Ca accumulation by the SR, the addition of 0.1 mM Ca induced only a gradual, smoothly sigmoidal rise in tension which attained and maintained its maximal value until the Ca was washed out. The slow rise in tension reflects the slow diffusion of Ca into the fiber which is caused by the presence of Ca binding sites other than those which had been eliminated when SR function was eliminated by Brij (Orentlicher et al., 1974).

In the final experiment of this series Ca was added when the fiber was

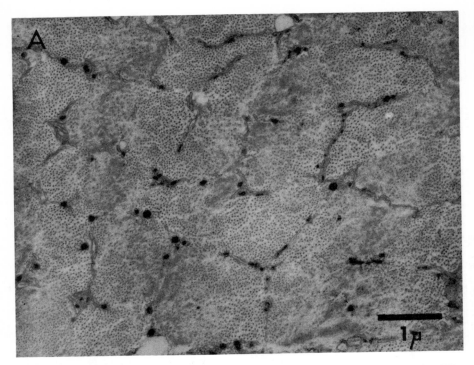

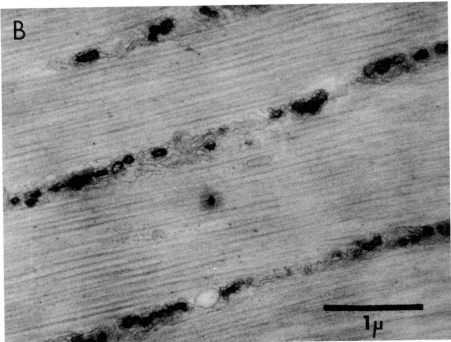

FIG. 4. Accumulation of Ca as electron dense deposits in the SR. A: Electron micrograph of cross section of fiber bathed for about 1 hr in a K-free isotonic saline containing 100 mM $CaCl_2$. The Ca is found in the SR which surrounds each myofibril. B: Longitudinal section from another similarly treated preparation. The Ca is deposited in the SR, which covered the myofibrils in the A-band region, but not in the membrane of the transverse tubules. (From Reuben et al., 1974.)

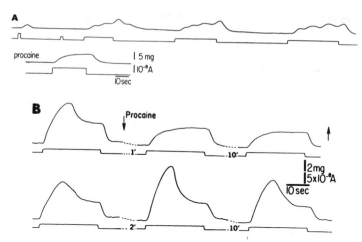

FIG. 5. A: Responses to prolonged ionophoretic injections that are only slightly above threshold for contraction. Currents monitored on the lower of each pair of traces. The first injection, by a 2-sec pulse, was well above threshold (ca. 8×10^{-9} A). It was reduced to half for the next stimulus which evoked a barely visible response. The duration of this stimulus was then increased to about 25, 35, and 45 sec for the next three stimulations. Note the oscillatory tension which continued to increase after the current was turned off. Procaine was then added and the current was again increased to about 8×10^{-9} A. The oscillations were abolished. The rate of relaxation was not altered by the drug. (From Reuben et al., 1974.) B: Responses to higher ionophoretic currents and the effect of procaine. Sequence of records made during an interval of about 30 min. Constant testing currents (about 2×10^{-9} A) were applied for about 25 sec. In the absence of procaine the tension attained a peak (overshoot) and then subsided to the steady state plateau. Procaine (10^{-3} g/ml) eliminated the overshoot, the tension rising slowly to the plateau. The effect was rapid, persisting as long as procaine was present. The procaine was then washed out (last arrow in upper set). The lower set of traces continues the sequence. A small overshoot developed immediately, was greater than in the control at 2 min, but returned to the control level after 10 min. Note that the rate of relaxation was not affected by the procaine. (From Reuben et al., 1974.)

maintained in procaine. The rise of tension was slowed still more, reflecting the presence of the Ca-binding sites of the SR. Now, however, the tension exhibited oscillations. The changes could not have been due to release of Ca from the SR and must represent a succession of phases in which Ca is depleted by its uptake in the SR alternating with periods when the uptake stops or is reduced. Thus, the initial experiment indicates that the addition of Ca while the SR was functionally intact led to a rapid release of Ca followed by enhanced uptake and a repetitive cycle of such alterations superimposed on the tension rise due to the gradual increase of free Ca within the myoplasm.

The data of Fig. 7A may be interpreted as suggesting that the initial response of the SR to the exogenous Ca was a dumping of Ca already stored in the SR. The experiments of Fig. 7B demonstrate, however, that dumping is preceded by an initial Ca uptake. On the left of Fig. 7B are three responses of a fiber with SR intact to additions respectively of 0.01; 0.05, and 0.1 mM Ca. On the right are repetitions of the same three experiments and on the same fiber, but after the SR had been inactivated by Brij. In the absence of the SR's activity addition of 0.01 mM Ca (PCA $= 6.3$) was without effect. With the SR intact, however, the fiber developed a sequence of "tension spikes," the first occurring only after a delay of about 30 sec and the subsequent ones following after intervals of about 1 min. Tension was almost 80% of maximal. Hence, it appears that release of stored Ca after addition of exogenous Ca had occurred only after the SR had first accumulated Ca from the medium. The amount taken up by, and the amount released from the SR must be large, since the released Ca is capable of eliciting nearly maximal tensions. Higher concentrations of exogenous Ca introduced the

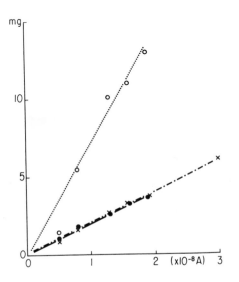

FIG. 6. Data from a similar experiment in which the ionophoretic current was increased with the fiber bathed in the K propionate saline before and after addition of procaine. (O) Peaks of the overshoots in the controls; (●) amplitudes of the plateaus following the overshoots; (X) procaine added. This phase of the response was not altered by addition of procaine, whereas the overshoots were abolished. (From Reuben et al., 1974).

additional complexity that the added Ca acts directly upon the contractile protein, but as was already seen in Fig. 7A, the cycle of oscillations resulting from activity of the SR is still in evidence. An initial uptake of Ca must still be present, but it is obscured by the rapidity of the uptake process which raises the Ca content of the SR quickly to the level for triggering contractions. The rapid uptake is, however, reflected in the rapid fall of the tension spikes during the oscillatory activity.

These experiments as well as other data on intact, Ca-loaded fibers (Reuben et al., 1974) have now led us to conclude that the primary response of the SR to exogenous Ca is not the regenerative release of SR Ca. We have suggested (Reuben and Brandt, 1972; Reuben et al., 1974; cf. Fabiato and Fabiato, 1972) that influx of exogenous Ca results in a rapid uptake to some "supernormal" level which induces inhibition of further uptake. The superloaded SR then releases Ca. Thus, the "trigger" for Ca release from the SR lies in the complex properties of the SR itself; its ability to take up Ca very rapidly to reach a superloaded state in which uptake is inhibited and gives

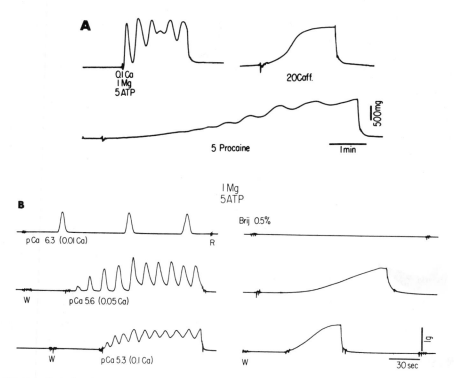

FIG. 7. A: Ca-induced tensions when the SR is in different functional states. Skinned muscle fiber bathed in medium containing 5 mM ATP and 1 mM Mg. Initial artifact on each trace denotes addition of 0.1 mM Ca. Relaxations were produced by removing the Ca. Top left: Both uptake and release functions of SR intact. Top right: Uptake by Ca by SR blocked with 20 mM caffeine. Bottom: Release of Ca from SR blocked with 5 mM procaine. Further description in text. (Unpublished records kindly supplied by Dr. J. P. Reuben.) B: Cyclic contractile activity resulting from accumulation by and release of Ca from the SR in a skinned fiber. Left: SR functionally intact. Right: After disruption of the SR with Brij-58. Upper records: 0.01 mM Ca was added to the bath, which already contained 1 mM Mg and 5 mM ATP. The free Ca in the solution ($P_{CA} = 6.3$) was below threshold for inducing contractile activity after the SR had been disrupted by Brij. With the SR functional, however, cyclic tension spikes of about 0.8 g were evoked in this solution. The delay before the first tension (ca 30 sec) and the intervals between the tensions, presumably represent the times during which the SR was accumulating Ca from the subthreshold concentration in the bath. The delay and the intervals between oscillatory spikes decreased with summation of tension as free Ca was increased when the SR was intact. The oscillatory response at $P_{CA} = 5.3$ resembles the responses to prolonged injections of Ca into intact muscle fibers (Fig. 5). Initial artifacts marked W indicate exchange of bathing medium. Responses were terminated by withdrawing Ca. (From Reuben et al., 1974.)

play to still another characteristic of the SR, its ability to release the Ca, also a rapid rate.

If the mechanism of ECC is, indeed, so complex a process a rigorous analysis of its quantitative aspect will require much further study. Participation of exogenous Ca in ECC is only one of the many roles that Ca influx plays in cell functioning. Among the best known is the initiation of secretory activity of the vesicles of gland cells and nerve terminals. Also well characterized is the mitochondrial uptake of Ca. Like the SR these membranous organelles derive from endoplasmic reticulum (ER) and the effects on them induced by raising intracellular Ca may reflect a general responsiveness of ER membranes to the cation. In this light the SR may be regarded as a modification which is specialized to accumulate Ca and, when exogenous Ca enters, responding with secretory release from its Ca store. The release is intracellular to meet the particular requirements of the contractile proteins. Secretory vesicles are specialized to accumulate and store various cellular products which, upon entry of Ca release these agents to the environment of the cells.

ACKNOWLEDGMENTS

The author's work was supported in part by grants from the Muscular Dystrophy Associations of America, Inc.; by Public Health Service Research Grant (NS 03728) and Training Grant (NS 05328) from the National Institute of Neurological Diseases and Stroke; and from a grant from the National Science Foundation (GB 31807X). The new work reported here was also supported in part by various grants to Drs. John P. Reuben and Philip W. Brandt. I thank them and various Fellows of the Laboratory of Neurophysiology for permission to present data to which my contribution was minimal.

REFERENCES

April, E. W., Brandt, P. W., Reuben, J. P., and Grundfest, H. (1968): Muscle contraction: the effect of ionic strength, Nature, 220:182–184.

Atwater, I., Rojas, E., and Vergara, J. (1974): Calcium influxes and tension development in perfused single barnacle muscle fibres under membrane potential control. J. Physiol (Lond.), 243:523–552.

Bianchi, C. P., and Bolton, T. C. (1967): Action of local anesthetics on coupling systems in muscle. J. Pharmacol. Exp. Ther., 157:388–405.

Brandt, P. W., and Grundfest, H. (1968): Sarcomere and myofilament changes accompanying local contractile activation in crayfish muscle fibers. Fed. Proc., 27:375.

Chiarandini, D. J., Reuben, J. P., Girardier, L., Katz, G. M., and Grundfest, H. (1970): Effects of caffeine on crayfish muscle fibers. II. Refractoriness and factors influencing recovery (repriming) of contractile responses. J. Gen. Physiol., 55:665–687.

Chiarandini, D. J., Reuben, J. P., Brandt, P. W., and Grundfest, H. (1970a): Effects of caffeine on crayfish muscle fibers. I. Activation of contraction and induction of Ca-spike electrogenesis. J. Gen. Physiol., 55:640–664.

Endo, M., Tanaka, M., and Ogawa, Y. (1970): Calcium-induced release of calcium from the sarcoplasmic reticulum of skinned skeletal muscle fibers. Nature, 228:34–36.

Fabiato, A., and Fabiato, F. (1972): Excitation–contraction coupling of isolated cardiac fibers with disrupted or closed sarcolemmas. Circ. Res., 31:293–307.

Fenn, W. O. (1923): A quantitative comparison between the energy liberated and the work performed by the isolated sartorius muscle of the frog. J. Physiol. (Lond.), 58:175.

Ford, L. E., and Podolsky, R. J. (1970): Regenerative calcium release within muscle cells. Science, 167:58–59.

Ford, L. E., and Podolsky, R. J. (1972): Intracellular calcium movements in skinned muscle fibres. *J. Physiol. (Lond.)*, 223:21–33.

Forssmann, W. G., Brandt, P. W., Reuben, J. P., and Grundfest, H. (1974): Reversible morphological changes in the contractile sphere of crayfish muscle fibers. *J. Mechanochem. Cell Motility*, 2:269–285.

Girardier, L., Reuben, J. P., Brandt, P. W., and Grundfest, H. (1963): Evidence for anion permselective membrane in crayfish muscle fibers and its possible role in excitation–contraction coupling. *J. Gen. Physiol.*, 47:189–214.

Needham, D. M. (1971): *Machina Carnis.* Cambridge.

Orentlicher, M., Reuben, J. P., and Brandt, P. W. (1972): Morphology and physiology of detergent-treated crayfish muscle fibers. *16th Ann. Mtg. Biophys. Soc.*, Abstr. p. 81a.

Orentlicher, M., Reuben, J. P., Grundfest, H., and Brandt, P. W. (1974): Calcium binding and tension development in detergent-treated muscle fibers. *J. Gen. Physiol.*, 63:168–186.

Reuben, J. P., and Brandt, P. W. (1972): Oscillatory tensions in skinned crayfish muscle fibers. *16th Ann. Mtg. Biophys. Soc.*, Abstr. p. 81a.

Reuben, J. P., Brandt, P. W., Garcia, H., and Grundfest, H. (1967): Excitation–contraction coupling in crayfish. *Am. Zool.*, 7:623–645.

Reuben, J. P., Brandt, P. W., and Grundfest, H. (1974): Regulation of myoplasmic calcium concentration in intact crayfish muscle fibers. *J. Mechanochem. Cell Motility*, 2:269–285.

Reuben, J. P., Brandt, P. W., Katz, G. M., and Grundfest, H. (1970): Augmentation of responses to calcium injection by agents that reduce calcium sequestration. *J. Gen. Physiol.*, 55:140.

Reuben, J. P., Katz, G. M., and Berman, M. (1969): Two phases of contractile activation induced by Ca injections in crayfish muscle fibers. *Fed. Proc.*, 28:711.

Sandow, A. (1965): Excitation–contraction coupling in skeletal muscle. *Pharmacol. Rev.*, 17:265–320.

Suarez-Kurtz, G., Reuben, J. P., Brandt, P. W., and Grundfest, H. (1972): Membrane calcium activation in excitation–contraction coupling. *J. Gen. Physiol.*, 59:676–688.

Uchitel, O., and Garcia, H. (1974): Muscle contraction during hyperpolarizing currents in the crab. *J. Gen. Physiol.*, 63:111–122.

Weber, A. (1971): Regulatory mechanisms of the Ca transport system of fragmented rabbit sarcoplasmic reticulum. II. Inhibition of out-flux in calcium-free media. *J. Gen. Physiol.*, 57:64–70.

Weber, A., and Herz, R. (1968): The relationship between caffeine contracture of intact muscle and the effect of caffeine on reticulum. *J. Gen. Physiol.*, 52:750–759.

Physiology of Smooth Muscle, edited by
E. Bülbring and M. F. Shuba.
Raven Press, New York © 1976.

Relationship Between Contraction and Transmembrane Ionic Currents in Voltage—Clamped Uterine Smooth Muscle

Jean Mironneau

*Laboratoire de Physiologie Animale, Faculté des Sciences, Université de Poitiers, 86022
Poitiers, France*

The application of the voltage-clamp technique to strips of uterus taken from rats has revealed that an inward current with slow kinetics carried by calcium and sodium ions underlies the development of the action potential (1–3).

A rise in intracellular calcium concentration is generally proposed as an activator of the contractile proteins in skeletal muscle (4), cardiac (5), and smooth muscle (6). In the present experiments, the relationships between contraction, membrane current, and imposed potential have been investigated on pregnant rat myometrium.

MATERIALS AND METHODS

Experiments were performed on small strips of pregnant rat myometrium (after 18 days of gestation). These muscular strips were 70 to 120 μm in diameter and 3 to 4 mm in length. As previously described by Rougier et al. (7) for cardiac trabeculae, the preparation was placed in a double sucrose-gap apparatus with an artificial node width of 100 μm for measurements of voltage and current. An optical method for measuring the contraction of the muscle strip in the central compartment was used (8). In the floor of the central compartment a small window allowed a beam of light to illuminate the preparation. The device for measuring light intensity consisted of a universal Zeiss microscope including a photomultiplier. The photomultiplier was used for visible light and the photoelectric current was displayed on oscilloscopes by means of a current–voltage transducer. The response time of the photomultiplier system to a change in light intensity was better than 50 μsec. When the muscle is observed to contract under these conditions it is accompanied by a change in the intensity of the light arriving at the photo-multiplier. Generally, movement is confined to the part of the muscle in the central compartment. The current–voltage clamps were applied at a rate of 0.5/min.

Physiologic solutions had the following compositions:

1. Reference solution (mM): NaCl 130; KCl 5.6; CaCl₂ 2.2; MgCl₂ 0.24; and glucose 11. The solution was aerated with O_2 and was buffered by Tris-HCl (8.3 mM) at pH 7.4.

2. The following inhibitors of permeability were used: manganese (5 mM), lanthanum (2.5 mM), and compound D 600 (α-isopropyl-α [(N-methyl-N-homoveratryl)-γ-aminopropyl]-3,4,5-trimethoxyphenylacetonitrile-HCl) (5×10^{-3} mM).

All solutions were maintained at $30 \pm 1°C$.

Differences in calcium concentration were obtained by increasing (6.5 mM) or decreasing (0.72 mM) the calcium without alteration of the other constituents. Ethyleneglycoltetraacetic acid (EGTA, added as 1 mM solution) was used to obtain a calcium-free solution. Choline was used as a substitute for sodium after the addition of 0.1 mM atropine.

RESULTS

Relation Between Contraction and Action Potential

Subthreshold depolarizations caused no change in the optical record. When an action potential was triggered a modification of light intensity oc-

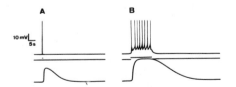

FIG. 1. Records of action potentials and contraction. A: Contraction triggered by an action potential. B: A fused tetanus in response to a train of action potentials provoked by a depolarizing current pulse. The tension is expressed in arbitrary units.

curred after a delay of 80 to 100 msec, and lasted 10 to 15 sec (Fig. 1A). The time course of the contraction recorded by the optical method was similar to the one observed with classical methods. Rhythmic activity can be triggered by long-lasting depolarizing current steps on uterine muscle (9). In these conditions, the amplitude of the contraction depends on the frequency of the action potentials. If the intervals between action potentials become short enough the tension does not revert to the resting level but summates to a fused tetanus (Fig. 1B). A similar tetanus which resembles that in striated muscle has been recorded previously on guinea pig taenia coli by Bülbring (10) and on other types of smooth muscle by West and Landa (11).

Effects of Duration, Size, and Frequency of Potential Steps upon Contraction

For step depolarizations between +20 and +60 mV a transient contraction is observed even when the step duration is only 10 msec. Increasing the duration of steps leads to an increase in the magnitude of contraction. The

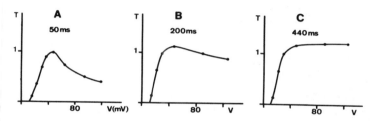

FIG. 2. Peak contraction–voltage relationship for steps of three different durations. A: 50 msec; B: 200 msec; C: 440 msec. The decrease of the maximum amplitude of the contraction observed for 50 msec steps higher than +50 mV becomes progressively less, rapidly developing into a plateau. The ordinate is expressed as a ratio of the maximum contraction obtained for 50 msec depolarizing steps.

duration of the contraction, which is approximately constant for depolarizing steps shorter than 100 msec, increases for longer depolarizations. However, a second phase of increasing tension distinct from the first one can be seen.

The curve of the maximum amplitude of the contraction plotted as a function of voltage shows two different phases in response to step durations between 50 and 500 msec. For a step lasting 50 msec, the contraction increases in voltage from +10 mV to reach a maximum between +45 and +50 mV (Fig. 2A), then decreases markedly for higher depolarizations. For durations of 200 and 440 msec, the first phase of the relation between voltage and contraction is similar for voltages between +10 and +50 mV but for higher depolarizations (from +50 to +120 mV) the second phase of decrease in contraction is flatter (Fig. 2B,C). These differences between the slopes of the second part of the curves suggest that the contractile response may be considered as resulting from two components. A similar hypothesis has been put forward by West et al. (12) from observations on intestinal smooth muscle in which the contraction has dynamic and tonic phases.

The effect of frequency has been recorded with trains of 50-msec depolarizations. Increasing the frequency progressively decreases the increments of the contraction while the final level of tension increases (Fig. 3). Up to 0.8 Hz, contractile responses after each depolarization are still distinct and repre-

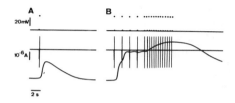

FIG. 3. Effect of inward current frequency on contraction: A: Response to a single 50 msec depolarization. B: Increasing the frequency decreased the increments of the contraction obtained in response to successive 50 msec steps, until a maximum level of tension was reached. Up to 0.8 Hz an incomplete tetanus can be observed. Over 0.8 Hz the contraction became a fused tetanus.

sent an incompletely fused tetanus. Over 0.8 Hz, they summate to a typical tetanus.

Dependence of the Contraction on the Inward Current

It has been suggested that the inward current of the uterine membrane is carried by both calcium and sodium ions (1,13). Several substances can abolish this current. Manganese ions are known to inhibit the slow inward current of the frog heart (7) and of uterine muscle (14,1,3). Application of

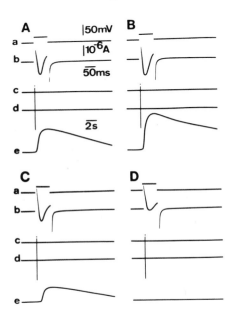

FIG. 4. Effect of different calcium concentrations on the inward current and the contraction. A: 2.16 mM Ca; B: 6.5 mM Ca; C: 0.72 mM Ca; D: without Ca. A high calcium solution was associated with an increase in the amplitude and the duration of the contraction; in the absence of calcium the contraction was abolished in spite of the persistence of a small inward current. The imposed clamp potentials are shown at a and c and the current records at b and d. Contractions are recorded at e. Time base: a, b: 50 msec; c, d, e: 2 sec.

manganese (5 mM) suppressed both inward current and contraction in uterine strips. Lanthanum ions are believed to block calcium regenerative systems (15,16) and have also been used to study the excitation–contraction coupling in rabbit aorta by Van Breemen et al. (17). Lanthanum ions (2.5 mM) inhibited completely the inward current and the contraction in uterine muscle for all values of depolarization. Finally, compound D 600 (5×10^{-3} mM) which is considered a highly selective substance for inhibition of calcium current (18) suppressed the contraction in uterine strips and strongly reduced the inward current. A small inward current persisted, however, which may have consisted of a sodium component.

The use of solutions containing different calcium concentrations modified the contraction (Fig. 4). A high-calcium concentration (6.5 mM) led to an increase in the maximum amplitude and duration of contraction for a given depolarizing step. In a low-calcium solution (0.72 mM) both amplitude and duration were reduced. In a calcium-free solution the contraction disappeared rapidly even with the highest depolarizations, in spite of the persistence of a small inward current, thus supporting the supposition that the latter was a sodium current unable to induce a mechanical response. For the different calcium concentrations, curves depicting the peak of the contraction and the maximum amplitude of the inward current (corrected for leak current) as a function of voltage show a correlation between the mechanical response and the inward current (Fig. 5). These results seem to support the fundamental role of the calcium current in the activation of contraction and in the maximum value of tension of uterine strips. For depolarizing steps over +50 mV

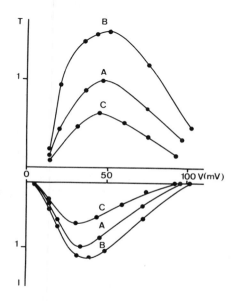

FIG. 5. Relation between the peak of contraction and the maximum inward current as a function of voltage in solutions with different calcium concentrations. A: 2.16 mM Ca; B: 6.5 mM Ca; C: 0.72 mM Ca. The inward currents were corrected for leak current. The ordinates are expressed as a ratio of the maximum contraction and of the maximum inward current obtained in 2.16 mM Ca.

the decrease in the inward current was associated with a decrease in tension, even though contractions were not abolished at the reversal potential of the total inward current (Ca + Na), especially in the high-calcium solution. However, the presence of sodium ions as a component of the inward current could be relevant here.

Removal of sodium ions caused a sustained contracture in uterine strips of the kind described for the frog heart by Lüttgau and Niedergerke (19) and for the mouse myometrium by Osa (20). In these circumstances, in spite of the occurrence of large inward currents during depolarizing steps, little or no contraction was observed. In solutions containing only 12.5 mM sodium ions, contractions associated with depolarizations were smaller because the resting tension was already slightly elevated, but the inward currents were hardly affected. This effect of reducing extracellular sodium on contraction could be explained as due to an increase in intracellular calcium concentration as a consequence of a decrease of calcium efflux, such as has been shown to occur in guinea pig auricle (21) and in squid giant axon (22).

Existence of Contraction in the Absence of Inward Current in Manganese Solution

In solutions containing manganese (5 mM), normal contractile responses to brief depolarizations cannot be elicited. Nevertheless, depolarizing steps longer than 200 msec and greater than +40 mV lead to the development of tension, where the onset of contraction is much delayed, often appearing only after 1 or 2 sec, and its amplitude is small. Under these conditions, there is no inward current. The slopes of the contraction and relaxation phases are less steep than those measured in manganese-free solution with depolarizing

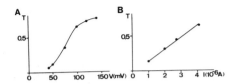

FIG. 6. A: Peak contractions in response to 440 msec depolarizations in manganese solution (5 mM) plotted against voltage. The relation has a sigmoid shape. B: Relation between the peak of contraction and the maximum outward current. A linear relation was observed. The ordinates are expressed as a ratio of the maximum contraction obtained in response to a 50-msec depolarization in the absence of manganese.

steps of 80 to 100 msec duration. Similar results were obtained with lanthanum ions and with compound D 600. This type of contraction resembles that described on frog auricle. The relation between peak tension and voltage has a sigmoid shape (Fig. 6A) as described for taenia coli smooth muscle by Imai and Takeda (23). Léoty and Raymond (24) plotted the peak tension developed by frog auricles as a function of the intensity of the outward current and observed a linear relationship. A similar relation between the maximum amplitude of the contraction and the outward current was obtained on rat myometrium (Fig. 6B), but other mechanisms may regulate the activation of the slow component of contraction, particularly the intracellular sodium concentration (25).

DISCUSSION

The use of a double sucrose-gap apparatus combined with an optical method permitted the simultaneous recording of both membrane currents and contraction of uterine strips taken from pregnant rats (26).

For depolarizing steps above threshold, contraction was recorded as a function of both inward and outward currents. Increasing the frequency of brief depolarizations demonstrated that the contractile responses of uterine smooth muscle would summate into a fused tetanus. This property has also been described for guinea pig taenia coli (10) and rabbit and rat uterus (27,28).

Experiments performed with different step durations and with specific inhibitors of ionic currents suggest that contractions can be activated by mechanisms of at least two distinct types: the first consisting of the inward current and the second depending on some other mechanism. The identity of the first mechanism and the inward current is strongly suggested in the experiments for the following reasons: (a) There is a relationship between the maximum amplitude of contraction and of the inward current, both being functions of potential. (b) Both contractions and inward currents in response to depolarizing steps of 80 to 100 msec duration are abolished in calcium-free, manganese, and lanthanum solutions. (c) Modifications of the calcium concentration are associated with alterations of peak tension developed in response to given depolarizations. These results suggest that the contraction is related to calcium inward current and agree with those obtained on other smooth muscles (29,20,17). The minimum intracellular concentration of calcium required to evoked tension is approximately 10^{-6} M according to estimates made for frog myoplasm by Ebashi and Endo (30). According to Casteels (31) the calcium concentration reached in the cell approximately 10^{-5} M during depolarization, could be sufficient for triggering contractions. On the other hand, several observations are not consistent with this hypothesis as it stands: (a) The peaks of maximum inward current and for contraction as a function of potential do not correspond. (b) There is an interval between the time at which depolarizations and contractions begin. (c) Contractions occur on depolarizations to the reversal potential for the total inward current (Ca + Na). It is possible to postulate that the calcium ions which flow into the cell as a current carrier may release a portion of the calcium ions bound to intracellular sites and that it is the released calcium which activates the contractile proteins. A similar conclusion was drawn by Sakamoto and Kuriyama (32) from experiments on muscle from the stomach of cats.

The tentative conclusion is that in response to depolarizations within the voltage range of the action potential the contraction of uterine muscle occurs when calcium flows into the cell. Influx of calcium could either act directly on the myofibrils or indirectly by activating intracellular stores from which calcium is released. This release is promoted either by the influx of calcium or by the increase in calcium conductance as proposed by New and Trautwein (33). Total replacement of sodium ions by choline or sucrose produces contracture, and it is generally believed that a rise in the intracellular calcium concentration is responsible. It is suggested that low extracellular sodium reduces calcium efflux. Thus calcium accumulation inside the muscle may result from impaired extrusion as well as from increased influx. Low external sodium has been associated with a reduction of calcium efflux in guinea pig auricle (21) and squid axon (22), and with an increase in calcium content of guinea pig taenia coli (34).

The second component of the mechanism activating contraction which appears with long depolarizing pulses is different from the first since (a) it is not suppressed by manganese; and (b) it is independent of inward current. It has been suggested that the diminution in amplitude and rate of rise of tension after the suppression of the inward current could be explained either by preventing the filling of intracellular stores with calcium or by entering the cells (35) and uncoupling the effect of calcium upon the contractile elements. Neither of these proposals, however, is consistent with the contraction which is observed in response to very long depolarizations in the presence of manganese. The existence of this second mechanism of activation (in the absence of inward current) is consistent with the hypothesis that calcium is released from intracellular stores from which it can be displaced. Since the structures of the sarcoplasmic reticulum and tubular system are poorly developed in the smooth muscle, the intracellular stores could be located in the sarcoplasmic membrane of uterine smooth muscle, which may have the properties of the sarcoplasmic reticulum of striated muscle, as suggested by Ito and Kuriyama (36) and Gabella (37).

All these results seem to support the existence of a dual source of activator calcium responsible for the activation of contractile proteins of uterine smooth muscle, as previously suggested for cardiac muscle and other smooth muscles.

ACKNOWLEDGMENT

This work was supported by a C.N.R.S. association, E.R.A. No. 111.

REFERENCES

1. Anderson, N. C., Ramon, F., and Snyder, A. (1971): *J. Gen. Physiol.,* 58:322–339.
2. Kao, C. Y. (1971): *Proc. Int. Congr. Physiol. Sci. Munich, 25th,* 9:288.
3. Mironneau, J., and Lenfant, J. (1971): *C. R. Acad. Sci. Paris,* 272:436–439.
4. Sandow, A. (1970): *Ann. Rev. Physiol.,* 32:87–138.
5. Katz, A. M. (1967): *J. Gen. Physiol.,* 50:185–195.
6. Somlyo, A. V., Vinall, P., and Somlyo, A. P. (1969): *Microvasc. Res.,* 1:354–373.
7. Rougier, O., Vassort, G., Garnier, D., Gargouïl, Y. M., and Coraboeuf, E. (1969): *Pflügers Arch.,* 308:91–110.
8. Gargouïl, Y. M., Léoty, C., Poindessault, J. P., and Raymond, G. (1969): *C. R. Acad. Sci. Paris,* 269:1686–1689.
9. Mironneau, J., and Lenfant, J. (1972): *C. R. Acad. Sci. Paris,* 274:3269–3272.
10. Bülbring, E. (1955): *J. Physiol. (Lond.),* 128:200–221.
11. West, T. C., and Landa, J. (1956): *Am. J. Physiol.,* 187:333–337.
12. West, T. C., Hadden, G., and Farah, A. (1951): *Am. J. Physiol.,* 164:565–572.
13. Mironneau, J. (1974): *Pflügers Arch.*
14. Abe, Y. (1969): *J. Physiol. (Lond.),* 200:1–2P.
15. Hagiwara, S., and Takahashi, K. (1967): *J. Gen. Physiol.,* 50:583–601.
16. Casteels, R., Van Breemen, C., and Mayer, C. J. (1972): *Arch. Int. Pharmacodyn. Ther.,* 199:193–194.

17. Van Breemen, C., Farinas, B. R., Gerba, P., and McNaughton, E. D. (1972): *Circ. Res.,* 30:44–54.
18. Tritthart, H., Grün, G., Byon, K. Y., and Fleckenstein, A. (1970): *Pflügers Arch.,* 319: R 117.
19. Lüttgau, H. C., and Niedergerke, R. (1958): *J. Physiol. (Lond.),* 143:486–505.
20. Osa, T. (1971): *Jap. J. Physiol.,* 21:607–625.
21. Reuter, H., and Seitz, H. (1968): *J. Physiol. (Lond.),* 195:451–470.
22. Blaustein, M. P., and Hodgkin, A. L. (1969): *J. Physiol. (Lond.),* 200:497–527.
23. Imai, S., and Takeda, K. (1967): *J. Physiol. (Lond.),* 190:155–169.
24. Léoty, C., and Raymond, G. (1972): *Pflügers Arch.,* 334:114–128.
25. Vassort, G. (1973): *Pflügers Arch.,* 339:225–240.
26. Mironneau, J. (1973): *J. Physiol. (Lond.),* 233:127–141.
27. Csapo, A. (1962): *Physiol. Rev.,* 42:7–33.
28. Casteels, R., and Kuriyama, H. (1965): *J. Physiol. (Lond.),* 177:263–287.
29. Tomita, T. (1970): In: *Smooth Muscle,* edited by E. Bülbring, A. F. Brading, A. W. Jones, and T. Tomita, pp. 198–243. Edward Arnold, London.
30. Ebashi, S., and Endo, M. (1968): *Progr. Biophys. Molc. Biol.,* 18:123–183.
31. Casteels, R. (1970): In: *Smooth Muscle,* edited by E. Bülbring, A. F. Brading, A. W. Jones, and T. Tomita, pp. 70–99. Edward Arnold, London.
32. Sakamoto, Y., and Kuriyama, H. (1970): *Jap. J. Physiol.,* 20:640–656.
33. New, W., and Trautwein, W. (1972): *Pflügers Arch.,* 334:24–38.
34. Bauer, H., Goodford, P. J., and Hüter, J. (1965): *J. Physiol. (Lond.),* 176:163–179.
35. Ochi, R. (1970): *Pflügers Arch.,* 316:81–94.
36. Ito, Y., and Kuriyama, H. (1971): *J. Gen. Physiol.,* 57:448–463.
37. Gabella, G. (1971): *J. Physiol. (Lond.),* 216:42–43 P.

Physiology of Smooth Muscle, edited by
E. Bülbring and M. F. Shuba.
Raven Press, New York © 1976.

Topical Differences in Excitation and Contraction Between Guinea Pig Stomach Muscles

H. Kuriyama, T. Osa, Y. Ito, H. Suzuki, and K. Mishima

Department of Physiology, Faculty of Dentistry, Kyushu University, Fukuoka 812, Japan

It has already been established that the contraction–relaxation cycle of striated muscle in physiological conditions is regulated by the intracellular free Ca ion concentration, which is under the control of the sarcoplasmic reticulum. In the absence of Ca ions, troponin, in collaboration with tropomyosin, exerts a particular effect on F-actin to depress its interaction with myosin; the Ca ion removes this depression through its binding to troponin. Thus, Ca ion behaves like a kind of depressor (21,8,33,9,10). In visceral smooth muscle, a similar mechanism is postulated, and the differences in mechanical responses of the various tissues in identical conditions are believed to be caused primarily by different requirements of Ca (5,25).

Recently, Ito and Kuriyama (12,13) and Ito et al. (14,15) observed the responses of the membrane and contractile machinery to caffeine and thymol, both of which are potent Ca releasers from the sarcoplasmic reticulum in striated muscle. They differed from one region in the alimentary canal to another, apparently because of differences in the role of sequestered Ca in the tissues. This is because the release of sequestered Ca in visceral smooth muscles may lead directly to changes in ionic permeability of the membrane in addition to changes of the mechanical activity.

The stomach has three functional regions which correspond to its anatomical divisions, namely, fundus, corpus, and antrum. The fundus and corpus both act as reservoirs. The major functions of the antrum are propelling, retropelling, and triturating the contents. Therefore, the activity of the stomach might be reflected in topical differences in muscle properties. Moreover, in the stomach, the longitudinal and circular muscle show different responses to electrical and chemical stimulation (18,24).

This chapter describes an investigation of the properties of the longitudinal and circular muscles of the stomach in relation to the role of Ca ion, and discusses the possible mechanism underlying the differences between the two tissues.

ELECTRICAL PROPERTIES OF THE LONGITUDINAL AND CIRCULAR MUSCLE MEMBRANE

Longitudinal muscle was taken from the fundus and corpus regions, and circular muscle from the antrum region. The membrane potentials recorded from the various stomach regions, including the longitudinal and circular muscles, were nearly the same (−55 to −60 mV) and the length constants were also the same (1.2 to 1.5 mm).

When the microelectrode penetrated the cell, slow potential changes with superimposed spikes could be recorded from the lower portions of the corpus and the pyloric region. However, no slow potentials and spikes were recorded from the fundus region (Fig. 1).

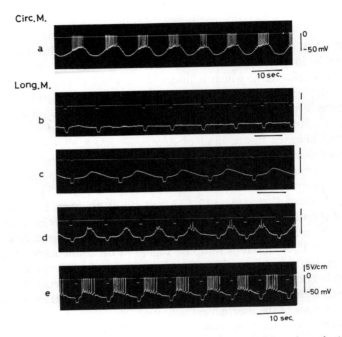

FIG. 1. Spontaneous membrane activity recorded from various regions of the guinea pig stomach. (a) Circular muscle fiber of the pyloric region; (b) longitudinal muscle fiber of fundus region; (c) longitudinal muscle fiber of corpus region; (d) longitudinal muscle fiber of lower corpus region; (e) longitudinal muscle fiber of pyloric region. (b–e) Inward current pulses were applied during the intervals between burst discharges.

Outward current pulses did not always trigger a spike. In some fibers, the spike frequency increased in proportion to current intensity. In other less excitable muscle fibers outward current pulses only depolarized the membrane, but the amplitude of the electrotonic potential did not reach the steady level expected from the cable properties of the muscle. During the course of the membrane depolarization by strong outward current, the membrane resistance gradually increased. When an outward current was applied for more than 30 sec, in most of the less excitable muscle fibers, a spike or abortive spikes could be elicited. As shown in Fig. 2, a spike was evoked with a very long latency after the generation of an oscillatory potential.

A spike with overshoot could also be recorded from the less excitable muscle fibers after treatment with tetraethylammonium (TEA), although the amplitude of the electrotonic potentials evoked by the inward current pulses remained nearly the same before and after TEA treatment. The electrically less excitable muscle fibers were mainly distributed in the upper region of the longitudinal muscle, and the electrically excitable muscle fibers were in the lower region of the longitudinal and circular muscles (pyloantral region). In Na-free Krebs solution spikes could be elicited in all regions of the stomach after treatment with TEA, but the slow potential change was suppressed. It is

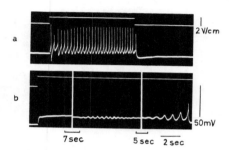

FIG. 2. Effects of outward current pulses on the stomach muscle fibers. (a) Electrically excitable fiber; (b) electrically less-excitable fiber. In (b) a spike was generated after 10 sec following oscillation of the membrane.

therefore postulated that spike generation results from influx of Ca ions during the active state of the membrane.

MECHANICAL PROPERTIES OF THE STOMACH MUSCLES IN RELATION TO ELECTRICAL PHENOMENA

In striated muscle as well as in the many visceral smooth muscle tissues, depolarization of the muscle fiber produces a mechanical response. In this respect, the stomach muscle is not exceptional, i.e., depolarization of the membrane induced a mechanical response irrespective of whether a spike was generated or not.

The effects of step depolarization on the mechanical response were studied in conventional voltage-clamp experiments with the double sucrose-gap method. An increase in membrane depolarization enlarged the mechanical response in proportion, and when an early transient inward current was evoked the amplitude became greater and the upstroke of the contraction became faster. The longitudinal muscle did not produce a transient inward current, but the amplitude of the mechanical response was also enlarged in proportion to the depolarization. The mechanical response of the circular muscle was larger than that of the longitudinal muscle, due to generation of the transient inward current, even when the membrane potentials were clamped at the same levels. It is concluded that both the depolarization of the membrane and the action current increase the Ca concentrations in the myoplasm.

Depolarization of the membrane with excess $[K]_o$ also produced a mechanical response (K-induced contracture). The contracture was composed of a phasic and a tonic response. The shape of the K-induced contracture differed according to the regions and muscle layers of the stomach, e.g., in the circular muscle, isotonic K-Krebs solution produced mainly a phasic response which relaxed rapidly to a level close to the resting level. In contrast, in the longitudinal muscle, the amplitude of tonic response often exceeded the amplitude of the phasic response (see Fig. 4a). The depolarizations induced by the

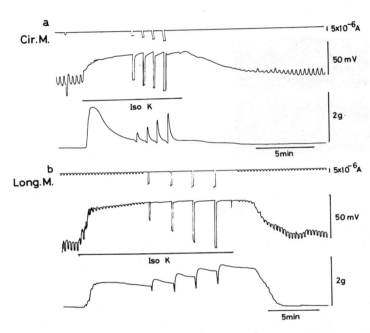

FIG. 3. Effects of displacements of the membrane potential by applications of inward current pulse on the mechanical response during the generation of K-induced contracture. Current-clamp method was used. (a) Circular muscle tissue; (b) longitudinal muscle tissue. Upper record: stimulus current; middle record: membrane potential; lower record: mechanical response.

isotonic K solution were, however, nearly the same in both tissues and were maintained throughout the exposure to isotonic K-Krebs solution.

The amplitude of the tonic response of the K-induced contraction was modified by various factors. For example, when atropine was added to isotonic K-Krebs solution, the amplitude of the tonic response was lower than the control; conversely, prostigmine enlarged the amplitude. However, neither drug affected the depolarization of the membrane. This phenomenon can be explained by the release of acetylcholine from the enteric plexus by isotonic K-Krebs solution.

The responses to strong inward current pulses during the K-induced contraction were recorded in both tissues. When strong inward current pulses were applied under current-clamp conditions, the membrane was repolarized to a certain level and the muscle relaxed (Fig. 3). The circular muscle (Fig. 3a), produced a rebound contraction on cessation of the inward current pulse. This rebound contraction rapidly declined to the level of the tonic response. In the longitudinal muscle (Fig. 3b), the rebound contracture declined very slowly. Therefore, after repetitive application of inward current pulses, the rebound contractures summed up to a higher level than the original tonic response. In both tissues, the Q_{10} value for the rate of the relaxation (3.2) was higher than that of the rebound contracture (2.6), and the generation of

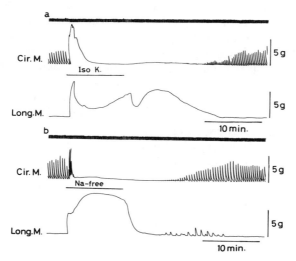

FIG. 4. Effects of Na-free sucrose solution and isotonic K-Krebs solution on the longitudinal and circular muscle tissues. Long bars indicate application of either Na-free solution or isotonic K-Krebs solution.

the relaxation required a longer pulse duration than the generation of the rebound contracture (20).

Responses of the two muscle layers also differed in Na-free Krebs solution. As shown in Fig. 4b, when Na was removed from the Krebs solution, the circular muscle developed only a transient mechanical response, whereas that of the longitudinal muscle was large and often exceeded the amplitude of the K-induced contracture. The changes in the membrane potential in Na-free solution were, however, nearly the same in both tissues. In the longitudinal muscle, more than two-thirds of the normal Na concentration was required to suppress the generation of contracture.

It has been demonstrated by Magaribuchi et al. (19), that lowering the temperature of Krebs solution below 15°C produces a contracture in the taenia coli and in the circular muscle of the stomach. The present experiments confirmed these observations. However, when the temperature was lowered in either isotonic K-Krebs solution or in Na-free solution, the cold-induced contracture developed in the circular muscle, but not in the longitudinal muscle, in which the contracture already produced by the above solutions was relaxed. Therefore the source of Ca to generate the contracture in the two tissues may not be the same.

Release of Ca Ion from the Sequestered Site

To investigate further the role of Ca ion in the K-induced contracture, the effects of Ca removal were observed. Removal of Ca ion from isotonic K-Krebs solution suppressed the K-induced contracture. Application of carbachol to the depolarized muscle produced a small contracture (carb-induced contracture), but after repeated application, carbachol could no longer evoke contracture. If it is assumed that carbachol releases sequestered Ca, the amplitude of the carb-induced contracture in Ca-free isotonic K-Krebs solution should indicate the existence and amount of sequestered Ca. Figure 5 shows such an experiment, i.e., after the exhaustion of sequestered Ca by repeated carbachol application (Fig. 5b), Ca-containing isotonic K-Krebs solution was applied for various durations (Fig. 5c). In the circular muscle, the application of Ca caused only a very slight tonic response, and when carbachol was applied after the Ca contracture had relaxed it caused a small contracture.

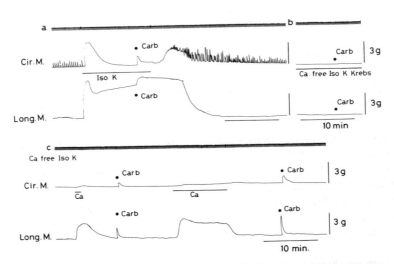

FIG. 5. Effects of carbachol (10^{-5} g/ml) on the circular and longitudinal muscles in isotonic K-Krebs solution. (a) Effects of carbachol on the K-induced contracture; (b) effects of carbachol after several previous applications of carbachol in Ca-free isotonic K-Krebs solution; (c) effects of carbachol after briefly readmitting Ca to Ca-free isotonic K-Krebs solution. Ca ion (2.2 mM) was added for 1 min and 10 min and carbachol was applied to the tissue after the contracture relaxed to the resting level in Ca-free isotonic K-Krebs solution.

The amplitude of the carb-induced contraction increased in proportion to the duration of the preceding exposure to Ca-containing solution. In contrast, in the longitudinal muscle, immersion in Ca-containing isotonic K-Krebs solution after treatment with Ca-free isotonic K-Krebs solution caused a large contracture, the amplitude of which was proportional to the duration of the immersion. The subsequent carb-induced contractures were affected in a way similar to that of circular muscle. These results indicate that in the circular muscle Ca might be mainly sequestered in storage sites, whereas in the longitudinal muscle, the influx of Ca ion might directly increase the amount of intracellular free Ca ions and only a part of the Ca ions might be sequestered.

It is also known that Cd and Mn ions suppress the influx of Ca ion in visceral smooth muscles and therefore suppress the development of tension. As shown in Fig. 6 the application of Cd ion (0.5 mM) suppressed the K-induced contracture without any marked change of the membrane depolarization level (Fig. 6A). Two minutes after addition of Cd ion to Ca-free isotonic K-Krebs solution, acetylcholine induced a small contracture (Fig. 6B). The amplitude of the contracture caused by successive application of acetylcholine

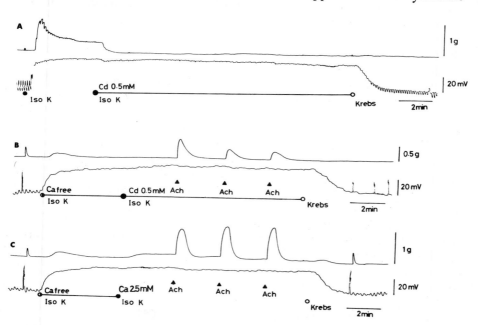

FIG. 6. Effects of acetylcholine (10^{-5} g/ml) and Cd on the K-induced contracture evoked in circular muscle of the stomach. (A) Effects of Cd ion (0.5 mM) during the generation of K-induced contracture. Inward current pulses (3 sec pulse duration) were continuously applied throughout. (B) Effects of acetylcholine in Cd-containing Ca-free isotonic K-Krebs solution. The tissue was depolarized by Ca-free isotonic K-Krebs solution, then 0.5 mM Cd was added. (C) Effects of acetylcholine in Ca-containing isotonic K-Krebs solution. The tissue was depolarized by Ca-free isotonic K-Krebs solution; then 2.5 mM Ca was added in the solution. In b and c, to evoke the electrical and mechanical responses outward current pulses were applied before and after application of isotonic K-Krebs solution.

gradually diminished. In contrast, on addition of Ca after pretreatment with Ca-free isotonic K-Krebs solution, the tonic response of the K contracture recovered and acetylcholine induced large contractures (Fig. 6C). These results support the view that carb-induced contracture was due to release of sequestered Ca.

Mechanical responses could also be evoked by electrical stimulation in both tissues. A single stimulus of 30-msec pulse duration produced a twitch and repetitive stimulation produced a tetanus. The duration of the twitch tension was shorter in the circular muscle than in the longitudinal muscle. The amplitude of the tetanus rapidly declined to a certain level in the circular muscle, but it was sustained at a high level in the longitudinal muscle. There was another marked difference between these muscles, in that the positive inotropic effect (staircase phenomenon) could be observed in the circular muscle but not in the longitudinal muscle. This means that repetitive generation of twitch tension mobilizes Ca from sequestered sites in the circular muscle, but this mechanism is less dominant in the longitudinal muscle (20). It is an interesting feature of stomach muscle fibers that the amplitude of the K-induced contracture is smaller than that of the tetanus tension in the circular muscle, while the reverse is observed in the longitudinal muscle. Presumably, the Ca uptake mechanism into sequestering sites might be well developed in the circular muscle, thereby suppressing the amplitude of the K-induced contracture.

Caffeine is known to suppress the mechanical response induced by electrical and chemical stimulation in various regions of the alimentary canal, with a few exceptions (12,23,15). In longitudinal and circular stomach muscle pretreatment with caffeine not only suppressed the amplitudes of the phasic and tonic responses of the K-induced contracture, but also reduced the carb-induced contracture during the K-induced depolarization. These phenomena might indicate that the suppression of K-induced and of carb-induced contractures by caffeine is due to removal of Ca ion from sequestered sites and also to extrusion into the extracellular space.

A similar effect on the mechanical response to that observed with caffeine was observed in the Ca-free Krebs solution on treatment with D 600 (10^{-5} g/ml), with verapamil, and with EGTA (3 mM). Presumably the sequestered Ca in both tissues is bound loosely and is easily released into the extracellular spaces.

DISCUSSION AND CONCLUSION

The electrical and mechanical properties observed in the longitudinal muscles (L) and circular muscles (C) are described. The differences between the two tissues can be summarized as follows: (a) The membrane potential and the length constant are the same in both tissues but the spike can only be evoked in C. However, treatment with TEA or prolonged electrical depolarization of the membrane activate the spike-generating mechanism in L. (b) Repetitive stimulation of the tissue produces the staircase phenomenon in C but not in L. A tetanus can be evoked in both tissues. During the tetanic stimulation, tension declines in C muscles but is sustained in L muscles. (c) Depolarization of the membrane and the action current (early transient inward current) both generate a mechanical response. The experiments indicate that depolarization induced by electrical displacement of the membrane potential releases sequestered Ca, whereas the depolarization induced by excess $[K]_o$ increases the influx of Ca and also releases sequestered Ca. In C muscles, the Ca that enters the cell is rapidly sequestered, but in L, the intracellular free Ca concentration rapidly increases. (d) In Na-free solution, contracture is evoked in L but not in C muscles. In isotonic K-Krebs solution, the tonic phase of the contracture is very marked in L muscles, but only slightly developed in C muscles. (e) When the temperature is lowered to below 15°C, a contracture is evoked in both tissues in Krebs solution. In C muscles, the contracture is still produced at low temperature in isotonic K-Krebs solution or in Na-free solution; however, in L muscles the contracture developed in these two solutions is partly relaxed by cooling.

From the above results, the role of Ca in stomach muscle activity, as observed in the two different tissues, is illustrated schematically in Fig. 7.

In excitable tissues, e.g., in cardiac muscle and nerve, Ca influx and efflux are partially dependent on the sodium gradient across the membrane and the possible involvement of a Na–Ca exchange system has been discussed (27,1,2,11). The existence of a similar mechanism has also been postulated to maintain a low Ca ion concentration in the myoplasm of visceral smooth muscles (28,22,26,16,30). On the other hand, it has been reported that the large transmembrane Ca gradient in smooth muscle depends on cellular ATP, but not on the Na gradient, since ATP depletion abolished the Ca gradient by increasing Ca influx (32).

It is difficult to assess the two alternative hypotheses concerning the Ca extrusion mechanism on the basis of the present experiments, because the observations on stomach muscle fit both mechanisms. In the circular muscle, a Ca pump may be the mechanism to reduce the Ca ions in the myoplasm, and in the longitudinal muscle a Na–Ca exchange mechanism may be more likely.

Removal of Ca from the myoplasm may occur mainly by uptake into the sequestered sites, which are presumably distributed just beneath the muscle membrane, either by a Na–Ca exchange mechanism or a Ca pump mechanism. On the basis of an electrically neutral (2 Na for 1 Ca) exchange, postulated from thermodynamics, the Na gradient across the muscle membrane cannot alone provide sufficient energy to reduce the myoplasmic Ca concentration to cause complete relaxation of the contractile protein (26).

It is known that the sarcoplasmic reticulum in many visceral smooth muscles is less developed than in skeletal muscle, and the stomach muscle is not exceptional. However, in the circular muscle, the vesicles and the sarco-

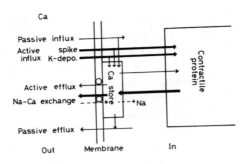

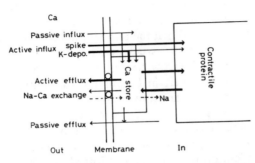

FIG. 7. Schematic illustration of the possible movements of Ca during the resting and active state of the membrane. Thick lines indicate more dominant path of Ca ion than that of thin lines. Top: Longitudinal muscle; bottom: circular muscle.

plasmic reticulum are more developed than in the longitudinal muscle (Yuko Nishio, *personal communication*). Therefore, the Ca pump mechanism may be dominant in the longitudinal tissue. In general, the Na–Ca exchange mechanism depends not only on the $[Na]_o$ and $[Ca]_i$, but also on the $[Na]_i$, which is modified by $[K]_o$, and by the Na–K electrogenic pump mechanism (29,17,31,3,6,7). Moreover, $[Ca]_i$ is thought to control the K conductance in the taenia coli (4).

It may be too speculative to propose that the Na–Ca exchange at the muscle membrane might have a causal relationship with the passive diffusion of monovalent cations, i.e., that an early increase of K permeability during the active state of membrane might suppress the spike generation in the longitudinal muscle tissue by release of Ca which is stored just beneath the cell membrane.

As described previously, the motility of the stomach differs according to the region. The longitudinal muscle in the upper region of stomach may only be activated by extension of the stomach wall through increased contents, since this region acts as a reservoir. It may not be necessary for this muscle to possess the properties required by the lower region of the stomach where there are various patterns of motility such as propelling, retropelling, and triturating of the contents. Hence, it is postulated that the activity of the stomach might be controlled not only by the activity of the autonomic nervous system and the enteric plexuses, but also by topical differences in the properties of the smooth muscle tissues themselves.

ACKNOWLEDGMENTS

The authors wish to record their sincere graditude to professor E. Bülbring for valuable discussions and her critical review of the manuscript. This work was supported in part by Educational Ministry of Japan.

REFERENCES

1. Baker, P. F., Blaustein, M. P., Hodgkin, A. L., and Steinhardt, R. A. (1969): *J. Physiol. (Lond.)*, 200:431–458.
2. Blaustein, M. P., and Hodgkin, A. L. (1969): *J. Physiol. (Lond.)*, 200:497–527.
3. Brading, A. F. (1973): *Philos. Trans. R. Soc. Lond. [Biol. Sci.]*, 265:35–46.
4. Brading, A. F., Bülbring, E., and Tomita, T. (1969): *J. Physiol. (Lond.)*, 200:637–654.
5. Bülbring, E., and Needham, D. M. (1973): *Philos. Trans. R. Soc. Lond. [Biol. Sci.]*, 265:1–231.
6. Casteels, R., Droogmans, G., and Hendrickx, H. (1971): *J. Physiol. (Lond.)*, 217:297–313.
7. Casteels, R., Droogmans, G., and Hendrickx, H. (1973): *J. Physiol. (Lond.)*, 228:733–748.
8. Ebashi, S., and Endo, M. (1968): *Prog. Biophys. Mol. Biol.*, 5:125–183.
9. Ebashi, S., Ohtsuki, I., and Mihashi, K. (1972): *Symp. Quant. Biol.*, 37:215–223.
10. Fuchs, F. (1974). *Annu. Rev. Physiol.*, 36:461–502.
11. Glitsch, H. G., Reuter, H., and Scholz, H. (1970): *J. Physiol. (Lond.)*, 209:25–44.
12. Ito, Y., and Kuriyama, H. (1971): *J. Gen. Physiol.*, 57:448–463.
13. Ito, Y., and Kuriyama, H. (1973): *J. Physiol. (Lond.)*, 231:455–470.
14. Ito, Y., Osa, T., and Kuriyama, H. (1973): *Jap. J. Physiol.*, 24:217–232.
15. Ito, Y., Osa, T., and Kuriyama, H. (1974): *Jap. J. Physiol.*, 24:343–357.
16. Katase, T., and Tomita, T. (1972): *J. Physiol. (Lond.)*, 224:389–500.
17. Kuriyama, H., Ohshima, K., and Sakamoto, Y. (1971): *J. Physiol. (Lond.)*, 217:179–199.
18. Kuriyama, H., Osa, T., and Tasaki, H. (1970): *J. Gen. Physiol.*, 55:48–62.
19. Magaribuchi, T., Ohbu, T., Sakamoto, Y., and Yamamoto, Y. (1972): *Jap. J. Physiol.*, 22:333–352.
20. Mishima, K., Suzuki, H., and Kuriyama, H. (1975): *In preparation.*
21. Ohtsuki, I., Masaki, T., Nonomura, Y., and Ebashi, S. (1967): *J. Biochem.*, 61:817–819.
22. Osa, T. (1971): *Jap. J. Physiol.*, 21:607.
23. Osa, T. (1973): *Jap. J. Physiol.*, 23:199.
24. Osa, T., and Kuriyama, H. (1970): *Jap. J. Physiol.*, 20:626.
25. Prosser, C. L. (1974): *Annu. Rev. Physiol.*, 34:503–535.
26. Reuter, H., Blaustein, M. P., and Haeusler, G. (1973): *Philos. Trans. R. Soc. Lond. [Biol. Sci.]*, 265:87.
27. Reuter, H., and Seitz, N. (1968): *J. Physiol. (Lond.)*, 195:451.
28. Sitrin, M. D., and Bohr, D. F. (1971): *Am. J. Physiol.*, 220:1124–1128.
29. Taylor, G. S., Paton, D. M., and Daniel, E. E. (1970): *J. Gen. Physiol.*, 56:360–375.

30. Tomita, T., and Watanabe, H. (1973). *Philos. Trans. R. Soc. Lond. [Biol. Sci.]*, 265:73–85.
31. Tomita, T., and Yamamoto, T. (1971): *J. Physiol. [Lond.]*, 212:851–868.
32. van Breemen, C., Farinas, B. R., Casteels, R., Gerba, P., Wuytack, F., and Deth, R. (1973): *Philos. Trans. R. Soc. Lond. [Biol. Sci.]*, 265:57–71.
33. Yasui, B., Fuchs, F., and Brigg, F. N. (1968): *J. Biol. Chem.*, 243:735–742.

Physiology of Smooth Muscle, edited by
E. Bülbring and M. F. Shuba.
Raven Press, New York © 1976.

Theory of P and T Systems for Calcium Activation in Smooth Muscle

K. Golenhofen

Department of Physiology, University of Marburg/Lahn, West Germany

CALCIUM ANTAGONISM

Fleckenstein et al. (1) found that a special group of substances is able to suppress the contraction of heart muscle without producing significant changes in the action potential. This effect was interpreted as an inhibition of calcium activation in the process of electromechanical coupling, and these substances were called "calcium antagonists." Further studies have shown that these drugs inhibit the calcium current during heart muscle activation (2). The main representatives in this group are verapamil, D 600 (methoxy-verapamil) (Knoll A. G., Ludwigshafen, West Germany), and nifedipine (BAY a 1040, Bayer, Leverkusen, West Germany).

DIFFERENTIATED SUPPRESSION OF CALCIUM ACTIVATION

These calcium antagonists are also effective inhibitors in smooth muscle (3-7) where they suppress not only contraction, but also the spike discharges (8). It had usually been assumed that their mechanism of action in smooth muscle was similar to that in heart muscle. However, we observed (Fig. 1) that during complete suppression of spike discharges with verapamil, in guinea pig portal vein, norepinephrine (NE) could still induce a contraction which was reduced to 40 to 50% of the control reaction. Such a "spike-free" activation could also be elicited in stomach smooth muscle by acetylcholine (10). This spike-free activation was as sensitive to calcium deprivation and to application of lanthanum as was the normal activation, which indicated that, in addition to the verapamil-sensitive "spike activation mechanism," another calcium activation mechanism exists in smooth muscle, which is similarly dependent on transmembrane calcium fluxes (10,11). The final proof for the dual nature of calcium activation in smooth muscle was provided by the observation that the verapamil- and D 600-resistant tonic activation of stomach fundus could be selectively blocked by another antagonist, sodium nitroprusside (12; Boev, Golenhofen, and Lukanow, *this volume*).

Comparative studies with other types of smooth muscle are now available, which allow more general statements to be made about smooth muscle activa-

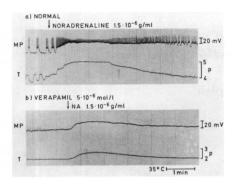

FIG. 1. Effect of norepinephrine (NE) application on membrane potential (MP), measured intracellularly, and tension development (T) of portal vein smooth muscle (guinea pig). (a) Under normal conditions; (b) under continuous treatment with verapamil. NE was given as a single application into the continuously perfused organ bath. Tension calibrated in pond. (From Ref. 9.)

tion (13). One experiment with vascular smooth muscle is shown in Fig. 2 in which aorta and portal vein were chosen as examples of tissues producing typically tonic and typically phasic mechanical activity, respectively. Norepinephrine (NE) produced a tonic activation of the aortic strip and a slight increase in amplitude of the phasic activity of the portal vein. Sodium nitroprusside (NP) inhibited selectively (or preferentially) the tonic activation of the aortic preparation, whereas D 600 suppressed the phasic activity of portal vein with little alteration of the NE-induced tonic activation of aorta. The effect of NP on the aortic reaction during D 600 treatment was the same as that on the normal aortic reaction. The aorta and portal vein of the guinea pig were less sensitive than were those of the rat to both NP and D 600. Experiments with the gallbladder were also included because of its pronounced tonic character, and a comparison with stomach preparations is shown in Fig. 3. The typical selective inhibition of the tonic activity in the fundus preparation by NP is visible, and also the selective inhibition of the phasic activity of the antrum by D 600. The cholecystokinin (CCK)-induced activation of the gallbladder, however, was resistant to both NP and D 600, either alone or in combination.

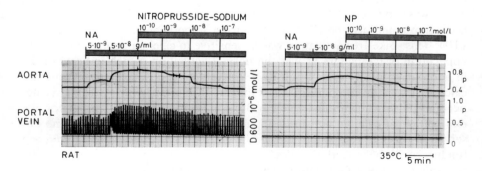

FIG. 2. Tension development of a spiral strip of aorta and of an isolated portal vein, both from the rat. Cumulative application of norepinephrine (NE) and sodium nitroprusside (NP) as indicated. D 600 10^{-6} mol/liter was applied 15 min before the second NE application.

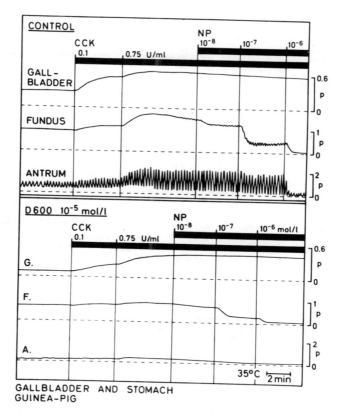

FIG. 3. Tension development of isolated strips from gallbladder and stomach, fundus and antrum region, all from the guinea pig. Cumulative application of cholecystokinin (CCK), in units per milliliter, and sodium nitroprusside (NP, in moles per liter) as indicated. Top: Under normal conditions; bottom: 15 min after application of D 600 10^{-5} mol/liter.

Uterine smooth muscle is a typical spike-producing and phasically active muscle. Even here a strong spike-free, tonic component could be unmasked in the oxytocin contraction by D 600 as shown in Fig. 4 (14). A pure tonic activation appeared under D 600 10^{-5} mol/liter, which was not significantly changed with further increase of the D 600 concentration to 10^{-4} mol/liter. The same tonic reaction appeared under D 600 10^{-6} mol/liter, with some phasic contractions superimposed. This observation suggests that a D 600-resistant tonic component is included in the normal oxytocin-induced activation. However, this tonic reaction of uterine smooth muscle, which can be clearly separated from the phasic activity by D 600, is not sensitive to NP (in contrast to stomach and aorta) (Fig. 5). Papaverine acts as an unspecific inhibitor of both phasic and tonic activations, in the uterus (Fig. 5) as well as in the stomach. The D 600-resistant tonic activation is, at least in some tis-

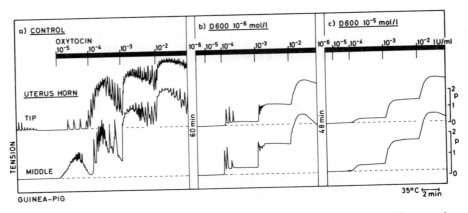

FIG. 4. Tension development of two isolated strips of guinea pig uterus, excised from different regions of the uterus horn. Cumulative application of oxytocin as indicated, (a) under normal conditions, (b) and (c) after application of D 600 in the indicated concentration. Application of oxytocin at least 15 min after application of D 600. (Golenhofen and Neuser, *unpublished*.)

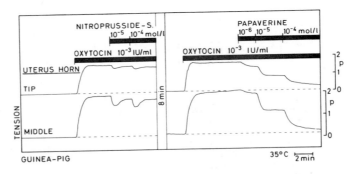

FIG. 5. Tension development of two isolated strips of guinea pig uterus, during continuous treatment with D 600 10^{-5} mol/liter. Details as in Fig. 5. Application of sodium nitroprusside and papaverine as indicated. (Golenhofen and Neuser, *unpublished*.)

sues, similarly dependent on extracellular calcium as is the phasic activity (10,11).

THEORY OF P AND T SYSTEMS

A satisfying interpretation of all results (from which only examples could be presented here) is possible on the basis of the following theory, which is illustrated in Fig. 6:

1. Two chemically different systems for calcium activation exist in the membrane of smooth muscle cells which are called P and T systems.

2. The P system is preferentially used for producing phasic activity, the T system preferentially for tonic activity.

3. The P system can be blocked selectively by a group of substances such as verapamil, D 600, and nifedipine, which may be called P antagonists. The T system is resistant to P antagonists. A universal T antagonist is not yet known; NP is a partial T antagonist which can inhibit the T systems of some tissues. The T systems appear to be more diverse than the P systems, and a further subdivision of the T system can be visualized.

4. The P system is usually responsible for spike discharges; the activation of the T system normally occurs with a spike-free depolarization.

5. P and T systems lie in parallel in the chain of events leading from the interaction of an agonist with its receptor to the activation of the contractile proteins. The action of P and T antagonists takes place beyond the level of re-

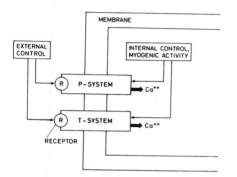

FIG. 6. Diagrammatic illustration of P and T systems for calcium activation in the membrane of smooth muscle cells.

ceptors, since it is independent of whether the cell is driven via adrenergic or cholinergic receptors or by internal myogenic pacemaker processes.

6. Special connections between one of the systems and one of the different pools of activator calcium may exist, and they may be different from one type of smooth muscle to another. However, in principle, the differentiation of calcium pools is independent of the chemical differentiation of the activator systems.

7. A differentiation between the cells of one type of tissue into more P and more T cells may exist, but this is not a necessary precondition for the P–T differentiation.

8. The differentiation of P and T systems expresses the tendency of the organism to control phasic and tonic processes selectively, the T system apparently being the more specialized and phylogenetically the younger type.

9. The degree of differentiation of calcium activation processes into P and T systems differs from species to species and from one type of smooth muscle to another.

10. P and T antagonists are different in their ability to block one of the systems selectively, i.e., they are different in their specificity.

Not all of these statements are equally well proved at the moment. They also contain hypotheses that have to be tested experimentally.

REFERENCES

1. Fleckenstein, A., Döring, H. J., and Kammermeier, H. (1968): *Klin. Wochenschr.*, 46:343–351.
2. Kohlhardt, M., Bauer, B., Krause, H., and Fleckenstein, A. (1972): *Pflügers Arch.*, 335:309–322.
3. Bilek, I., Laven, R., Peiper, U., and Regnat, K. (1974): *Microvasc. Res.*, 7:181–189.
4. Fleckenstein, A., Grün, G., Tritthart, H., and Byon, K. (1971): *Klin. Wochenschr.*, 49:32–41.
5. Grün, G., and Fleckenstein, A. (1972): *Arzneim. Forsch.*, 22:334–344.
6. Haeusler, G. (1972): *J. Pharmacol. Exp. Ther.*, 180:672–682.
7. Peiper, U., Griebel, L., and Wende, W. (1971): *Pflügers Arch.*, 330:74–89.
8. Golenhofen, K., and Lammel, E. (1972): *Pflügers Arch.*, 331:233–243.
9. Golenhofen, K., Hermstein, N., and Lammel, E. (1973): *Microvasc. Res.*, 5:73–80.
10. Golenhofen, K., and Wegner, K. (1975): *Pflügers Arch.*, 354:29–37.
11. Golenhofen, K., and Hermstein, N. (1975): *Blood Vessels*, 12:21–37.
12. Boev, K., Golenhofen, K., and Lukanow, J. (1973): *Pflügers Arch.*, 343:R56.
13. Golenhofen, K. (1973): *Pflügers Arch.*, 343:R57.
14. Golenhofen, K., and Neuser, G. (1974): *Pflügers Arch.*, 347:R15.

Physiology of Smooth Muscle, edited by
E. Bülbring and M. F. Shuba.
Raven Press, New York © 1976.

Selective Suppression of Phasic and Tonic Activation Mechanisms in Stomach Smooth Muscle

K. Boev,* K. Golenhofen, and J. Lukanow*

Department of Physiology, University of Marburg/Lahn, West Germany

As early as 1898, Cannon (1) distinguished between two types of motor activity in the stomach: the active (tonic) reservoir function of the fundus and the (phasic) peristaltic activity of the pyloric portion. With modern electrophysiological techniques it has been shown that slow fluctuations of the membrane potential, with an amplitude of 30 to 40 mV, are associated with the stomach peristalsis (2–7). The absence of these potential waves in the fundus muscle seemed to provide an electrophysiological basis for the regional differences in stomach motility. Recently Golenhofen and Wegner (8), using circular muscle strips of the corpus/antrum region of guinea pig stomach, showed that phasic, peristaltic activity could be blocked with verapamil, while tonic activity persisted. Spike discharges were also suppressed by this procedure, but the slow basic potential waves were not significantly altered (8,9). Based on these studies the activation mechanisms of stomach smooth muscle have now been investigated in more detail (10).

METHODS

Electrical and mechanical activity were measured in circular strips of stomach smooth muscle from 20 guinea pigs and 16 cats. In addition, in nine experiments the activity of a taenia coli preparation (guinea pig) was measured. The preparations were usually excised from the middle part of fundus, corpus, and antrum of the stomach, and preparations from all regions were mounted in a conventional thermostatically controlled organ bath with modified Krebs solution for the simultaneous recording of mechanical activity. Additional preparations were usually excised for the simultaneous recording of electrical and mechanical activity. Electrical activity was measured either extracellularly with platinum wire electrodes (11), or intracellularly (4), or with the sucrose-gap technique (12).

---

* Present address: Bulgarian Academy of Sciences, Institute of Physiology, Ul. 36 Bl. 1, Sofia, Bulgaria.

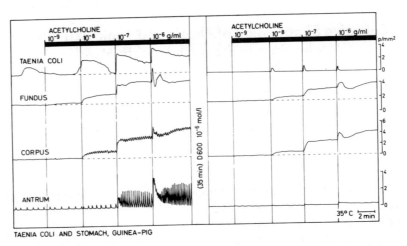

TAENIA COLI AND STOMACH, GUINEA-PIG

FIG. 1. Simultaneous recording of tension development of a taenia coli preparation and three circular muscle strips from different regions of a guinea pig stomach. Application of acetylcholine cumulatively as indicated, under control conditions (left) and in the presence of D 600 10^{-6} mol/liter (right). Tension calibrated in pond per square millimeter.

RESULTS

As shown in Fig. 1, acetylcholine (ACh) produced an increase in the amplitude and frequency of phasic contractions in the antrum preparation, a more tonic activation in the fundus strip, and an intermediate type of reaction in the corpus preparation. In taenia coli the spontaneous phasic activity, which is slower than in antrum preparations and is of the minute-rhythm type, was intensified, and a more continuous activation often appeared with higher concentrations of ACh. In the presence of D 600 (methoxyverapamil; Knoll, A. G., Ludwigshafen, West Germany), applied in concentrations of 10^{-6} to 10^{-5} mol/liter, at least 15 min before the next ACh application, all phasic components of the activity in the stomach preparations were suppressed, while the tonic reaction of the fundus strip persisted, only slightly reduced in size. The intermediate phasic–tonic activation of the corpus was converted to a purely tonic activation. Spontaneous and ACh-induced activity was completely suppressed in taenia coli, with no tonic activation appearing in the presence of D 600. The small and short-duration contractions seen with each increase of the ACh concentration in Fig. 1 are negligible. These results are in good agreement with those of Golenhofen and Wegner (8) who only used preparations from the distal corpus and the proximal antrum regions of guinea pig stomach and therefore usually observed responses of a type intermediate between those of the corpus and antrum in Fig. 1. Verapamil and D 600 were qualitatively similar in their effects but D 600 was more effective and more specific and was therefore preferred.

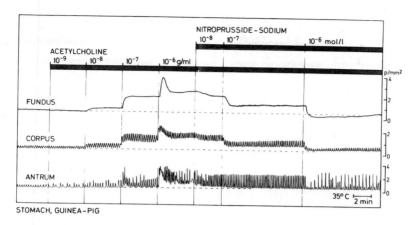

FIG. 2. Tension development of three circular muscle preparations of guinea pig stomach, as described in Fig. 1. Cumulative application of sodium nitroprusside in addition to ACh.

In the search for a drug able to suppress tonic activation selectively we tried sodium nitroprusside (NP), which is highly effective in blocking activation of aortic smooth muscle, but has little effect on intestinal smooth muscle (tested with duodenum) and uterine smooth muscle (13). Because aortic contraction is similar to that of the fundus in its tonic character and in its low sensitivity to verapamil and D 600 (14), the expectation that NP might be able to suppress tonic activity in fundus preparations was justified. A typical result is shown in Fig. 2. Threshold concentrations of NP (10^{-8} mol/ liter) reduced fundus tone and the tonic component in the corpus, without significantly affecting the phasic contractions in corpus and antrum. This effect was augmented at 10^{-7} mol/liter, and small phasic contractions became visible in the fundus record. With further increase of the NP concentration to 10^{-6} mol/liter, the fundus tension declined below the level of its initial spontaneous tone, and corpus tone was completely suppressed. Phasic contractions, however, were also slightly inhibited, at least transiently, in a manner similar to that described for other phasic contractions by Kreye (13).

Figure 3 shows the effect of a combination of NP and D 600 and, in addition, the similarity of the reactions in cat stomach to those in guinea pig stomach. Both drugs acted more selectively in cat stomach than in the guinea pig, which is also seen in Fig. 3. Tonic reactions of cat fundus often reached the same size under D 600 treatment as under control conditions, and the effect of NP 10^{-6} mol/liter on phasic contractions was usually less pronounced in cat preparations. As shown in Fig. 3, the effect of NP on the ACh response after D 600 application was not significantly different from the effect under control conditions.

In adequate concentrations, therefore, D 600 (and verapamil) are able to suppress selectively phasic contractions, and NP is able to suppress selectively

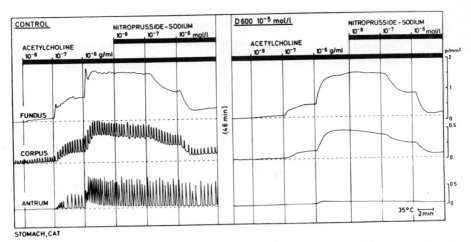

STOMACH, CAT

FIG. 3. Tension development of three circular muscle preparations from different regions of a cat stomach. Details as described in Figs. 1 and 2.

tonic activation. Such observations can be taken as final proof that two different mechanisms, which can be designated as the P and T mechanisms, respectively, are responsible for these two types of activity.

Tetrodotoxin (up to $3 \cdot 10^{-6}$ g/ml) often augmented the spontaneous activity in preparations of cat stomach, but had no influence on the described reactions. Spontaneous tonic activity was particularly pronounced in fundus preparations (see Figs. 2 and 4) and was very sensitive to NP.

The dependence of the T mechanism on extracellular calcium is shown in Fig. 4. Tension declines in Ca-free solution, and an initial rapid phase can be distinguished from a second slower phase. The full response is rapidly reestablished after readmission of calcium. There is no doubt that the T activation is dependent on the presence of external calcium. The electrical behavior of a corpus/antrum preparation during T activation is shown in Fig.

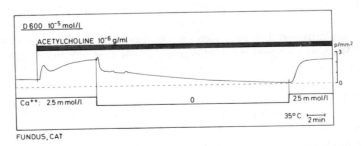

FUNDUS, CAT

FIG. 4. Tension development of a fundus strip of cat stomach during P blockade by D 600 10^{-5} mol/liter. ACh-induced tonic activation, and the effects of calcium removal and calcium readmission during this activation. Note the spontaneous tonic activity before ACh application.

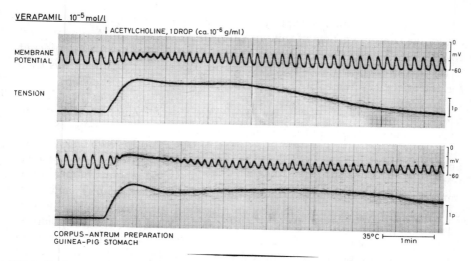

VERAPAMIL 10⁻⁵mol/l

FIG. 5. Simultaneous recording of membrane potential (above, measured intracellularly) and tension (below) in a corpus/antrum preparation of guinea pig stomach. P system blocked by verapamil 10^{-5} mol/liter. Application of ACh into a continuously perfused organ bath, with rapid washout of ACh. Both parts are sections of a continuous recording; 2 drops of ACh were applied in the lower part.

5. As described by Golenhofen and Lammel (9), the spontaneous phasic contractions and the spike discharges were blocked by verapamil, but the basic slow waves of the membrane potential persisted without much alteration. Under these conditions, ACh induced a dose-dependent depolarization combined with contraction (comparable results in portal vein smooth muscle: see ref.

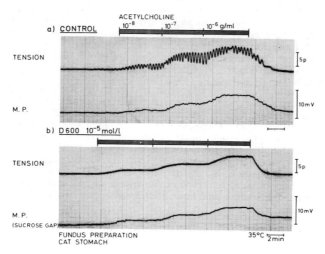

FIG. 6. Simultaneous recording of tension development (above) and membrane potential (below, with the sucrose-gap technique) of a fundus preparation of cat stomach. ACh-induced activation, under control conditions (a) and in the presence of D 600 10^{-5} mol/liter (b). Both are sections of the same experiment.

15). These effects rapidly disappeared because ACh was progressively washed out by the continuous flow of solution through the organ bath. A similar good correlation between the depolarization and the tonic reaction was observed in the fundus preparation of Fig. 6 where the sucrose-gap technique was used for potential measurement.

More quantitative measurements of the electrical responses need to be made, but some conclusions can be drawn. The depolarization combined with tonic response was, except for very high ACh concentrations, smaller than the spontaneous potential waves of corpus/antrum preparations which were not able to trigger tension development after P blockade. Furthermore, T activation could also be induced by ACh in K^+-depolarized preparations (120 mmol/liter K^+). The depolarization normally combined with T activation cannot, therefore, be interpreted as the causal factor for producing the mechanical activation. On the other hand, it cannot be excluded that depolarization plays a definite part in the activation processes. We therefore hesitate to call the T activation a "nonelectrical activation," as did Bohr and Uchida (16) in their description of a similar activation in vascular smooth muscle.

REFERENCES

1. Cannon, W. B. (1898): *Am. J. Physiol.*, 1:359–382.
2. Daniel, E. E., and Irwin, J. (1968): Electrical activity of gastric musculature. In: *Handbook of Physiology*, Sect. 6, Vol. IV, pp. 1969–1984. American Physiological Society, Washington, D.C.
3. Golenhofen, K. (1971): Intrinsic rhythms of the gastrointestinal tract. In: *Gastrointestinal Motility*, edited by L. Demling and R. Ottenjann. Academic Press, New York, pp. 71–81.
4. Golenhofen, K., v.Loh, D., and Milenov, K. (1970): *Pflügers Arch.*, 315:336–356.
5. Papasova, M. (1970): *An Electrophysiological Study of the Motor Activity of the Stomach*. Publishing House of the Bulgarian Academy of Sciences, Sofia.
6. Sakamoto, Y., and Kuriyama, H. (1970): *Jap. J. Physiol.*, 20:640–656.
7. Sarna, S. K., and Daniel, E. E. (1973): *Am. J. Physiol.*, 225:125–131.
8. Golenhofen, K., and Wegner, H. (1975): *Pflügers Arch.*, 354:29–37.
9. Golenhofen, K., and Lammel, E. (1972): *Pflügers Arch.*, 331:233–243.
10. Boev, K., Golenhofen, K., and Lukanow, J. (1973): *Pflügers Arch.*, 343:R56.
11. Golenhofen, K., and v.Loh, D. (1970): *Pflügers Arch.*, 314:312–328.
12. Boev, K., and Golenhofen, K. (1974): *Pflügers Arch.*, 349:277–283.
13. Kreye, V. A. W. (1974): Untersuchungen zum zellulären Wirkungsmechanismus vasodilatatorischer Pharmaka unter besonderer Berücksichtigung von Natrium-Nitro-prussid. Habilitationsschrift, Heidelberg.
14. Golenhofen, K., and Hermstein, N. (1975): *Blood Vessels*, 12:21–37.
15. Golenhofen, K., Hermstein, N., and Lammel, E. (1973): *Microvasc. Res.*, 5:73–80.
16. Bohr, D. F., and Uchida, E. (1969): Activation of vascular smooth muscle. In: *The Pulmonary Circulation and Interstitial Space*, pp. 133–143. University of Chicago Press, Chicago, Ill.

Physiology of Smooth Muscle, edited by
E. Bülbring and M. F. Shuba.
Raven Press, New York © 1976.

The Slow Potential and Its Relationship to the Gastric Smooth Muscle Contraction

M. Papasova and K. Boev

Institute of Physiology, Bulgarian Academy of Sciences, Sofia, Bulgaria

The problem of the correlation between electrical and contractile activities of gastric smooth muscle attracted the attention of the pioneering workers in this field. First, there was the problem of the role of the slow waves recorded from the wall of the gastrointestinal tract, and, second, the problem of whether the spike potentials were an expression of the action potentials. The slow wave in the small intestine has an almost sinusoidal shape, whereas in the stomach of the cat, dog, and man, a complex configuration may be observed that resembles the plateau-type potential recorded from heart muscle. In chronic experiments on cats and dogs, with electrodes and strain gauges on the gastric wall, Papasova et al. (6) established that the slow wave generated spontaneously in the gastric smooth muscle of dog was always accompanied by contractions whether it was associated with spike potentials or not. Rhythmic slow waves followed by low-amplitude contractions were always recorded from the gastric wall during the so-called relative quiescence of the stomach of a starved animal (Fig. 1A). The normal occurrence of stronger motor activity of the stomach was accompanied by the appearance of spike potentials (Fig. 1B). The amplitude of the contractions of the gastric wall depended on both frequency and amplitude of spike potentials. The fact that the slow wave recorded from gastric smooth muscle was always accompanied by contractions, as well as by some changes in the slow potential related to changes in the ions and temperature in the nutrient medium in *in vitro* experiments, formed the basis on which to speak of "slow potentials in the stomach" in contrast to "slow waves" in the small intestine, which did not always lead to contractions.

The results reported here have been obtained with gastric preparations mainly from cat, and partly from dog and man, by using the sucrose-gap method and pressure electrodes. The slow potential from the gastric smooth muscle of dog, cat, and man resembles the plateau-type potential characteristic for heart muscle. However, slow potentials of almost sinusoidal shape are recorded from the gastric smooth muscle of rat and guinea pig. This was demonstrated by Bogach et al. (7) and by Golenhofen et al. (2). The detailed studies of Milenov and Boev (3) and Boev (1) of the configuration of the gastric slow potential in these two species indicated that it was not a

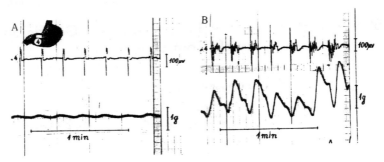

FIG. 1. Dependence of the gastric wall contraction force on the pattern of bioelectrical activity. (*Upper traces*) Bioelectrical activity. (*Lower traces*) Contraction (O) Electrode and (□) strain gauge chronically implanted in the dog stomach. For description see text.

matter of errors in method, as some authors considered, but of some special property of the smooth muscle cells in the gastric wall of these animals.

It has been shown by Papasova et al. (5) that the slow potential from cat stomach consists of two components—an initial fast component, which is an expression of the initial depolarization of the cell membrane, and a second component that is an expression of the repolarization. The first component depends on the entry of Na ions although it is not sensitive to tetrodotoxin and is reduced or eliminated by ouabain. With respect to its responses to low concentrations of sodium, tetrodotoxin, and ouabain, the first fast component of the action potential from cat stomach is very similar to the sinusoidal slow-wave characteristic for duodenum and jejunum. The second component decreases in amplitude and frequency with the decrease of Ca ions. It is sensitive to manganese and epinephrine. By these responses, the second component resembles the potentials characteristic of taenia coli, as was shown by Nonomura et al. (4).

The duration of the slow potential is within the range of 5 to 10 sec. The amplitude of the first component (measured by the sucrose gap) is 13.5 ± 0.04 mV, and its duration is 0.77 ± 0.01 sec. The amplitude of the second component measured at the plateau maximum is 3.5 mV.

The second component is variable, and its duration and amplitude determine the pattern of muscle contractions. Whereas the slow wave in the gastric smooth muscle of rat and guinea pig is followed by contractions only when it is accompanied by spike potentials, there is a continuous correlation between slow potential and contraction, each slow potential being followed by contraction of the gastric smooth muscle from cat, dog, and man.

The slow wave from guinea pig stomach has a clear-cut sinusoidal shape identical with that of the slow wave recorded from the small intestine of the cat. The rate of increase of the initial depolarization is about 5 mV/sec and is almost equal to the repolarization. Contractions appear only in the presence of spike potentials. Moreover, the muscle contraction begins not

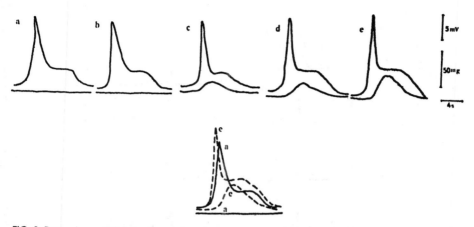

FIG. 2. Dependence of the smooth muscle contraction on the speed of the initial depolarizing phase of the slow potential. (a) Rate of increase: 16 mV/sec; (c) rate of increase: 25 mV/sec.

at the onset of the slow wave but just at the occurrence of the first spike. A different correlation exists in the gastric smooth muscle of the cat. Both components of the slow potential are necessary to initiate contraction. Slow potentials without a distinct second component are not followed by contractions. Here the rate of depolarization is relatively slow. It is not until the plateau-type potential is well defined, and the second component is clearly observable, that the contraction appears. The delay between the onset of the slow potential and the onset of contraction is 678 ± 25.8 msec, whereas the time of the initial depolarizing phase is 772 ± 15.5 msec. However, in some rare instances, the contraction is lacking even in the presence of the second component. The reason for this is the relatively slow rate of the initial depolarizing phase (Fig. 2). For instance, the rate of the initial depolarizing phase in Fig. 2a, b does not exceed 15 mV/sec, whereas the rate of depolarization in Fig. 2c, exceeding 25 mV/sec, produces contraction. This is clearly seen when the potentials are superimposed. With equal amplitude and relatively equal rate of rise of the initial depolarizing phase the contraction force depends on the amplitude of the slow-potential plateau. When spike potentials appear on the plateau an increase in the force of contraction may be observed. Indentation corresponding to the spike potentials is then seen in the mechanogram.

The dependence of the contraction on the configuration of the slow potential is particularly well demonstrated in a potential evoked by electrical stimulation. Application of suprathreshold cathodal stimulation, 1 μA, provokes the appearance of a potential of comparatively high amplitude and long duration (Fig. 3). The amplitude of the contraction is higher too. Evoked potentials are also obtained on switching off anodal stimulation. As a rule, anodal stimulation inhibits spontaneously occurring slow potentials,

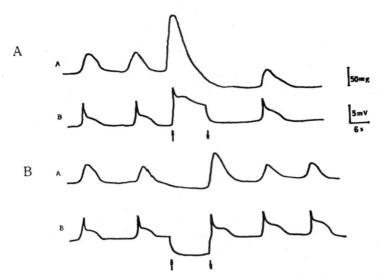

FIG. 3. Effect of electrical stimulation on the slow potential from the gastric smooth muscle. (A) Cathodal stimulation; (B) anodal stimulation. (a) Contractions; (b) potentials. (Arrows) Current on and off.

while switching off initiates potentials and contractions. During the development of the initial depolarizing phase the gastric smooth muscle excitability is reduced to zero, i.e., we observe an absolute refractory period, while during the repolarizing phase the excitability increases as the resting potential returns to its initial level. When cathodal stimulation with a current intensity of 1 μA is applied repeatedly during the latter part of the slow potential, a local response appears, which increases in amplitude with the termination of repolarization.

The spontaneous slow potential of the cat stomach is rather sensitive to changes in temperature and ionic composition of the medium. Its frequency increases with a rise in temperature. This is associated with a decrease in duration, as well as a shortening of the interval between two slow potentials. The effects of temperature on the various parameters are illustrated in Fig. 4. It is evident that the parameters that determine the duration of the separate phases, as well as the entire complex undergo insignificant changes in the temperature range between 33° and 39° C. However, their values increase sharply with lower temperatures. The frequency changes show a reverse relationship, i.e., an increase of frequency with a rise in temperature, a stabilization in the range of 33° to 37° C, and a sharp decrease with cooling. The amplitude of the separate phases of the slow potential shows a wide range of stability between 25° and 37° C. A change of the temperature outside this range causes a gradual reduction of the amplitude. Both frequency and amplitude of slow potentials change exponentially with the temperature at $Q = 10$.

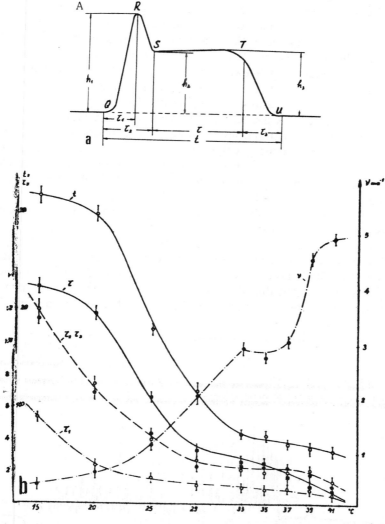

FIG. 4. Effect of temperature on the slow potential. (a) diagram showing the separate phases of the slow potential (b) changes in the separate phases of the slow potential by changes in temperature of the medium.

The rates of increase or decrease in potential of the separate phases are also temperature-sensitive, particularly those of the first component of the slow potential. There is a sharp drop in the rate of increase and decrease of the first component at a temperature of about 29° C. These changes differ from those in the descending phase of the second component. There are also changes in the propagation velocity of the slow potential, which reaches a maximum in the range of 33° to 38° C.

Compensation for the changes in the slow potential frequency at lower temperatures could be achieved by simultaneously increasing the Ca^{2+} concentration in the medium. Within the temperature range of 25° to 30° C, the relationship of the slow potential frequency to the Ca^{2+} concentration is linear. This dependence is likewise linear at 35° and 37° C, but only up to 50% increase of $[Ca^{2+}]_o$ after which there occurs a rapid rise in frequency.

The $[Na^+]_o$: $[Ca^{2+}]_o$ ratio is of importance for the pattern of slow potentials in cat stomach. A decrease of Na ions by sucrose substitution leads to a rise in slow potential frequency, i.e., the muscle reacts as if the Ca ions were increased. For instance, if the Na concentration is only 75.5 mM, the slow potential frequency rises sharply during the first 20 min, followed by the establishment of a relatively high frequency level. At the same time, both amplitude and duration of the slow potential decrease. When the Na concentration is reduced to 37 mM the ensuing changes in the muscle are irreversible.

The slow potential from the cat stomach displays some selectivity with regard to bivalent ions. If the $[Ca^{2+}]_o$: $[Na^+]_o$ ratio is kept constant but both are reduced by 50%, no essential changes are observed in the slow potential frequency. It turned out that Ba in low concentrations (0.2 to 0.8 mM) can substitute for Ca in its generation of spike potentials and contractions. How-

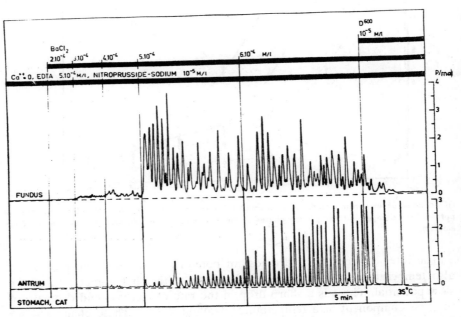

FIG. 5. Activation of phasic contractions of gastric smooth muscles by Ba^{2+}. The designations above the bottom solid line show the experimental conditions. The designations above the middle and top solid lines show the substances added to the medium. The vertical lines show the moment when the substances were added.

ever, it is interesting that in Ca-free medium Mn in low concentrations can also maintain the generation of slow potentials and contractions. A worthwhile peculiarity is observed here. While in Ca-free medium, Ba leads to restoration of the phasic contractions (Fig. 5), Mn potentiates the tonic contractions characteristic of the smooth muscles of both the corpus and fundus of the stomach.

This selectivity between Ba and Mn ions in substituting for Ca ions in excitation–contraction coupling in gastric smooth muscle is particularly distinct when the tonic contractions are blocked by sodium nitroprusside, or the phasic contractions are blocked by D_{600}. In a Ca-free medium the addition of Ba leads to generation of slow potentials with spikes accompanied by large muscle contractions (Fig. 6). The addition of D_{600} to this solution

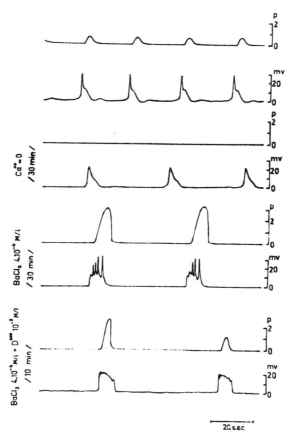

FIG. 6. Changes in the shape of the slow potential after Ca removal and after substituting Ca^{2+} by Ba^{2+}. (*Upper traces*) Contraction. (*lower record*) Bioelectrical activity (double sucrose gap). For description see text.

disturbs excitation–contraction coupling and diminishes muscle contractions, and gradually the spike, the slow potentials, and contraction disappear. The selective effects of Ca substitution by Ba for the phasic, and by Mn for the tonic contraction poses the question of two different mechanisms which bring about the two types of contractions in gastric smooth muscle. The same specificities as described for the cat were also observed in the gastric smooth muscles of dog and man.

Finally, a conclusion emerges that the spontaneously generated plateau-type potentials, characteristic for gastric smooth muscles of cat, dog, and man, determine the pattern of muscle contractions. Each potential with clearly manifested first and second components leads to contraction. The responses to changes in temperature, external sodium, and calcium ions, as well as the dependence of the muscle excitability on the different phases of the slow potential, suggest that the slow potential of gastric smooth muscles of cat, dog, and man is very similar to the plateau-type action potential of heart muscle.

REFERENCES

1. Boev, K. (1972): *Compt. r. Ac. Bulg. Sc.*, 25:541–544.
2. Golenhofen, D. v. Loh, and Milenov, K. (1969), *Pflüegers Arch.*, 315:336–356.
3. Milenov, K., and Boev, K. (1972): *Bull. Inst. Physiol. Bulg. Acad. Sci.*, 14:151–161.
4. Nonomura, I., Hotta, I., and Ohashi, H. (1966): *Science,* 97–98.
5. Papasova, M., Nagai, T., and Prosser, L. (1968), *Am. J. Physiol.*, 695–702.
6. Papasova, M., Boev, K., Milenov, K., and Atanassova, E. (1966), *Bull. Inst. Physiol. Bulg. Ac. Sci.*, 10:15–24.
7. Bogach, P. T., Kaplunenko, M. A., and Milenov, M. T. (1969): *Proc. III Ukrainian Symp. Physiol. Pathol. Digestion.* Jan. 16–17, Odessa.

Physiology of Smooth Muscle, edited by
E. Bülbring and M. F. Shuba.
Raven Press, New York © 1976.

Mechanical Properties of Contractile Elements of Smooth Muscle

Emil Bozler

Department of Physiology, Ohio State University, Columbus, Ohio 43210

Information on the nature of the contractile elements of smooth muscle can be obtained by studying the mechanical properties of the muscle at rest. Many years ago (1930) I reported (1) that stress relaxation of a snail retractor was much slower than that of skeletal muscle and, except for its beginning, closely paralleled isometric relaxation (Fig. 1A). Later the same result was obtained in uterine muscle (2,3). Here it is important that this muscle becomes electrically excitable and can conduct only under the influence of estrogens, as first shown by recording action potentials (4). After treatment with estrogens for 4 days, there are usually no spontaneous contractions in the uterus of the cat, but an electric shock produces a conducted response. In this muscle and in the guinea pig uterus during estrus, the time

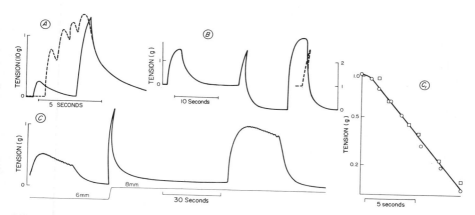

FIG. 1. Stress relaxation of relaxed muscles and isometric relaxation. In each graph first isometric twitch at the short length, then stretch and stretch relaxation, finally isometric contraction at the greater length. Tension was zero at the beginning of each experiment. A: Snail retractor of pharynx. (Graph redrawn from Fig. 7 in Ref. 1.) After first twitch and stretch relaxation recording drum was turned back so that relaxation of subsequent contraction was superimposed on stress relaxation. B: Guinea pig uterus. (Graph was redrawn from Fig. 1B in Ref. 2.) Muscle was stretched by moving the muscle lever, thereby shifting the baseline downwards, 34°. Interrupted line is stress relaxation superimposed on isometric relaxation. C: Pregnant rat uterus. Lower line: length of muscle. After stretch baseline dropped. Spontaneous contractions were preceded by a slow rise in tension, 28.5°. C_1: Semilogarithmic plot of relaxation of second twitch (◯), and stress relaxation (▢) of Fig. 1C are superimposed.

course of isometric relaxation agreed closely with that of stress relaxation (Fig. 1B).

However, because only one other author (5) has come to the same conclusion, while others (6–8) considered my results to be fallacious, I wish to report recent experiments in which my earlier work was repeated and extended. Experiments on the snail muscle have completely confirmed previous results, but I can now also sympathize with those who failed to confirm them. Unless nervous connections of the muscle with the nervous system are interrupted about a second after opening the animal, the muscle shortens from the length of more than 4 cm in a crawling animal to 0.6 cm, and then remains permanently in a state of weak contracture, gives irregular spontaneous contractions, and is unsuitable for that reason. In recent work I succeeded with the dissection only after considerable practice and only with modifications of the technique, such as cooling the animal.

If we wish to correlate mechanical properties with mechanical responses, the following conditions must be fulfilled. The preparations must be available in a completely relaxed condition and should give all-or-none, rapidly conducted responses, so that the whole muscle is rapidly and uniformly activated. This is rarely true of smooth muscle. The only vertebrate smooth muscle I could find that meets the criteria mentioned is the uterus, provided it is in the proper hormonal state.

I have recently carried out experiments with the 20- and 21-day pregnant guinea pig uterus. The spontaneous isometric contractions of short pieces of this muscle have waves of the same frequency as the spike potentials observed by previous investigators (Fig. 1C), showing that the contractions are incomplete tetani. That single contractions can be discerned is important because it shows that there is rapid conduction over the whole muscle, a condition essential for a comparison between stress relaxation and isometric relaxation. The preparation used is also favorable because immediately after a contraction it is in a strongly refractory state, so that stretching usually did not induce a response and sufficient time elapsed between contradictions, for complete relaxation to be possible. Not all preparations were equally favorable. In some, the intervals between contractions were too short or a tonic contraction was present which disappeared only briefly after a contraction. That no tonic contraction was present in a favorable preparation was shown by the fact that tension did not rise after a quick release and tension did not drop below the resting level in a Ca-free solution.

As shown in Fig $1C_1$, in which stress relaxation and isometric relaxation of Fig. 1C were plotted semilogarithmically, the time course of these processes are similar, but there are two deviations. At the beginning, the former was faster than the latter, as was true also in the muscles that were previously studied (1,2). This deviation was larger the faster the stretch, and therefore, is probably caused by a viscous resistance. Furthermore, toward the end, stress relaxation was slower than isometric relaxation, and continued at a diminishing rate for more than 2 min if the interval between contractions was long enough, producing a drop in the baseline for the next contraction. This effect was the larger the higher the permanent tension. The speed of this drop was similar to that observed in experiments of Axelsson et al. (9) on the taenia coli immersed in Ca-free solution and, therefore, was incapable of contraction. A slow drop in tension is also found in relaxed striated muscle and is generally considered to be caused by stress relaxation of the noncontractile part of the muscle. Because of the shape of the resting length-tension diagram this effect is small at short lengths of the muscle and is virtually eliminated if, before the stretch, the muscle is slightly slack and the muscle is stretched near the end of a preliminary twitch. Moreover, during a state of tonus or contracture stress relaxation also is slow.

The mechanical properties of smooth muscle have often been studied by recording the extension under a load. The slowness of extension ("creep") has been explained as being due to internal friction similar to that in plastics and has been described in terms of "viscous" properties and "plastic tonus" (cf. refs 9,10). The treatment of smooth muscle as a viscoelastic system may have some merit for the purpose of description, but cannot give information on the mechanisms involved, because it is uncertain to what extent length changes are determined by the properties of noncontractile structures or by the passive properties or activity of the contractile elements. Because the first phase of stress relaxation is faster than isometric relaxation, as mentioned above, the more so the faster the stretch, a frictional resistance seems to be present in relaxed smooth muscle, but the force so produced is of very short duration, and, therefore, cannot explain the slowness of extension under a load.

The results of previous investigators who disagree with these conclusions have in common that the drop in tension after stretch, except for the beginning, is much slower than isometric relaxation and that there is a considerable permanent increase in tension after the stretch. This is particularly true in the experiments of Greven (7). In the experiments of Abbott and Lowy on the Helix retractor the time course of stress relaxation was similar to that of isometric relaxation at short lengths, but the former became much slower at greater lengths. This deviation can be explained as being due to the presence of tonus or stress relaxation of the noncontractile component of muscle or a combination of both. Actually, Lowy confirmed my findings in later experiments (5, and *personal communication*).

Information on the mechanical properties of the contractile elements can be obtained only in relaxed muscles and at short lengths because the complicating factors mentioned are then minimized. The agreement between stress relaxation and isometric relaxation observed under these conditions shows that stress relaxation is mainly controlled by the contractile elements and that relaxation is a passive process. During isometric relaxation of a twitch, the muscle is in the same state as resting muscle after stretch. Therefore, stress relaxation must be assumed to be due to the breaking of bonds between actin and myosin filaments.

Stretching of resting striated muscle, in contrast to smooth muscle, does not produce any tension that can be attributed to the contractile elements. A quick stretch of resting striated muscle produces only a small tension, which disappears much more rapidly than that produced during an isometric twitch. This can be shown strikingly in the retractor penis of the turtle (Fig. 2), a striated muscle that has a small resting tension within a wide range of lengths and that relaxes very slowly (11). The difference in mechanical properties between the two types of muscle indicates an important difference in the contractile mechanism. The high extensibility of the contractile mechanism of striated muscle is explained by assuming complete dissociation of the thin and thick filaments. For smooth muscle, however, it must be assumed that dissociation of the filaments is incomplete at rest. This is perhaps the

FIG. 2. Isometric twitch of retractor penis of turtle. At *arrow* muscle was stretched quickly 10%. Downward movement indicates rise in tension, 20°.

most distinctive characteristic of smooth muscle, common to vertebrate and invertebrate smooth muscle. The assumption of incomplete dissociation conflicts with generally accepted ideas on the nature of the activity of crossbridges, based on studies on striated muscle, according to which bridges are either in an on- or off-position. However, in view of the large differences in submicroscopic structure and in the properties of myosin, it does not seem unexpected to find significant differences in the interaction of the filaments in the two types of muscle.

A high viscosity, that is, resistance to a change in length in both directions, would not be of any benefit to the muscle. However, as suggested above, the slowness of extension is chiefly due to a resistance encountered in the passive sliding of the filaments and thus is due to a force opposing only extension, not active shortening. This property largely explains the great economy of smooth muscle in maintaining tension. As pointed out by Rüegg (10) the slowness of the rising phase of contraction, presumably due to slow cycling of crossbridges, also contributes to the economy smooth muscle and may be more important than relaxation during tetanic contractions induced by high-frequency stimulation. When, however, activation of the fibers has a low frequency, as is likely to be true in the intact animal, the slowness of relaxation becomes much more important because of the asymmetric shape of the twitch.

In adductor muscles of mollusks the resistance to extension is further increased many times by various factors: increased CO_2 tension, prolonged stimulation (12), and drugs. Extension can become particularly slow in the anterior byssus retractor of *Mytilus*. It would be most satisfactory to assume that the differences between various muscles are only quantitative differences in the strength of the crosslinks, but it has been suggested that a distinct mechanism, the catch mechanism, involving the paramyosin fraction of the thick filaments, may be responsible for the change in extensibility of the muscle just mentioned (cf. 10). It is doubtful whether such a mechanism exists in vertebrate smooth muscle. On the contrary, in the muscles which I have studied the speed of relaxation did not change at all during an experiment and was not influenced after a prolonged tetanic contraction or by an increase in CO_2 (2).

In summary, stress relaxation of the contractile elements of smooth muscle at rest is slow, indicating that the crosslinks between the filaments are not completely broken. The mechanical force required to break these bonds determines the time course of relaxation. These properties are in sharp contrast with those of striated muscle and explain the large economy of smooth muscle in maintaining tension.

ACKNOWLEDGMENT

This work was supported by United States Public Service Grant 9 R01 HE-14548 from the National Heart and Lung Institute.

REFERENCES

1. Bozler, E. (1930): *Z. Vergl. Physiol.*, 12:579–602.
2. Bozler, E. (1941): *J. Cell. Physiol.*, 18:385–391.
3. Bozler, E. (1953): *Experientia*, 9:1–6.
4. Bozler, E. (1938): *Am. J. Physiol.*, 124:502–510.
5. Lowy, J., and Millman, B. M. (1963): *Philos. Trans. R. Soc. Lond. [Biol. Sci.]*, 246:105–148.
6. Abbott, B. C., and J. Lowy (1958): *J. Physiol. (Lond.)*, 141:398–407.
7. Greven, K. (1951): *Z. Biol.*, 105:373–383.
8. Schatzmann, H. J. (1964): *Ergeb. Physiol.*, 55:29–130.
9. Axelsson, J. (1970): In: *Smooth Muscle*, edited by E. Bülbring, A. F. Brading, A. W. Jones, and T. Tomita. Arnold, London.
10. Rüegg, J. C. (1971): *Physiol. Rev.*, 51:201–247.
11. Bozler, E. (1936): *J. Cell. Physiol.*, 8:419–438.
12. Bozler, E. (1930): *J. Physiol. (Lond.)*, 69:443–462.

Physiology of Smooth Muscle, edited by
E. Bülbring and M. F. Shuba.
Raven Press, New York © 1976.

The Time Course of Creep and Stress Relaxation in Relaxed and Contracted Smooth Muscles (Taenia coli of the Guinea Pig)

K. Greven

Abteilung für Allgemeine Physiologie im Zentrum der Physiologie, J. W. Goethe Universität Frankfurt/M, Germany

Creep and stress relaxation are useful mechanisms for adapting the volumes of hollow organs with smooth muscle walls to increasing contents without significant increase of pressure. The time course of these phenomena has been investigated by many physiologists since the earlier work of Jordan (1) and Bozler (2,3). For physical interpretation, Voigt and Maxwell elements with additional springs in series or in parallel, i.e., the standard physical models representing elastic forces with internal friction by combinations of springs and dashpots are usually employed (Fig. 1). If the initial purely

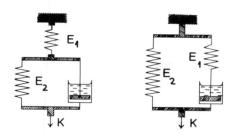

FIG. 1. Left: Voigt element with additional spring E_1 in series. Right: Maxwell element with additional spring E_2 in parallel. K = force or stretch.

elastic reaction at the beginning of the creep is neglected, and the decrease of stress during the relaxation is taken positively, the time course in the models can be given by

$$N \text{ or } R = C\,[1 - \exp(-t/\tau)] \tag{1}$$

where N is the elongation during creep after constant loading, R the decrease (positively taken) of stress during relaxation after constant stretching, C the final elongation or final stress at t_∞, and τ a time constant (4,5). The background of the above formula is the usual equation of motion given here by

$$a' = b'\,(N \text{ or } R) + \mu\,\frac{d(N \text{ or } R)}{dt} \qquad a', b', \mu = \text{const} \tag{2}$$

223

where a' is a force in the case of creep, or a stretch in the case of stress relaxation. b' and μ represent, in the most simple case of creep by a Voigt model, an elastic modulus and a coefficient of friction. Generally speaking, the form of Eq. (2) is valid for all movements of Voigt and Maxwell models representing creep or stress relaxation, the interpretation of parameters b' and μ being different in special cases. The equation is even valid for combinations of the models in parallel (only in the case of creep) or in series (only in the case of stress relaxation) (5). Thus the applicability of the models or of their combinations for representing our biologic phenomena

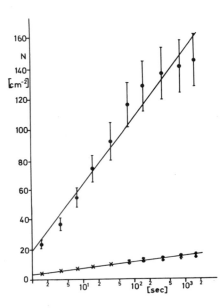

FIG. 2. Time course of creep (N) in the relaxed ($\otimes$), lower curve, Ca^{2+}-free Krebs solution with verapamil) and in the contracted ($\bullet$), upper curve, isotonic solution with K_2SO_4 in excess state. Standardization with respect to volume units (cm/cm³).

can be tested: $d(N \text{ or } R)/dt$ should be a linear function of N or R at least either during creep or during stress relaxation if the models apply.

However, this could not be confirmed in our experiments with the taenia coli (see Figs. 3, 4, and 5). According to some investigators (6,4,7) a linear relationship is often found between N or R and the logarithm of time. Figure 2 gives some support to this concept. Unfortunately it is very difficult to look for a physical or a physiologic interpretation of a function like

$$N \text{ or } R = c' + c \ln t \tag{3}$$

We therefore tried to adapt the first mentioned Eq. (2) which is founded on physical reasoning to the experimental data by replacing the parameter μ by a function of N or R. For reasons that will be discussed below, b' must remain constant. It could be shown that by merely inserting a linear function

$$\mu = a + bN \quad \text{or} \quad \mu = a + bR \tag{4}$$

in Eq. (2) the resulting equation which represents a hyperbola

$$\frac{a' - b'(N \text{ or } R)}{a + b(N \text{ or } R)} = \frac{d(N \text{ or } R)}{dt} \tag{5}$$

did agree rather well with the experimental results. On the other hand, if the logarithmic Eq. (3) is differentiated, an exponential function

$$d(N \text{ or } R)/dt = k \exp(-(N \text{ or } R)/c) \tag{6}$$

is obtained. Equation (6) gives a similar plot of N or R versus $d(N$ or $R)/dt$ as Eq. (5) does (Figs. 3 and 4, dashed and unbroken line). This may be the reason why Eq. (3), although without clear physical or physiologic meaning, often covers the experimental data.

Figures 3, 4, and 5 demonstrate how well these experimental data fit the hyperbola calculated by Eq. (5). Each figure is based on the results of 18 experiments (work done with B. Hohorst and U. Kröhnert) with pieces of taenia the initial length of which varied from 5 to 15 mm. To enable a better comparison of the calculated and the experimental results a rectified hyperbola is also inserted in Figs. 3 and 4. The rectifying function is given by

$$Y = f(N) = A + BN = \frac{N - N_{(c)}}{dN/dt - dN/dt_{(c)}} \tag{7}$$

where N and dN/dt denote of course variable pairs of experimental data. A and B are constant parameters. $N_{(c)}$ and $dN/dt_{(c)}$ characterize a distinct constant pair of values which can be chosen at random from among the

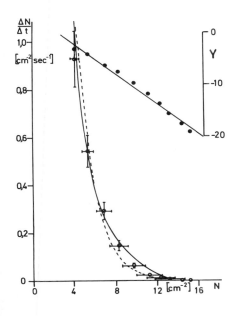

FIG. 3. Hyperbolic relation between the extent of creep (N) and the corresponding derivative dN/dt (substituted as usual by $\Delta N/\Delta t$). Relaxed muscles (Ca^{2+}-free Krebs solution with verapamil). Inserted below are the means of 18 experiments with SEM. Solid line: Calculated hyperbola [see Eq. (5)]. Dashed line: Calculated exponential function [see Eq. (6)]. Inserted above on the right side: rectified hyperbola with the same experimental data [see Eq. (7)]. Standardization with respect to volume units (cm/cm^3).

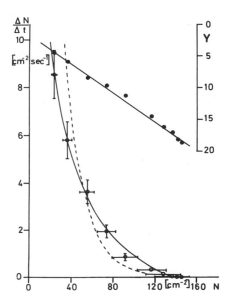

FIG. 4. Hyperbolic relation between the extent of creep (N) and the corresponding derivative dN/dt. Contracted muscles (isotonic solution with K_2SO_4 in excess). Means of 18 experiments. For further explanation, see Fig. 3. Note the difference in scale between Figs. 3 and 4.

experimental data (8). Creep was recorded during 30 min (see Fig. 2). Stress relaxation which occurs far more quickly was recorded during 10 sec. The changes in length (Figs. 2, 3, and 4) as well as those of stress (Fig. 5) were standardized with respect to volume (cm/cm³ resp. dyn/cm³) for reasons mentioned in an earlier paper (5). Load during creep = 9,928 dyn, stretch during relaxation = 2.5 mm.

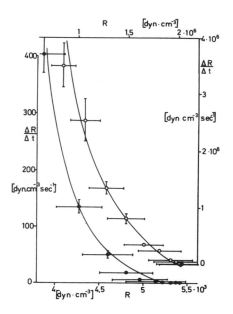

FIG. 5. Hyperbolic relation between the extent of stress relaxation (R) and the corresponding derivative dR/dt (substituted as usual by $\Delta R/\Delta t$) Left curve: Means of 18 experiments (SEM inserted) with relaxed muscles (Ca^{2+}-free Krebs solution with verapamil), scale on the left and below. Right curve: Means of 18 experiments with contracted muscles (isotonic solution with K_2SO_4 in excess), scale on the right and above. Standardization with respect to volume units (dyn/cm³). Note the difference between the corresponding scales.

As mentioned above, the parameter b' of Eq. (5) must remain constant. The reason is because Eq. (5) is valid for creep as well as for stress relaxation. A more complete presentation of the equations of the Voigt and Maxwell models (5) shows that the factor of friction in the dashpots of the models may be taken as a function of N or R, but that the parameters of the elastic forces may not be a function of N or R, if we require the *relations of N to dN/dt and those of R to dR/dt* to take *the same mathematical form.* Thus, the parameter b' cannot be transformed into a function of N or R without neglecting the empirical facts given by the figures.

All the figures demonstrate that changes in length or strain are significantly greater with contracted (Krebs solution with K_2SO_4 in excess) than with the relaxed (Krebs solution, Ca^{2+}-free with verapamil) muscle. As a consequence of this difference the calculated factor b in Eq. (5) which represents a factor of friction in the case of creep, is about 100 times smaller in the contracted than in the relaxed state. The factor b', which stands for an elastic modulus in the case of creep decreases only by about one-tenth if we compare the relaxed with the contracted state.

In our later experiments we always used K_2SO_4 instead of KCl for depolarization because, according to the paper by Jones et al. (9), KCl seems to damage the internal structures of the preparation. In our work, creep after depolarization and contracture by K_2SO_4 was significantly more pronounced than after treatment with KCl, which probably causes swelling of the cells, thus preventing a more pronounced change in the shape of the sample imposed by the load.

Equation (4), which replaces the factor μ by a linear function of N or R, can be interpreted as follows: (a) Because the volume of the preparation remains approximately constant, N is inversely proportional to the cross section, and it seems not unreasonable to think that friction would increase with decreasing cross section, i.e., $\mu = a + b/Q$, (where Q = cross section), as the internal structures approach each other. However, this is not the case with stress relaxation because the cross section is not altered under these circumstances. (b) To cover both cases another approach may be preferred. If μ is written as

$$\mu = \frac{k_1}{k_2 + k_3(dN/dt \text{ or } dR/dt)} \qquad k_1, k_2, k_3 = \text{const.} \qquad (8)$$

the former Eq. (2) is also transferred into a hyperbola formally the same as Eq. (5), i.e.,

$$\frac{a'k_2 - b'k_2(N \text{ or } R)}{(k_1 - a'k_3) + b'k_3(N \text{ or } R)} = \frac{d(N \text{ or } R)}{dt} \qquad (9)$$

The derivatives dN/dt as well as dR/dt are proportional to the velocity of the plunger in the dashpots of the Voigt and Maxwell models. Such motions may be interpreted in the case of the muscle as movements of the internal structures relative to one other by means of which transverse bindings may be broken. In this view, which supports that of Bozler (10), without any resort to distinct structures, resistence to friction is nothing but a force to break bonds which are permanently formed during the sliding of internal structures. The formation of such transverse bonds takes a certain time, and the time of contact Δt when such bindings can take place will be the longer the slower the relative velocity of the structures one to the other. Hence, if n is the number of bindings that can be formed with a certain probability during Δt, one may define a function $f(n) = \Delta t$ where $df(n)/dn > 0$. Consequently $1/(dN/dt)$ or $1/(dR/dt)$ in Eq. (8) may be replaced by $f(n)$ and a formally similar hyperbolic equation which is nothing but a transcription of Eqs. (2), (5), and (8).

$$K_1 - K_2(N \text{ or } R) = \frac{d(N \text{ or } R)}{dt} \frac{k_1 f(n)}{k_2 f(n) + k_3} \qquad K_1, K_2 = \text{const.} \qquad (10)$$

is obtained, i.e., the factor of friction is a function of the number of transversal bindings.

In this way the hyperbolic relations of N or R to their derivatives in time may perhaps be explained. It may be remarked also that the last term on the right side which stands for μ in Eq. (2), increases with $f(n)$ in a hyperbolic form, beginning with $f(n) = 0$, $\mu = 0$, and ending with $f(n) = \infty$, $\mu = k_1/k_2$.

REFERENCES

1. Jordan, H. (1918): *Ergeb. Physiol.*, 16:87–227.
2. Bozler, E. (1936): *Cold Spring Harbor Symp. Quant. Biol.* 6:260–266.
3. Bozler, E. (1941): *J. Cell. Physiol.*, 18:385–391.
4. Buchthal, F., and Kaiser, E. (1951): *The Rheology of the Cross-Striated Muscle Fibre.* Munksgaard, Copenhagen.
5. Greven, K., Gotthardt, H., and Hancke, E. (1973): *Pflügers Arch.*, 344:245–260.
6. Greven, K. (1950): *Z. Biol.*, 103:139–166.
7. Axelsson, J. (1970): In: *Smooth Muscle*, edited by E. Bülbring, pp. 289–315. Arnold, London.
8. Bronstein, I. N., and Semendjajew (1971): *Taschenbuch der Mathematik*, p. 523. H. Deutsch, Zürich and Frankfurt.
9. Jones, A. W., Somlyo, A. P., and Somlyo, A. V. (1973): *J. Physiol. (Lond.)*, 232:247–273.
10. Bozler, E. (1952): *Am. J. Physiol.*, 171:359–364.

Physiology of Smooth Muscle, edited by
E. Bülbring and M. F. Shuba.
Raven Press, New York © 1976.

Laser Raman Spectroscopy: A New Probe of the Molecular Conformations of Intact Muscle and Its Components

I. M. Asher, E. B. Carew, and H. E. Stanley

Harvard—MIT Program in Health Sciences and Technology, Massachusetts Institute of Technology, Cambridge, Massachusetts, 02139

The macroscopic, mechanical motion of muscle may ultimately derive from the coupling between the contractile proteins actin and myosin, accompanied by conformational changes in the latter (1,2). The proposed involvement of intramolecular changes in the contraction process extends the concern of muscular physiology to the submicroscopic level.

Conventional techniques, although useful, are often limited in their ability to investigate the molecular correlates of well-characterized physiologic states. For example, electron microscopy presently reveals the overall shape of myosin at the 20 to 50 Å level, but provides little detailed information on its internal structure or intramolecular bonding patterns. Furthermore, a dehydrated, "fixed" sample is utilized, in place of the highly hydrated, functioning system of interest to the physiologist. This problem of biological relevance also applies to high-resolution X-ray (or neutron) diffraction structural determinations of isolated muscle proteins in the crystalline state, although the size of myosin is too large ($\sim$ 480,000 daltons) to encourage such an undertaking. Optical rotatory dispersion and circular dichroism presently provide useful but rather general information, such as the helical content of protein preparations. Proton- or ^{13}C-nuclear magnetic resonance spectra of large, heterogeneous molecules, such as myosin, would probably consist of a semicontinuous superposition of peaks too dense to be readily analyzed. Finally, infrared absorption spectroscopy is severely limited by the opacity of water to infrared radiation.

In this chapter, we suggest that laser Raman spectroscopy is a technique capable of probing certain aspects of the structure of muscle proteins (e.g., myosin), both in aqueous solution and in the highly structured matrix of functionally intact muscle. Whereas some of the observed spectral peaks correspond to the localized vibrations of individual chemical groups (e.g., phenylalanine residues), others (e.g., the "amide modes") provide information on overall conformation. These considerations, and our preliminary investigations presented below, encourage us in the belief that laser Raman spectroscopy may prove to be a valuable complement to the more familiar laboratory techniques mentioned above, especially in those applications of greatest interest to the muscle physiologist.

LASER RAMAN SPECTROSCOPY

We begin with a simplified review of Raman spectroscopy, emphasizing its differences from traditional absorption spectroscopy. As indicated in Fig. 1a, infrared spectroscopy proceeds by illuminating a sample with infrared radiation of various frequencies, and then plotting the measured transmission intensity versus incident frequency. A dip in the transmission indicates the direct absorption of infrared quanta of that frequency, resulting in the excitation of a vibrational normal mode of the molecule with the same frequency. In laser Raman spectroscopy (Fig. 1b) a single highly monochromatic beam of visible light is incident on the sample. Most is transmitted, and some is elastically scattered with no change in frequency. A far smaller fraction is *inelastically* scattered with a definite shift in frequency. The physical origin of this inelastic "Raman" scattering is discussed elsewhere (3,4). Significantly, the frequency *shifts* of the scattered light correspond to the frequencies of the vibrational normal modes of the molecule. (These shifts are tradi-

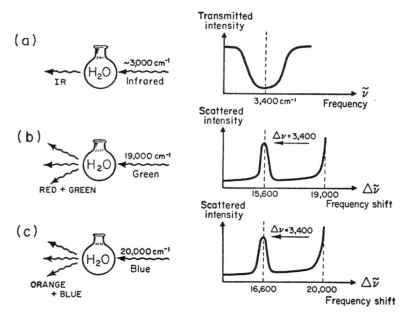

FIG. 1. A comparison of infrared absorption (a) and Raman scattering (b, c) spectroscopy. Raman scattered light displays a series of constant frequency *shifts* ($\Delta\tilde{v}$) from the frequency of the incident light. Both incident and scattered light can be visible, even if $\Delta\tilde{v}$ corresponds in frequency to infrared light; this permits observations in aqueous solution which is largely opaque to infrared radiation. The shifts $\Delta\tilde{v}$ arise from the inelastic excitation of corresponding vibrational states in the molecules of the sample. See text.

tionally called Raman "frequencies" and are measured in reciprocal centi-meters, where 1 cm^{-1} = 3 × 10^{10} Hz.)

In the simple example of Fig. 1b, the inelastic scattering of green light (~ 19,000 cm^{-1}) by water produces small amounts of red light (~ 15,600 cm^{-1}). The 3,400 cm^{-1} frequency shift corresponds to the ~ 3,400 cm^{-1} vibration frequency of water. With blue light (~ 20,000 cm^{-1}) incident, a different inelastically scattered frequency (~ 16,600 cm^{-1}) is observed, but the ~ 3,400 cm^{-1} Raman shift is maintained. Notice that both the incident and detected beams are visible light, which permits the use of aqueous solutions. In fact, water is an ideal Raman solvent; it has only two intense peaks, both of which are moved hundreds of reciprocal centimeters in D$_2$O. Deuteration also changes the vibration frequencies of the NH and OH groups of proteins in a useful characteristic manner. In small molecules of high symmetry, infrared absorption (associated with the polarization vector) and Raman scattering (associated with the polarizability tensor) often probe different vibrational modes; this complementarity is somewhat less important in large molecules of low symmetry.

METHODS AND MATERIALS

In a typical Raman apparatus (Fig. 2) an incident 10 to 1,000 mW beam from a gas laser is focused on the sample, which is often confined in a glass capillary tube suspended perpendicular to the scattering plane. Most of the light is transmitted; however, a small fraction is scattered and collected by a large lens placed to one side. This light is analyzed by a double (or triple) grating monochromater and detected by a phototube; the resulting spectrum is plotted by a chart recorder. Our own system (5) utilizes a SPEX Ramalog 4 spectrometer and a Spectra-Physics Model 124 argon ion laser. This system has been used extensively in Raman studies of ion-specific antibiotics (5–9).

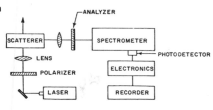

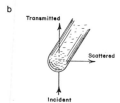

FIG. 2. A schematic diagram of a laser Raman system (a) and a simple scattering geometry for samples mounted in a glass capillary tube (b).

Myosin samples were prepared from rabbit psoas muscle by Prof. Carew using a modification of the method of Nauss, Kitagawa, and Gergely (9) in which a tenfold dilution in 0.6 M KCl was substituted for $(NH_4)_2SO_4$ precipitation. This myosin was typically isolated in $\sim$ 10 mg/ml concentrations and was then concentrated 2 to 5 times by dialysis against starch or sucrose followed by dialysis overnight against 0.6 KCl at pH 7.0. The resulting solution was spun at 100,000 $\times$ g for 1 hr to remove any remaining actomyosin, polymyosin, or denatured myosin. Samples strictly prepared in this manner gave reproducible results.

The myosin subfragments were prepared by J. C. Seidel at the Boston Biomedical Research Institute using the methods of Refs. 10 and 11; light meromyosin (LMM) was concentrated to 38 mg/ml and subfragment 1 (S-1) to 24 mg/ml. Whole rabbit psoas muscle fibers were washed and investigated in physiologic saline immediately after excision, without further preparation.

RAMAN INVESTIGATIONS OF MYOSIN

Our initial questions were: (a) Can one obtain useful Raman spectra of large, structurally heterogeneous proteins like myosin? (b) Can these results be interpreted with the aid of protein subfragments and the constituent amino acids? (c) Can such techniques be extended to investigate functionally intact whole muscle in physiologically distinct states? Figures 3, 4, 5, and 6, abstracted from a series of forthcoming papers (12–13), speak to these issues.

As Fig. 3 indicates, detailed laser Raman spectra of myosin can indeed be obtained; over 20 distinct peaks can be observed and identified (Table 1). The upper spectrum shows the Raman scattering intensity of a $\sim$ 60 μM aqueous solution of myosin (0.6 M KCl, pH 7.0) plotted against the observed frequency shift from the laser line. For comparison, the three lower spectra represent aqueous solutions of the component amino acids of myosin based on the chemical analysis of myosin by Lowey and Cohen (14) in which tryptophan is not reported. Figure 3b represents, in the appropriate proportions, the five most prevalent amino acids of myosin (Glu, Lys, Asp, Leu, Ala); Fig. 3c represents those amino acids which predominate in the globular head region of the molecule (Gly, Phe, Met, Pro, Tyr); and Fig. 3d represents the remaining seven amino acids. Such comparisons can be a great aid to spectral analysis. For example, the myosin peaks near 1,004, 1,033, and 1,209 cm^{-1} can be readily assigned to ring vibrations of the phenylalanine and tyrosine residues.

As seen in Fig. 3, some features, such as the pH independent CH bend modes near 1,450 cm^{-1}, are common to both amino acid and myosin spectra. Others, such as the pH dependent COO$^-$ stretch mode near 1,405 cm^{-1}, are intense only in the spectra of the ionized amino acids (which have terminal

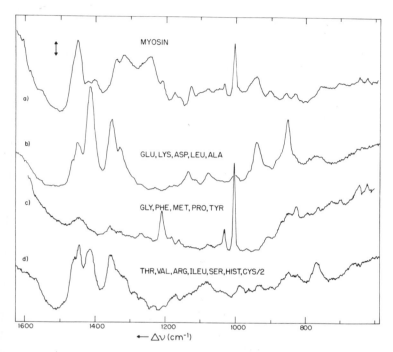

FIG. 3. Raman spectra of aqueous solutions of myosin (a) and its constituent amino acids (b–d). The relative concentrations of the amino acids follow the chemical analysis of myosin in Ref. 14. The five most prevalent amino acids appear in (b); those in (c) are most common in the globular head portions of the myosin molecule. Myosin at 35 mg/ml in 0.6 KCl, pH 7.0; spectra b, c, d at pH 6.0, 3.0, 11.0, respectively (to aid solution). Spectral resolution 5 cm⁻¹, scanning speed 30 cm⁻¹/min. Vertical arrow represents 300 counts/sec in (a) and 3,000 counts/sec in (b–d). Laser power 600 mW in (a) and 200 mW in (b–d). (From Ref. 12.)

COO⁻ groups). Conversely, the intense activity in the 1,240 to 1,310 cm⁻¹ region of the myosin spectrum is lacking in spectra of the constituent amino acids, suggesting that much of this activity must be associated with vibrations involving the peptide linkages. In fact these particular vibrations (called "amide III" modes) are known to contain considerable contributions from simultaneous CN stretching and NH in-plane bending motions (15,16).

The Raman frequencies of the amide III modes are sensitive to the changes in hydrogen bonding, skeletal bending, and local environment that accompany conformational change. For example the 1,311 cm⁻¹ amide III frequency of α-helical poly-L-lysine (17) shifts to 1,240 cm⁻¹ in the pleated β-chain conformation. In general, α-helical molecules have amide III frequencies near 1,265 cm⁻¹ [e.g. α-glucagon (18), lysozyme (19)], or 1,310 cm⁻¹ [e.g., α-poly-L-lysine (15)], whereas β-chain frequencies cluster between 1,230–1,250 cm⁻¹ [e.g., β-poly-L-lysine (17), polyglycine I]. Unfortunately random-coil conformations also have frequencies in the latter region. For example, the β-structure portion of denatured insulin gives rise to a doublet

TABLE 1. Raman frequencies (cm^{-1}) of myosin

Peaks	Assignment
622	CC ring twist (Phe)
645	CC ring twist (Tyr)
660–695	CS vibrations
704	CS stretch (Met)
755	Amide V; residue vibrations (Val, Ileu)
780	—
830	CC symmetric ring stretch (Tyr)
855	CH_2 residue rock
(882)	—
901	CC residue stretch
940	CC residue stretch (Lys, Asp); CH_3 symmetric stretch (Leu, Val)
(962)sh	CH_3 symmetric stretch
1,004	CC ring stretch (Phe); weak CC stretch (Lys, Glu)
1,033	CC ring bend (Phe)
(1,044)?	CH_2 twist (Glu, Arg, Pro)
(1,058)?	CH_2 twist (Lys, Arg); CH, CC stretch (?)
1,081	CN, CC skeletal stretch
1,104	CN, CC skeletal stretch
1,128	Isopropyl residue antisymmetric stretch; CN stretch (?)
1,160	CH_3 antisymmetric rock (Leu, Va); CH rock (Phe,Tyr)
1,175	CH_3 antisymmetric rock (Leu, Va); CH rock (Phe,Tyr)
1,209	Tyr, Phe modes
1,244	Amide III (β-chain + random coil)
(1,265)sl	Amide III; $C_\alpha H$ bend (Leu, Asp, Glu, Tyr, Pro)
(1,304)sl	Amide III (α-helix); $C_\alpha H$ bend, CH_2 twist
1,320	Amide III (α-helix); $C_\alpha H$ bend, CH_2 twist
1,342	CH bend (esp. residue)
1,402	COO^- symmetric stretch (Asp, Glu)
1,423	Residue vibration (Asp, Glu, Lys)
1,451	CH_3 (antisymmetric), CH_2, CH bend
1,553	Amide II; COO^- antisymmetric stretch (Asp)
1,587	Phe, Arg vibrations
1,607	Phe, Tyr ring vibrations
1,617	Tyr, Phe ring vibrations
1,650	H_2O, amide I regions

B, broad; sl, slant; sh, shoulder; ?, existence uncertain; parentheses, exact frequency uncertain.

at 1,227 and 1,252 cm^{-1}, whereas its random coil portion gives rise to a 1,239 cm^{-1} band (20). The proper identification of this region thus requires considerable care (21).

Figure 4 summarizes our tentative assignment of this region in myosin. The shoulders near 1,265 cm^{-1} represent α-helical structure, whereas the broad 1,244 cm^{-1} band represents local β-chain and random-coil conformations. Deuteration of the NH groups of myosin greatly reduces the Raman activity of the 1,200 to 1,300 cm^{-1} region; the missing peaks are thereby firmly identified as amide vibrations. (Deuterated amide III modes are shifted to the 875 to 975 cm^{-1} region.) Only part of the 1,304 cm^{-1} peak disappears upon deuteration, indicating that only a portion of the activity

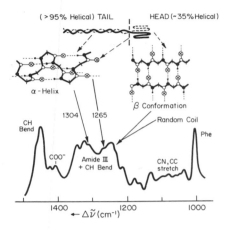

FIG. 4. Proposed analysis of the Raman spectrum of myosin in which several spectral features in the amide III region (1,220 to 1,320 cm^{-1}) are associated with specific protein structures (α-helix, β-chain, random coil) in specific portions (globular heads, helical tail) of the molecule. See text for further explanations. (From ref. 12.)

in this region represents an amide III mode; the remainder may represent CH bending vibrations and undeuterated NH groups. Interestingly, a similar portion of this peak is found to be polarization sensitive (i.e., it disappears in measurements on Raman scattered light with polarization perpendicular to that of the incident light).

RAMAN INVESTIGATION OF MYOSIN SUBFRAGMENTS

Additional insight can be obtained by examining the LMM and S-1 subfragments of myosin, which are, respectively, characteristic of the helical

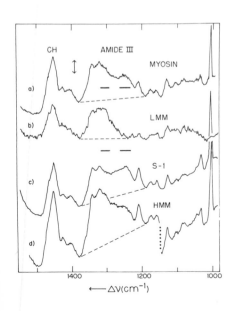

FIG. 5. A comparison of the Raman spectrum of myosin (a) with that of its tryptic subfragments (b–d): light meromyosin (LMM), subfragment 1 (S-1), and heavy meromyosin (HMM). Notice that the spectrum of LMM (> 95% α-helical; a coiled coil) displays considerable Raman activity near 1,305 cm^{-1}, but lacks the band near 1,250 cm^{-1} characteristic of HMM, S-1, and whole myosin. Such comparisons tend to support the assignments suggested in Fig. 4. (Samples from Boston Biomedical Research Center; from ref. 13.)

tail and globular head portions of the molecule (Fig. 5). The LMM sub-fragment consists of two α-helical chains wound into a rodlike, closely associated supercoil. In contrast, the S-1 subfragment is believed to contain only 35% α-helical structure [as measured by optical rotatory dispersion (1)] perhaps in small, dispersed regions. It is interesting that the amide III frequencies of these subfragments are quite different, and in fact comple-mentary. The Raman spectrum of LMM (Fig. 5b) contains a prominent peak near 1,306 cm^{-1}, but lacks resolvable Raman activity near 1,265 or 1,244 cm^{-1}. Conversely, Raman spectra of S-1 (Fig. 5c) lack a recognizable peak near 1,306 cm^{-1}, but display considerable activity between 1,240 and 1,280 cm^{-1}. These observations not only support our previous assignments of the amide III vibrations (e.g., β-chain and random coil can exist only in S-1), but also suggest that there are two structurally and spatially distinct α-helical regions in the myosin molecule. The 1,304 cm^{-1} shoulder of myosin appears to be associated with the coiled helices of the fibrous tail portion, whereas the 1,265 cm^{-1} shoulder appears to be associated with α-helical portions of the globular heads.

Such assignments permit the systematic investigation of conformation changes in different parts of the myosin molecule. For example, ~ 6 hr exposure to 100 μM MgCl$_2$ in 0.6 KCl greatly reduces the 1,244 cm^{-1} peak of the β-chain (or random coil) regions in the globular head portion of myosin (12). Further exposure (~ 18 hr) leads to the appearance of a prominent peak near 1,310 cm^{-1}. Thermal denaturation of myosin also leads to interesting spectral changes in the 1,230 to 1,310 cm^{-1} (amide III) and 1,040 to 1,120 cm^{-1} (CN and CC stretch) regions (12), which are expected to be sensitive to changes in the backbone conformation. Similar experi-ments have been performed with myosin subfragments (13).

RAMAN SPECTROSCOPY OF WHOLE MUSCLE

We now turn to our final question: Can useful Raman spectra be obtained from such complex systems as physiologically intact whole muscle? At the bottom of Fig. 6 we present the first Raman spectrum of whole-rabbit psoas muscle, obtained with no other preparation than excision and suspension in a capillary tube of physiologic saline. Considerable detail is present even in this preliminary spectrum. Comparison with the Raman spectrum of myosin (Fig. 6a) reveals numerous similarities—e.g., the phenylalanine ring mode near 1,004 cm^{-1}, valyl residue mode near 1,130 cm^{-1}, and CH bend modes near 1,450 cm^{-1}. Myosin accounts for 50% of the dry weight of muscle; the other components—actin, tropomyosin, and the troponins—are presently under investigation.

How will the addition of chemical agents (e.g., Ca^{2+}, ATP, drugs, prosta-glandins) to the surrounding saline affect the Raman spectrum of whole muscle? Will normal and pathological muscle display interpretable spectral differences? Can spectral changes be observed during electrically induced

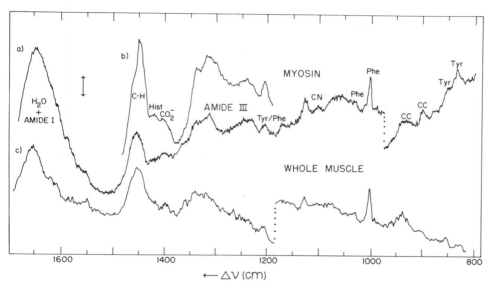

FIG. 6. A comparison of the Raman spectrum of myosin (a, b) with a preliminary spectrum of freshly excised rabbit psoas muscle (c). The presence of recognizable spectral features in the latter encourages further Raman studies of physiological intact muscle under various conditions. (From E. B. Carew, *unpublished data.*)

contraction? Such questions are just beginning to be investigated, but it already appears likely that laser Raman spectroscopy will be an important new technique for future investigations in this area.

ACKNOWLEDGMENTS

The possibility of using laser Raman spectroscopy to study muscle and contractile proteins was first explored by Prof. E. B. Carew at the University of Michigan at Dearborn and in the laboratory of Prof. R. C. Lord, M.I.T. Our collaboration has been recently expanded to include workers at the Boston Biomedical Research Institute (BBRI), including Dr. J. Seidel, Dr. J. Potter, and Dr. J. Gergely, from whom the myosin subfragments of Fig. 5 were obtained. We also wish to thank A. Hewitt and G. D. J. Phillies for their generous assistance and R. C. Lord and K. J. Rothschild for valuable discussions. Our work is supported by the Research Corporation and the National Heart and Lung Institute (Grant HL 14322–03, R. W. Mann, Principal Investigator).

REFERENCES

1. Tonomura, Y. (1972): *Muscle Proteins, Muscle Contraction and Cation Transport.* University of Tokyo Press, Tokyo.
2. Seidel, J. C., and Gergely, J. (1973): *Arch. Biochem. Biophys.*, 158:853.

3. Raman, C. V., and Krishnan, K. S. (1928): *Nature,* 121:501.
4. Tobin, M. C. (1971): *Laser Raman Spectroscopy.* Wiley-Interscience, New York.
5. Asher, I. M., Rothschild, K. J., and Stanley, H. E. (1974): *J. Mol. Biol.,* 89:205.
6. Rothschild, K. J., Asher, I. M., and Anastassakis, E., and Stanley, H. E. (1973): *Science,* 182:384.
7. Asher, I. M., Phillies, G. D. J., and Stanley, H. E. (1974): *Biochem. Biophys. Res. Commun.,* 61:1306.
8. Rothschild, K. J., and Stanley, H. E. (1974), *Science,* 185:616.
9. Nauss, K. M., Kitagawa, S., and Gergely, J. (1969): *J. Biol. Chem.,* 244:755.
10. Gergely, J., Gouvea, M. A., and Karibian, D. (1955): *J. Biol. Chem.,* 212:165.
11. Nauss, K. M., and Gergely, J. (1967): *Fed. Proc.* 26:727.
12. Carew, E. B., Asher, I. M., and Stanley, H. E. (1974): *Science,* 188:933.
13. Carew, E. B., Asher, I. M., Seidel, J. C., Hewitt, A., Potter, J., Stanley, H. E., and Gergely, J. (1974): In preparation.
14. Lowey, S., and Cohen, C. (1969): *J. Mol. Biol.* 4:293.
15. Miyazawa, T., Shimanouchi, T., and Mizushima, S. (1958): *J. Chem. Phys.,* 29:611.
16. Fanconi, B., Small, E., and Peticolas, W. L. (1971): *Biopolymers,* 10:1277.
17. Yu, T. J., Lippert, J. L., and Peticolas, W. L. (1973): *Biopolymers,* 12:2161.
18. Yu, N. T., Liu, C. S., and O'Shea, D. C. (1972): *J. Mol. Biol.,* 70:117.
19. Lord, R. C., and Yu, N. T. (1970): *J. Mol. Biol.,* 51:203.
20. Lord, R. C., and Yu, N. T. (1970): *J. Mol. Biol.* 50:509.
21. Lippert, J. L. (1974): Submitted for publication.

Physiology of Smooth Muscle, edited by
E. Bülbring and M. F. Shuba.
Raven Press, New York © 1976.

Biophysical and Physico-chemical Properties of Actomyosin and Myosin of Smooth Muscles

P. G. Bogach, V. L. Zyma, V. M. Danilova, V. Yu. Sokolova,
V. N. Dubonos, and D. S. Yankovsky

Department of Biophysics, T. G. Shevchenko State University, Kiev, USSR

Skeletal and smooth muscles differ essentially in their mechanical properties, the character of neuromuscular transmission, the relationship between membrane processes and contraction, as well as the nature of contractile proteins. Much attention has been focused on the study of the structural organization and the interaction between the contractile proteins. A great number of papers has been devoted to the contractile proteins obtained from different organs (uterus, arteries, stomach, and chicken gizzard) (1–5). The ATPase activity of actomyosin complexes of skeletal and smooth muscles are differently affected by environmental changes (temperature, pH, cofactors, and ionic strength). These differences may be connected with the protein structure.

This chapter is a comparative study of some biophysical and biochemical properties of actomyosin and myosin in the two muscle types.

MATERIALS AND METHODS

Skeletal muscle actomyosin as well as intestinal smooth muscle actomyosin complex were obtained from rabbit according to Bàràny et al. (2,3). The preparations were fractionated with ammonium sulfate and purified on the DEAE-cellulose column.

Highly purified preparations of skeletal muscle myosin were obtained by Kielley and Bradley (6). The proteins were salted out with ammonium sulfate in two stages and were then subjected to dialysis and centrifugation for 2 hr at $105,000 \times g$. Rabbit intestinal smooth muscle myosin preparations were obtained according to Yamaguchi et al. (7). The preliminary purification up to 55% salt saturation was carried out with ammonium sulfate, and further exhaustive dialysis and repeated ultracentrifugation for 2 hr at $105,000 \times g$. Both types of myosin preparations were purified. Nucleic acids and protein contaminations were removed from myosin by chromatography on DEAE-cellulose column, using a KCl gradient.

Actomyosin and myosin ATPase were estimated by measuring free inorganic phosphate. Actomyosin ATPase activity was measured under the following conditions: 0.03 or 0.5 M KCl, 0.05 M Tris-HCl buffer (pH 7.1), protein concentration was 0.5 mg/ml, 1 mM ATP, 5 mM $CaCl_2$, or 5 mM $MgCl_2$; for myosin 0.08 M or 0.5 M KCl, 20 mM histidine buffer (pH 7.6), protein concentration 0.14 mg/ml, 1 mM ATP, 5 mM $CaCl_2$. Incubation lasted for 10 min at 37°C.

The actomyosin superprecipitation was followed by the changes in absorbance at 660 nm (D_{660}) (8).

The absorption spectra of myosin solutions were recorded on a spectrophotometer SF-4A.

The fluorescence spectra were measured with an instrument described previously (9). The excitation of protein ultraviolet fluorescence was carried out at $\lambda = 280$ nm and the dye fluorescence 1-anilino-8-naphtalene sulfonate (ANS) at $\lambda = 366$ nm. The fluorescence spectra obtained were corrected for the spectral sensitivity of the instrument.

Infrared absorption spectra of proteins were recorded on a spectrometer IKS-14.

RESULTS AND DISCUSSION

It has been established that under the same conditions the smooth muscle actomyosin yield from gastrointestinal digestive tract is well below that of skeletal muscle. Ca^{2+}-activated ATPase activity of smooth muscle actomyosin measured at low ionic strength was seven to eight times lower than that obtained for actomyosin of skeletal muscle. Higher ionic strength resulted in an increased Ca^{2+}–ATPase activity of smooth muscle actomyosin; however, it did not reach the level of that from skeletal muscle. In the presence of Mg^{2+} ions the smooth muscle actomyosin, unlike the skeletal one, was characterized by low ATPase activity whose values were often within possible measuring errors.

Superprecipitation was studied in a reaction mixture containing 0.22 mg/ml actomyosin in the presence of 0.1 M KCl, 1 mM $MgCl_2$ at pH 6.8 and $t = 25$°C after addition of 0.1 mM ATP. The D_{660} value for skeletal muscle myosin increased more rapidly and reached the maximum in 10 sec. The superprecipitation rate for actomyosin of smooth muscle was lower, the maximum D_{660} value was reached after 360 sec. Thus, the superprecipitation for smooth muscle actomyosin was 36 times slower than that of skeletal muscle. The behavior of such a simple contractile model indicates the differences in structural organization of actomyosin complexes between the two types of muscle.

According to the literature (10,11) actin strands are similar in different types of muscle, and one can suggest that the myosin molecules are responsible for differences in functional behavior of muscles. To study this, highly purified samples of myosin have been used. The smooth muscle myosin has been found to aggregate more rapidly than that of skeletal muscle in a solution of equal protein concentration. The increase in absorption at 320 nm indicated the presence of aggregates. Figure 1 represents the absorption spectra for myosin solutions of the two muscle types. The absorption spectrum of skeletal muscle myosin (Fig. 1, curve 1) is characterized by $D_{280}/D_{260} = 1.7$ to 1.8 and also by the absence of absorption in the region of 320 nm. The smooth muscle myosin showed additional absorption at 320 nm indicating the presence of aggregates in solution. As a result, the D_{280}/D_{260} ratio decreased to the value of 1.2. The light-scattering component in

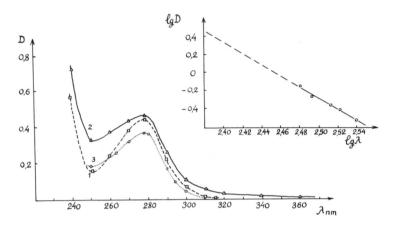

FIG. 1. Absorption spectra of myosin molecule solutions: (1) skeletal muscle; (2) smooth muscle; (3) smooth muscle after light scattering correction. Insert: Graphic calculation of the light scattering. Experimental conditions: 0.5 M KCl, 0.005 M Tris-HCl buffer (pH 7.5); $t = 20°C$; D, optical density of solution; λ, wavelength of light.

the ultraviolet range was calculated by using the Rayleigh equation $D = A/\lambda^n$, where A and n are constant (12). Beyond the protein absorption spectrum region $\log D = f (\log \lambda)$ is represented in the linear dependence. An example of light-scattering correction of myosin aggregate is shown in the inset of Fig. 1. Curve 3 demonstrates the corrected spectrum in which D_{280}/D_{260} reach 1.5.

ATPase activity of the smooth muscle myosin in the presence of Ca^{2+} and 0.5 M KCl was about two times less than that of skeletal muscle. Ca^{2+}–ATPase activity of smooth muscle myosin and actomyosin increased with the ionic strength of the solution, while that of skeletal muscle myosin decreased.

Applying hydrodynamic methods and electron microscopy, some authors (13–15) came to the conclusion that the myosin molecules of skeletal and smooth muscle were similar in shape, sedimentation coefficients, and molecular weights. The data obtained in this work and in other papers (7,15) show essential differences in ATPase activity between the two types of myosin. Thus, it follows that differences in functional properties of the myosin molecules resulted from differences in structural organization of molecules and their active sites.

For studying myosin conformational states we applied the fluorescence method. Excitation of myosin solutions at $\lambda = 280$ nm produces a fluorescence spectrum with the maximum at 336 nm for skeletal and at 334 nm for smooth muscle (Fig. 2). This position of the fluorescence spectrum maxi-

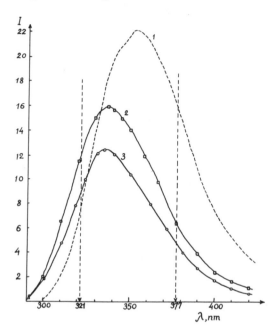

FIG. 2. Ultraviolet fluorescence spectra of solutions: (1) tryptophan; (2) skeletal muscle myosin; (3) smooth muscle myosin. The solutions were prepared in Tris-HCl buffer (pH 7.5), 0.5 M KCl; $D_{280} = 0.4$; $t = 17°C$; I, fluorescence intensity.

mum (λ_m) of myosin molecules in comparison with $\lambda = 354$ nm for tryptophan in water (Fig. 2, curve 1) indicates that most of the tryptophan residues emit from hydrophobic regions of the molecule (9). The λ_m fluorescence intensity in smooth muscle myosin was 1.3 times less than that in skeletal muscle.

It is known that tryptophan fluorescence parameters are sensitive to conformation changes. For this reason we have studied the effect of temperature and pH on the conformational state of myosin molecules. Shifts in the fluorescence spectrum, measured with a sensitive biwave method (16), indicated the myosin conformational state. The parameter $B = I_{321}/I_{377}$ was calculated from the intensities of fluorescence obtained at 321 and 377 nm. In Fig. 2 these regions are indicated with arrows.

In Fig. 3, where B is plotted versus temperature, the curve for myosin molecules has some parts with constant values. On the other hand, this curve clearly shows other temperature ranges at which B decreases with a rise of temperature. The decrease of B points to the red shift of λ_m as well as to the structural transitions between the various conformational states. We observed two reversible conformation transitions for skeletal muscle myosin (Fig. 3, curve 1) at 19° and 37°C. The B change can be produced by local changes in the environment of some tryptophan residues in myosin.

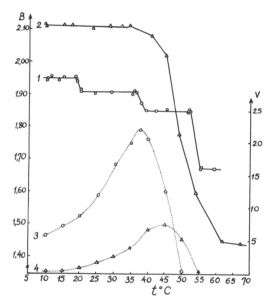

FIG. 3. Temperature dependence of B parameter (curves 1 and 2), and Ca^{2+}–ATPase activity (curves 3 and 4): 1 and 3 for skeletal muscle myosin; 2 and 4 for smooth muscle myosin; V, Ca^{2+}–ATPase activity is expressed $\mu g\ P_i/mg$-min.

It should be noted that the transition occurs in a rather narrow temperature range (1° to 2°C). This points to the high cooperativity of these changes.

The study of ATPase activity of skeletal muscle myosin showed (Fig. 3, curve 3) that enzyme activity increased with rising temperature up to 37°C and then rapidly slowed with further rise in temperature. Thus, one may suggest that the conformational changes at 37°C affect the active site structure of the enzyme, which results in a change of enzyme activity. A significant decrease of B was observed at 52°C. The fluorescence spectrum of a myosin solution heated to 52°C had a red shift and it approached that of free tryptophan. This behavior of B indicated significant conformational changes affecting the whole myosin molecule. The cooling of the solution to room temperature did not restore the initial B values, indicating the irreversibility of such structural transitions. Complete enzyme inactivation corresponded to this transition. According to the papers (17,18), the Arrhenius plots for ATPase activity at 16° to 18°C showed departure from the straight line. Our results indicate that within the 10° to 35°C temperature range the skeletal muscle myosin is characterized by two discrete conformational states and the transition between them takes place at 19°C.

The B plot versus temperature for smooth muscle myosin is somewhat different (Fig. 3, curve 2). The value of B remains constant up to 35°C.

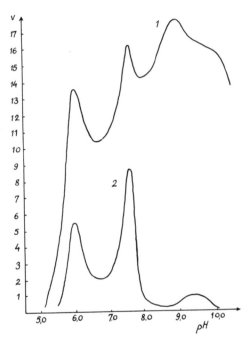

FIG. 4. pH dependence of Ca^{2+}–ATPase activity of skeletal muscle myosin (1) and smooth muscle myosin (2). Experimental conditions: 0.5 M KCl, histidine buffer (pH 7.6), 1 mM ATP, 5 mM $CaCl_2$; $t = 37$°C.

The reversible decrease of B took place in the 35° to 40°C range. The heating above 40°C decreased B and in the 45° to 62°C range we observed irreversible decrease of B. ATPase activity of the smooth muscle myosin molecules (Fig. 3, curve 4) increased with temperature rise to the maximum value of 45°C. Perhaps the increase of ATPase activity of smooth muscle myosin is determined by the conformational changes that are observed at 35° to 45°C. Complete inactivation of the smooth muscle myosin ATPase activity was observed at 55°C.

Figure 4 represents the pH dependence of Ca^{2+}–ATPase activity of the myosin molecules of the two muscle types. The first maximum on the curve is found at pH 6.0 and it coincides for both types of myosin; the second maximum of ATPase activity appears at pH 7.6 to 7.8 and its value increases considerably after careful purification of the column. Essential differences in behavior between the two types of myosin are observed at more alkaline pH values. ATPase activity of skeletal muscle myosin increases gradually and at pH 9.0 we have the third and the largest maximum of activity, whereas the smooth muscle myosin ATPase activity decreases with a negligible elevation at pH 9.5.

Our next step was to study the change of the intrinsic myosin fluorescence and the fluorescence of the dye ANS bound to myosin at different pH. The freshly prepared nonaggregated myosin preparations were used for binding to ANS, because we have shown that ANS can be absorbed to myosin aggregates distorting the interpretation of results. The addition of 1 to 1.5 10^{-5}M ANS to the myosin solution is accompanied by additional fluorescence of the ligand in the visible spectrum region. It is known that ANS absorbs on the hydrophobic sites of the molecules which frequently correspond to the functionally important centers. The parameter A has been found from the spectrum fluorescence of ANS, where $A = I_{447}/I_{525}$ and I_{447} and I_{525} are the intensity of fluorescence of ANS at $\lambda = 447$ and 525 nm. In Fig. 5, the pH dependence of parameters B and A (curves 1 and 3) and also the maximum fluorescence intensity of ANS (curve 2) are shown. The B parameter decrease of smooth muscle myosin (Fig. 5b) begins at pH 8.5, while for the skeletal muscle myosin (Fig. 5a) the analogous change occurs at pH 10.0. The red shift of the ultraviolet fluorescence corresponding to the B decrease points to the "uncoiling" of myosin molecules. It is probable that these structural changes take place in the heavy meromyosin region (HMM), which is the site of enzymatic activity. As shown in Fig. 4, the rate of ATPase activity for smooth muscle myosin decreases considerably at pH 8.5, whereas for skeletal muscle myosin it decreases at pH 10.0. These results indicate lower stability of the HMM of smooth muscle on transition to alkaline pH.

The examination of the ANS fluorescence parameters (Fig. 5) shows that the pH change from 7.0 to 10.0 for myosin molecules of both muscle types increases the values of I_m and A. The A parameter increase is related to the blue maximum shift of fluorescence. The fact that the ultraviolet shift of the fluorescence spectrum goes parallel with its rise in intensity implies that ANS is in a more hydrophobic microenvironment on the surface of the myosin molecule (19). It is of great interest to consider the character of the A parameter change (Fig. 5, curve 3). Its value undergoes several discrete transitions. This points to the fact that on transition to the alkaline pH the ligand molecule finds itself in several different microenvironments. The latter are created on the surface of the myosin molecule at the expense of conformational changes caused by ionization of the proper side groups of amino acid residues. We observed two conformational transitions in the 7.0 to 11.0 pH region. The regions of these transitions correspond to maxima of ATPase activity: for skeletal muscle myosin at pH 7.8 and pH 9.5; for smooth muscle myosin at pH 7.6 and pH 9.5.

It is of interest to consider the A parameter change for skeletal muscle myosin on both sides of pH 7.0. The character of change of both ATPase activity and A parameter during the pH shifts toward acidity or alkalinity are rather similar. We observed some differences in behavior of ANS bound to myosin molecules of the two muscle types. Changes in pH altered the value of the A parameter. The associated shift of the fluorescence spectrum of ANS bound to skeletal muscle myosin was greater than that of smooth muscle (Fig. 5, curve 3). With pH increase the fluorescence intensity of the ANS–myosin complex for skeletal muscle increased 3 times and for smooth muscle only 1.5 to 2 times (see curves 2). A slower rate of change of A and I_m for smooth muscle ligand proves that on binding with smooth muscle myosin at pH 7.0 ANS finds itself in a certain region of the molecule that is more hydrophobic compared with that of skeletal muscle myosin. This is proved when the higher value of $A = 0.92$ for smooth muscle myosin is compared with that of skeletal muscle $A = 0.78$. In view of the fact that the change of ANS fluorescence correlates well with the behavior of ATPase activity at various pH one may assume that ANS is bound at the region of the active site of myosin. As shown by Cheung and Morales (20), the binding of ANS takes place at the HMM region and never on the surface of light meromyosin (LMM). The structural organization of the active site of myosin is determined by the spatial arrangement of heavy and light chains or subunits (21). In such a structure the weak bonds (hydrogen bonds, electrostatic and hydrophobic interactions) play the main part in stabilization. On transition to pH 11.0 the subunits undergo dissociation accompanied by a complete loss of myosin ATPase activity. The results of the present study show that in the preinactivation zone of pH 8.0 to 10.0

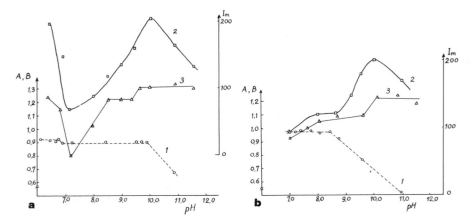

FIG. 5. pH dependence of fluorescence parameters for skeletal (a) and smooth (b) muscle myosins: (1) parameter B for ultraviolet fluorescence; (2) intensity at the fluorescence maximum of ANS (I_m); (3) parameter A for ANS fluorescence. Experimental conditions: myosin concentration 1 mg/ml, ANS concentration 1.0×10^{-5} M; 0.5 M KCl; $t = 17°C$.

certain changes take place in the active site structure, that bring about an increase of ATPase activity. One possible cause for the structural rearrangement of the active site is the change of electrostatic interactions between the heavy and light subunits. This results in a partial unfolding of the hydrophobic locus on the surface of myosin, which includes bound ANS and causes the increase of I_m and A. In the myosin molecule of smooth and skeletal muscle such changes manifest themselves differently, which proves the variations of tertiary and quaternary structures of these molecules.

The infrared absorption spectra of myosin molecules in the 2,700 to 3,500 cm^{-1} region showed definite differences in the secondary protein structure. Figure 6 represents the results of hydrogen–deuterium exchange (H–D) in myosin films, precipitated on CaF$_2$ at 70% relative humidity. The analysis of the residual absorption bands of amide A and B showed that the amide A of skeletal muscle myosin is 3,292 cm^{-1} and its half-width has a value of 89 cm^{-1} and for smooth muscle myosin these parameters are 3,298 cm^{-1} and 135 cm^{-1}, respectively. Amide B of skeletal muscle myosin is at 3,064 cm^{-1} and that of smooth muscle myosin at 3,077 cm^{-1}.

In conclusion, the data obtained show the differences in structural organization of smooth and skeletal muscle myosin molecules, which result in

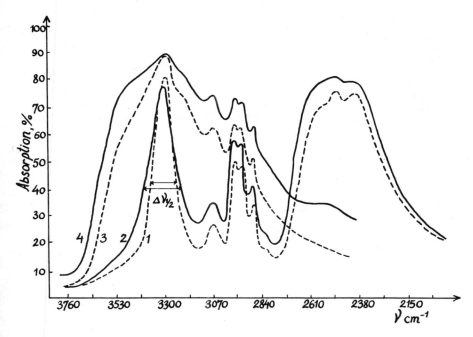

FIG. 6. Infrared absorption spectra of myosin molecules in amide A and amide B bands: (1 and 3) skeletal muscle after and before H–D exchange; (2 and 4) for smooth muscle after and before H–D exchange. v, Reciprocal wavelength; $\Delta\nu_\frac{1}{2}$, half-width of the spectrum band.

different ATPase activity and probably determine the rate of superprecipitation of actomyosin complexes. The actomyosin superprecipitation rate of smooth muscle is $\frac{1}{36}$th of that of skeletal muscle. This correlates with the contraction rate *in vivo*.

REFERENCES

1. Needham, D. M., and Shoenberg, C. F. (1968): In: *Handbook of Physiology*, Sec. 6, Vol. 4, pp. 1793–1810. American Physiological Society, Washington, D.C.
2. Bàràny, M., Bàràny, K., Geatjens, E., and Bailin, G. (1966): *Arch. Biochem. Biophys.*, 113:205–222.
3. Bàràny, M., and Bàràny, K. (1966): *Biochem. Z.*, 345:37–56.
4. Rüegg, J. C., Strassner, E., and Shirmer, R. H. (1965): *Biochem. Z.*, 343:70–76.
5. Bogach, P. G., Sokolova, V. Yu., Zyma, V. L., Danilova, V. M., and Yankovsky, D. S. (1972): *Biophysical Fundamentals and Regulation of Muscle Contraction Process*, pp. 133–141. Puschino-na-Oke.
6. Kielley, W. W., and Bradley, L. B. (1956): *J. Biol. Chem.*, 218:653–659.
7. Yamaguchi, M., Miyazawa, G., and Sekine, T. (1970): *Biochem. Biophys. Acta*, 216:411–421.
8. Ebashi, S. (1961): *J. Biochem. (Tokyo)*, 50:236–243.
9. Zyma, V. L., Demchenko, O. P., and Varetska, T. V. (1972): *Ukr. Biochim. Zh.*, 44:10–15.
10. Squire, J. M. (1971): *Nature*, 233:457–462.
11. Tonomura, Y., and Oosawa, F. (1972): *Annu. Rev. Biophys. Bioeng.*, 1:159–190.
12. Windor, A. F., and Gent, W. L. (1971): *Biopolymers*, 10:1243–1251.
13. Stracher, A. (1969): *Biochem. Biophys. Res. Commun.*, 35:519–525.
14. Weeds, A. G. (1969): *Nature*, 223:1362–1364.
15. Wachsberger, P., and Kaldor, G. (1971): *Arch. Biochem. Biophys.*, 143:127–137.
16. Turoverov, K. K., and Shchelchkov, B. V. (1970): *Biophysica*, 15:965–972.
17. Kominz, D. R. (1970): *Biochemistry*, 9:1792–1800.
18. Levy, H. M., Sharon, N. S., Ryon, E. M., and Koshland, D. E. (1962): *Biochim. Biophys. Acta*, 56:118–126.
19. Stryer, L. (1968): *Science*, 162:526–533.
20. Cheung, H. C., and Morales, M. F. (1969): *Biochemistry*, 8:2177–2182.
21. Gershman, L. C., Stracher, A., and Dreizen, P. (1969): *J. Biol. Chem.*, 244:2726–2736.

Physiology of Smooth Muscle, edited by
E. Bülbring and M. F. Shuba.
Raven Press, New York © 1976.

Contractile Properties of Isolated Smooth Muscle Cells

Fredric S. Fay,* Peter H. Cooke,† and Peter G. Canaday*

*Department of Physiology, University of Massachusetts Medical Center, Worcester, Mass. 01605 and
†Department of Physiology and Cell Biology, University of Kansas, Lawrence, Kansas 66044

Studies of the physiology of smooth muscle have been largely performed on intact tissues. Although considerable information about the function of these tissues has been developed, results obtained on intact tissue may often reflect the consequences of a complex organization of smooth muscle cells into a multicellular unit rather than the properties of the contractile cells themselves. Hence in the last few years several investigators (Bagby et al., 1971; Bagby and Fisher, 1972, 1973; Fay and Delise, 1972, 1973; Purves et al., 1973, 1974; Singer and Fay, 1974; Small, 1974) have attempted to investigate the physiology of smooth muscle utilizing single isolated smooth muscle cells.

The studies reported here represent an attempt to characterize the contractile process in isolated smooth muscle cells and from these studies to derive an understanding of mechanisms underlying the development of force in smooth muscle. The studies were performed largely on single smooth muscle cells obtained from the stomach muscularis of *Bufo marinus* by a modification (Fay and Delise, 1973) of the technique originally described by Bagby et al. (1971).

CELL VIABILITY

The purpose of studies such as these is eventually to use the information obtained from single smooth muscle cells to understand physiological processes that involve the contraction of smooth muscle in whole tissue. Hence, it is important that one be certain that the single smooth muscle cells have not been altered significantly by the isolation procedure. A brief review of the features of the isolated cells, which suggest that they are basically unaltered, is therefore appropriate.

A. The cell membrane appears to be intact and functioning in a normal manner as judged by

  1. The ability of the cells to exclude dyes such as trypan blue (Bagby et al., 1971).

  2. The behavior of the cells as "perfect osmometers" when exposed to

solutions exerting between 0.33 x and 2.5 x, the osmotic pressure of amphibian Ringer (Fay and Delise, 1973).

3. The measurement of transmembrane potentials between -30 and -60 mV (Fay and Singer, *unpublished observations*), which are in line with those obtained from smooth muscle cells within intact tissue (Casteels, 1970).

4. The presence of membrane ultrastructure (plasma membrane, dense bodies, caveolae, and a continuous trilaminar membrane) (Fay and Delise, 1973) characteristic of smooth muscle cells within intact tissues (Burnstock, 1970).

5. The presence of cholinergic receptors the properties of which are similar to those in intact tissue. Specifically the cells contract in response to cholinergic agonists and the response is blocked by atropine. Furthermore, values for the affinity of cholinergic receptors for atropine are almost identical in isolated cells and intact tissue (Fay and Singer, 1974).

6. The presence of β-adrenergic receptors whose properties are similar to those found in intact tissues. Specifically, isoproterenol inhibits the contractile response to various stimuli and this effect is antagonized by propranolol. In addition, the response to isoproterenol is accompanied by an increase in the level of cyclic adenosine 3',5'-monophosphate (Honeyman and Fay, *unpublished observations*). The sensitivity of the isolated cells to isoproterenol ($ED_{50} = 3 \times 10^{-7}$ M) is similar to that seen in intact visceral smooth muscles (Lee, 1970).

B. Several lines of evidence indicate that the contractile mechanism is unaltered by the isolation procedure.

1. The isolated cells contract in response to K⁺, ACh, and electrical stimulation and these responses, like those in the intact tissue, are inhibited after removal of Ca^{2+} from the medium (Fay and Delise, 1973).

2. Ultrastructural analysis reveals no qualitative or quantitative differences between the intact tissue and isolated cells with respect to filaments, dense bodies, or mitochondria (Fay and Delise, 1973).

3. The initial velocity of shortening in response to electrical stimulation of the isolated cells (0.13 ± 0.01 muscle lengths/sec) is comparable to V_{max} reported in intact strips of smooth muscles at 20°C (Gordon and Siegman, 1971).

C. The metabolic and growth properties of these cells appear unaltered by the isolation procedure as

1. The rate of lactate production in the isolated cells (0.3 μmole/g cells-hr) is comparable to that obtained in intact tissue from the same animals (0.2 μmole/g tisue-hr.) (T. Honeyman, *unpublished observation*).

2. The isolated cells can be sustained under culture conditions for more than 14 days, during which they retain their contractile ability and incorporate ³H-amino acids (Fay and Taber, *unpublished observation*).

The foregoing properties of the isolated cells indicate that they are indeed maintained in a normal functional state through the isolation procedure, and are thus appropriate for a study of the physiology of smooth muscle.

CONTRACTILE RESPONSE TO ELECTRICAL STIMULATION

Our attention was first focused on a description, at the level of the light microscope, of the contractile response of isolated smooth muscle cells to extracellular electrical stimulation. Observations were made using Nomarski optics. As may be seen from Fig. 1, depicting a typical contractile response to

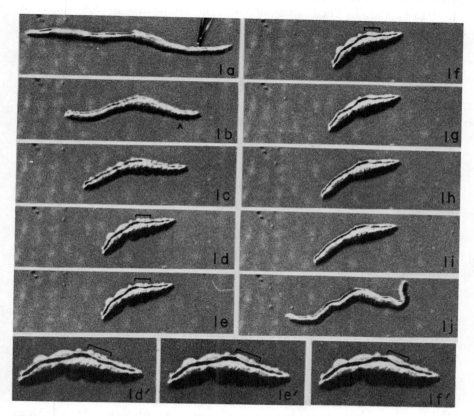

FIG. 1. Ten frames from a film record showing the contractile response of an isolated smooth muscle cell to brief electrical stimulation. Time intervals: (a) just prior to stimulation; (b) 1 sec after initiation of contraction, cell at 66% of its initial length (L_i); (c) 2 sec after initiation of contraction, 51% L_i; (d, d') 9 sec after initiation of contraction, 38% L_i; (e, e') 10 sec after initiation of contraction, 38% L_i; (f, f') 11 sec after initiation of contraction, 38% L_i—cell maximally contracted; (g) 20 sec after initiation of contraction, 40% L_i; (h) 26 sec after initiation of contraction, 44% L_i; (i) 31 sec after initiation of contraction, 45% L_i; (j) 10 min after initiation of contraction, 74% L_i. Note the appearance of evaginations after shortening to 66% L_i (b, dart), whereas the loss of such evaginations during relaxation is almost complete at 45% L_i. In frames d, d', e, e', f, and f' three small evaginations (brackets) are seen to fuse in a stepwise manner. a–i, ×200; d', e', f', ×295.

currents greater than 10^{-4} A, the cell contracts in a synchronous uniform manner. During contraction numerous evaginations appear along the cell surface. These surface irregularities are lost, however, following relaxation and reextension of the cell. Typically cells contracted to $35 \pm 9\%$ (mean $\pm$ SD, $N = 39$ cells) of their resting length. Within 15 min after the cessation of electrical stimulation, relaxation had progressed to $67 \pm 14\%$ (mean $\pm$ SD, $N = 39$ cells) of the prestimulation resting length. One other capability of the isolated smooth muscle cells became evident when it was found that cells stimulated by current intensities of less than 10^{-5} A, exhibited a partial contraction localized to the area near the stimulating electrode (Fay and Delise, 1973). The same cell could be stimulated to contract in a synchronous uniform manner when the current intensity was increased, indicating that the highly localized contraction was not the result of an impairment of the contractile mechanism in one part of the cell.

ARRANGEMENT OF CONTRACTILE ELEMENTS WITHIN SMOOTH MUSCLE

The model for the arrangement of contractile elements within the smooth muscle cell shown in Fig. 2 has been proposed (Fay and Delise, 1973) to

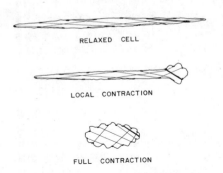

RELAXED CELL

LOCAL CONTRACTION

FULL CONTRACTION

FIG. 2. Schematic representation of a smooth muscle cell showing how contractile units are attached to the cell surface. The densities along the cell membrane represent plasma membrane dense bodies and the lines connecting then represent the contractile units. One of these lines has been widened to facilitate identification in the three different contractile states. Note that contractile units run for only relatively short lengths within the cell and that the angle between contractile units and the long axis of the cell increases during contraction.

explain both the observed ability of the isolated cells to contract locally and the observed formation of evaginations during contraction. The model represents an amplification of the general suggestions of Pease and Molinari (1960) that plasma membrane dense bodies act as anchoring sites for contractile units within the smooth muscle cell. Two important additional features of the model shown in Fig. 2 are that contractile units run for only relatively short lengths within the cell and the angle between contractile units and the long axis of the cell increases with a decrease in length. According to this model evaginations are the consequence of two opposing tendencies: (a) the need to increase cellular diameter in order to accommodate the volume displaced by cell shortening, and (b) the vectorial component of force of the contractile units pulling in on the cell membrane at the points of attachment to the cell membrane, i.e., the plasma membrane dense bodies.

Several specific predictions follow from this scheme: (a) evaginations ought to be formed from areas of the cell membrane not anchoring filaments and thus free of plasma membrane dense bodies, and the nonevaginated portions of the cell surface ought to be sites of filament attachment; (b) shortening of the cells during contraction ought to be associated with an increase in the angle of contractile elements relative to the long axis of the cell; (c) as activation of the contractile elements results in inwardly directed forces being applied over the entire extent of the cell surface, an increase in intracellular pressure and a consequent loss of volume might be expected to be associated with contraction. The volume loss might be expected to reverse as the force of contractile elements decreases during relaxation.

An alternative explanation for the formation of evaginations during contraction has been put forward by Lane (1965) and Kelly and Rice (1969) who noted outpocketings in cells within intact strips of smooth muscle during isotonic contraction. They propose that evaginations merely represent a means of accommodating excess surface membrane in going from a cylindrical to a more spherical geometry upon contraction. A specific prediction that follows from their interpretation is that the appearance of evaginations should be determined solely by cell length, regardless of contractile state. On the other hand, if evaginations are formed as we suggest, one might expect them to appear early during the contraction, but they should be lost following slight reextension of the cell during relaxation as both the inwardly directed forces and the need to accommodate cellular volume in the evaginations diminish.

ULTRASTRUCTURE OF CONTRACTED AND RELAXED CELLS

The predictions were tested by further analysis of cinematographic records of contractile responses of isolated cells and by ultrastructural observations on single cells whose contractile state was precisely defined. As may be seen from Fig. 3a,b, the plasma membrane of the relaxed smooth muscle cell consists of two types of areas that alternate over the entire length of the cell: one contains amorphous dense material subtending the membrane proper (plasma membrane dense bodies) and the other contains numerous micropinocytotic vesicles. In some thin sections, bundles of myofilaments appear to merge with the plasma membrane densities (Fig. 5). The surface of maximally contracted cells has numerous evaginations (Fig. 3c). The plasma membrane in this state has three characteristic areas. One is found at the base of evaginations and contains dense amorphous material subtending the membrane (Fig. 3d); the others, not associated with such densities, form either the surface of the evaginations or micropinocytotic vesicles (Fig. 3d). The cytoplasm enclosed by the evaginations is finely granular and does not appear to contain myofilaments (Fig. 3d). Similar outpocketings to those seen in the contracted isolated cells have been reported in isotonically contracted guinea pig ileum (Lane, 1965) and guinea pig taenia coli (Kelly and

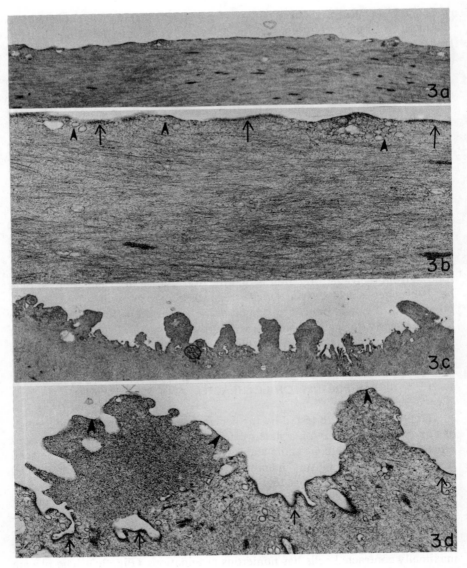

FIG. 3. (a) Electron micrograph showing the smooth surface contour of a relaxed cell. (b) The plasma membrane of a relaxed cell consists of areas that are either subtended by amorphous dense material (arrows) or differentiated into micropinocytotic vesicles (darts). (c) Electron micrograph showing the rough surface contour of a contracted cell. (d) The plasma membrane of a contracted cell is subtended by amorphous dense material along the bases and "neck regions" of the surface evaginations (arrows). The membrane covering the evaginations proper (darts) is not subtended by these dense areas, and in general it does not have micropinocytotic vesicles which characterize the plasma membrane of relaxed cells. a, c ×7,740; b, d ×21,635.

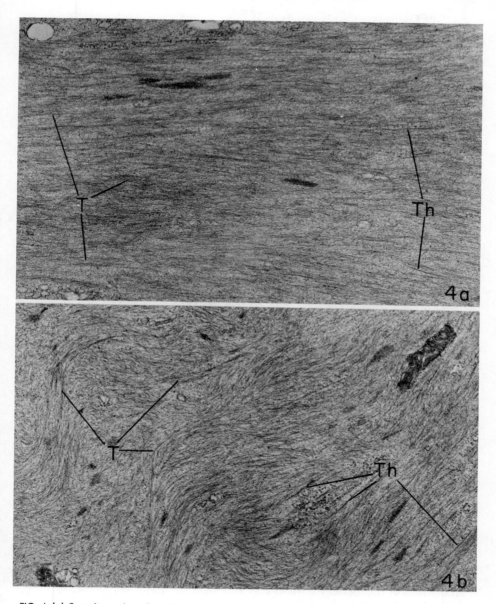

FIG. 4. (a) Cytoplasm of a relaxed cell showing the nearly uniform orientation of the thick (Th) and thin (T) myofilaments parallel to the long axis of the cell. (b) Cytoplasm of a contracted cell showing the non-parallel orientation of thick (Th) and thin (T) myofilaments at various angles to the long axis of the cells ×24,575.

Rice, 1969). Such structures have also been observed in rabbit colon and guinea pig urinary bladder (Nagasawa and Suzuki, 1967), in vascular smooth muscle of the dog (Ashford et al., 1966), and in guinea pig vas deferens (Merrilles, 1968), but have not always been ascribed to contraction. However, it is possible that such outpocketings reflect contraction of the cells under study, as the incidence of such evaginations is increased in hemorrhagic shock or after the tissues are traumatized.

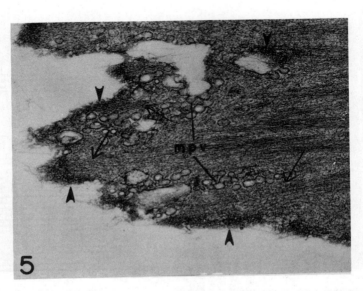

FIG. 5. Tangentially sectioned surface of a relaxed cell showing the relationship between the amorphous dense areas subtending the plasma membrane (*darts*) and bundles of myofilaments (*arrows*). The myofilaments appear to be continuous with the dense areas of the membrane but not with those areas showing micropinocytotic vesicles (mpv). ×29,020.

In relaxed isolated smooth muscle cells thick and thin myofilaments are generally parallel to the long axis (Fig. 4a), whereas in contracted cells thick and thin myofilaments are oriented in a nonparallel manner at considerably larger angles relative to the long axis (Fig. 4b). Hence, the ultrastructural appearance of the evaginated and nonevaginated portions of the cell fit the predictions of the model shown in Fig. 2.

APPEARANCE OF EVAGINATIONS DURING CONTRACTION AND RELAXATION

In order to determine when evaginations appear during contraction and when they are lost during relaxation, cinematographic records of the contractile responses of 19 cells to electrical stimulation were analyzed. Evaginations covered the entire cell surface after cells had shortened to

$57 \pm 9\%$ (mean $\pm$ SD) of their initial length. Evaginations were first evident on the cell shown in Fig. 1 after it contracted to 66% of its initial length (Fig. 1b) and its entire surface exhibited these irregularities after contraction had proceeded to 51% (Fig. 1c). Often, during contraction, several smaller evaginations fused to form large protrusions as shown in Fig. 1d,e,f. In this cell, evaginations disappeared (Fig. 1g–j) following re-extension of the cell to 45% of its initial length, i.e., a change of 7% from its initial length prior to initiation of relaxation. Analysis of the movie records of 14 cells revealed that the surface irregularities which characterize

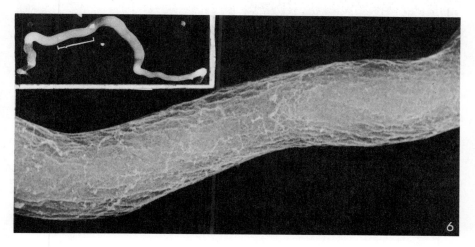

FIG. 6. Scanning electron micrograph of an isolated, unstimulated smooth muscle cell fixed at its resting length. Note the generally smooth surface of this cell. ×5,780. Inset at upper left shows entire cell. ×580.

contracting smooth muscle cells were lost after they had relaxed to $44 \pm 5\%$ (mean $\pm$ SD) of their initial lengths. When the electrical stimulation of a cell was maintained it did not relax and evaginations did not disappear.

To obtain further insight into the appearance and disappearance of the surface evaginations during the contraction/relaxation cycle, cells were fixed at various stages of the cycle and observed with the scanning electron microscope. The relaxed smooth muscle cell (Fig. 6) was characterized by a smooth surface, although occasionally tufts and strands of material were found in association with the cell surface. These probably represent remnants of the extracellular matrix. The generally smooth surface of the relaxed cell contrasts with the numerous evaginations present on the fully contracted cell in Fig. 7. This highly irregular surface was observed in all 12 of the fully contracted cells viewed with the scanning electron microscope. The evaginations in many of the cells were bulbous as exemplified by the cell in Fig. 7; these evaginations have a narrow stalk and a spherical head. Although most

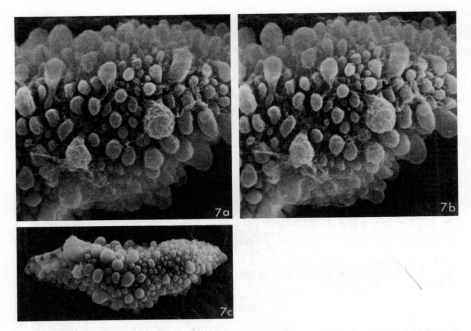

FIG. 7. Scanning electron micrograph of an isolated smooth muscle cell fixed after maximal shortening in response to brief electrical stimulation Cell had contracted to 35% L_i at the time of fixation. Part (c) shows entire cell, whereas micrographs a and b are a stereo pair of left end of cell. Stereo pair may be viewed with binocular viewer (available from E. F. Fullam, Inc., Schenectady, N.Y.) to obtain three-dimensional image of cell. Note that most evaginations are connected to main body of cell by slender neck; occasionally evaginations appear connected by multiple stalks (*darts*). a, ×1,496; b, c, ×5,280.

evaginations have a single narrow stalk connecting to the main body of the cell, occasionally protrusions are found which appear to be connected to the main body of the cell by more than one stalk (Fig. 7). It may be that these larger evaginations with several stalklike connections result from the fusion of several small evaginations as noted during contraction of living smooth muscle cells (Fig. 1d,e,f). A few of the fully contracted cells were found to contain predominantly moundlike evaginations with only occasional bulbous protrusions. Figure 8 shows a cell fixed after contracting to 72% of its original length. This degree of shortening represents approximately the half-maximal extent usually obtained. The surface of the cell shown in Fig. 8 is covered by small but definite evaginations which presumably increase in size to form larger protrusions characteristic of fully contracted cells (Fig. 7). The appearance of the cell in Fig. 8 was typical of all eight cells which were fixed after contracting to between 60 and 80% of their initial length. Small evaginations were also noted in cells studied after contraction to between 80 and 90% of their initial length, but such protrusions were less numerous than found in more contracted cells.

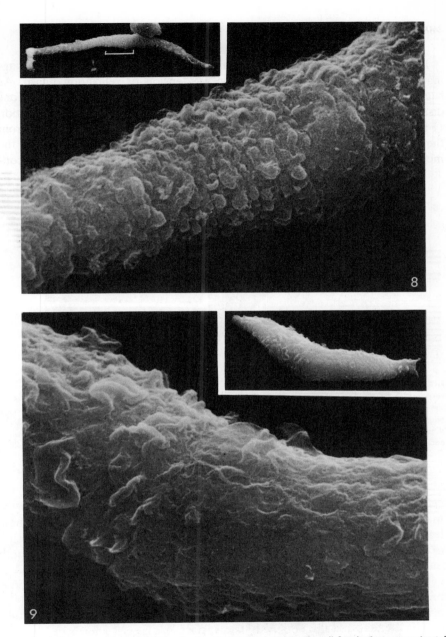

FIG. 8 (*top*). Scanning electron micrograph of an isolated smooth muscle cell fixed after approximately half-maximal shortening in response to brief electrical stimulation. Cell had contracted to 72% L_i at time of fixation. Note that the cell surface is completely covered by small evaginations. ×6,000. The inset at the upper left shows the entire cell. Spherical structure (partially shown) in inset is adherent debris. ×500.

FIG. 9 (*bottom*). Scanning electron micrograph of an isolated smooth muscle cell fixed following slight relaxation after contractile response to brief electrical stimulation. Cell length at time of fixation, 47% L_i; cell length at point of maximum shortening, 40% L_i. Note that the bulbous and moundlike evaginations characteristic during contraction are largely absent in this cell fixed during the early part of relaxation. Occasional foldings of the cell surface are evident, however. ×6,000. Inset at upper right shows entire cell. ×1,120.

In order to understand the mechanism underlying the formation of evaginations, cells were studied during the reextension or relaxation phase. As suggested by light-microscope observations (Fig. 1g,h) the evaginations disappear almost coincidentally with the earliest indications of relaxation (Fig. 9). The surface of cells fixed after relaxing approximately 10% from the fully shortened state were largely free of evaginations. Folding of the membrane was occasionally found, but evaginations of the type seen in con-

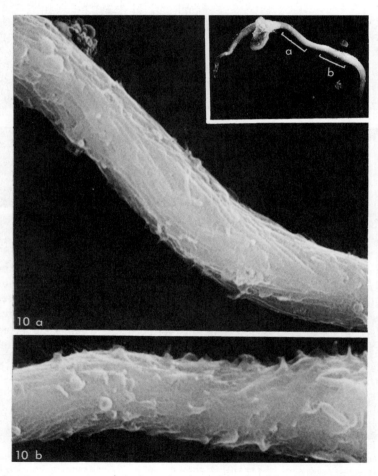

FIG. 10. Scanning electron micrograph of an isolated smooth muscle cell fixed during relaxation following the contractile response to brief electrical stimulation. Cell length at time of fixation, 79% L_i; cell length at point of maximal shortening, 37% L_i. (a) and (b) show portions of the cell shown in the inset at upper right. Note that large portions of the cell surface typified by (a) were free of evaginations; other regions of the cell displayed occasional slight folding of the cell membrane. A small amount of debris adherent to the cell is shown in the upper portion of the inset. Inset, ×430; a, b, ×5,390.

tracting cells were absent. The cell shown in Fig. 9 is typical of the six cells which were studied after relaxing by about 10% from their fully contracted state. The percentage relaxation reflects the ratio of the observed length change relative to the initial length of the cell. Cells fixed after relaxing for 15 min, although typically still somewhat shortened, had smooth surfaces (Fig. 10a,b). The cell shown in Fig. 10 relaxed to 79% of its initial length and its surface, typical of the seven cells studied at this stage, is basically smooth, although occasionally folds in the membrane were still apparent. In all cells fixed during intermediate stages of relaxation occasional folding of the cell membrane was noted, but the nature of these irregularities in surface contour were markedly different from the bulbous evaginations that characterize the contracting cell. These folds presumably accommodate excess surface membrane, as similar folding was noted in cells shrunk by hyperosmotic solutions. Thus the light and scanning microscopic evidence indicates that evaginations are a consequence of active force development and can not be viewed simply as a means of accommodating excess surface membrane resulting from shortening of the cells.

CELL VOLUME DURING CONTRACTION AND RELAXATION

As volume changes during the contraction/relaxation cycle are predicted by the model, cinematographic records of 40 cells which contracted and subsequently relaxed in response to electrical stimulation were analyzed. The analysis of possible volume changes required measurement of the lengths and areas of the two-dimensional projections of each cell as determined from cinematographic records (Fay and Delise, 1973) and rested on the assumption that the ratio of height and width of any cell remained constant. The validity of this approach was verified first by its application to records of cells exposed to a series of solutions of different osmotic pressures. As predicted, a linear van't Hoff relationship between volume and (osmotic pressure)$^{-1}$ was obtained. Subsequent application of this method of analysis to cells during their contraction/relaxation cycle revealed a progressive volume loss as the cells shortened (Fig. 11); upon maximum shortening to 30% of their initial length the measured cell volume decreased to almost 80% of its initial value. The lost volume was regained following relaxation of the cells. The force driving volume back into the cell during the relaxation phase might be an osmotic gradient. Such a gradient would exist if the fluid lost during contraction were an ultrafiltrate of the cell cytoplasm.

All data presently in hand thus support the model for the arrangement of contractile elements which has been discussed. The recent demonstration by Fisher and Bagby (1974) of diminished birefringence associated with the long axis of isolated smooth muscle cells during contraction also supports the proposed arrangement of contractile elements.

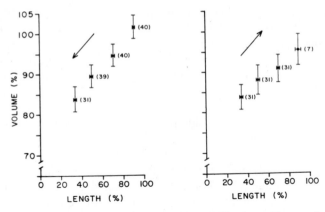

FIG. 11. The variation of volume of isolated smooth muscle cells with length during a single contraction–relaxation cycle. Volumes and lengths of each cell were compared to those at rest, which were designated as 100%. The data are grouped into length intervals of 20%. The points represent the mean length and volume for each interval. The vertical and horizontal bars indicate the SE and the number in parentheses indicates the number of observations. The panel to the left shows the relation of length and volume during shortening and that to the right during reextension.

MECHANICAL PROPERTIES OF SINGLE SMOOTH MUSCLE CELLS

As pointed out by Rosenbluth (1965) the arrangement of contractile elements in an oblique manner as in the model in Fig. 2, rather than in an end-to-end series arrangement as in striated muscle cells, would result in profound differences in such mechanical properties, as maximum force/cm^2 as well as maximum velocity and extent of shortening. Furthermore, the observed reorientation of filaments that has been observed in the present studies would also be expected to have a marked influence on the length–tension and force–velocity relationship. We have therefore undertaken studies on isolated single cells in the hope of clearly defining the relationship between cellular and subcellular changes in structure and mechanical activity. Numerous problems are posed for such a study on single smooth muscle cells, as the isolated cells do not possess a tendon which can be used to anchor the cells for force measurements as has been done in single striated muscle fibers. Furthermore, because of the smaller diameter of smooth muscle fibers (10 μm) the forces expected from a single smooth muscle cell are at least 2 orders of magnitude smaller than those measured in single striated muscle cells. Nonetheless, attempts have been made to record the force of contraction of individual smooth muscle cells and quite recently these efforts have begun to generate the first positive results. Two typical responses of brief electrical stimulation may be seen in Fig. 12. The time course of a typical contractile response to brief electrical stimulation is shown in Fig. 12a. Force rises to a peak within 4.0 sec and declines with a $t\frac{1}{2} = 3.0$ sec. As can be seen from Fig. 12b the latency between a single shock of 1-msec duration and the first detectable increase in force is approximately 0.6 sec. Observations on four cells revealed

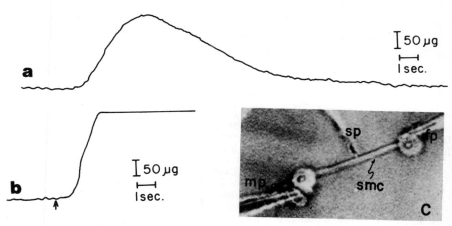

FIG. 12. Isometric recordings of the contractile response of single isolated smooth muscle cells to electrical stimulation. The smooth muscle cell (*smc*) was wrapped at each end around a glass probe as depicted in the photomicrograph ($\times$276) in part (c). One probe (*fp*) was attached to a force transducer, while the other probe (*mp*) was controlled by a micromanipulator. Each cell was suspended at 20°C in amphibian Ringer containing isoproterenol (10^{-5} M) to reduce spontaneous activity. (a) Isometric force recording obtained during stimulation of a single smooth muscle cell with a 1-sec burst of 1-msec pulses (10^{-5} A; cathode; 10 pulses/sec) applied extracellularly via a 3 M NaCl filled microelectrode (sp, c) set several tenths of a micrometer from the cell surface. Peak force achieved during contraction was 235 μg. The $t\frac{1}{2}$ for relaxation was 3.0 sec. (b) Isometric force recording obtained following electrical stimulation of a single smooth muscle cell different from that shown in a. Stimulation was as in a, but a single pulse only was applied with the stimulating electrode in this case applied directly to the cell surface. The arrow indicates the time of stimulation. A latency of 0.6 sec between the time of stimulation and the first noticeable increase in force was observed. The flat portion of the record indicates saturation of this recording channel. Peak force recorded during this contraction was 1470 μg, or 0.57 kg/cm². Force transducer characteristics: sensitivity, 600 mV/mg; compliance, 1.5 μm/mg; resonant frequency, 100 Hz; minimum detectable force applied as a square pulse, 10 μg.

an average delay between the electrical stimulus and a detectable increase in force of 0.70 $\pm$ 0.24 sec (mean $\pm$ SD). The cell shown in Fig. 12b developed a peak force of 1470 μg or 0.57 kg/cm² when normalized for cross-sectional area. The largest peak force/cm² detected during the contractile response of a single isolated smooth muscle cell to electrical stimulation was 2.0 kg/cm², but may not be the highest achievable force as a systematic study of the effects of the stimulus characteristics and cell length has not been completed.

These characteristics of the contractile response to electrical stimulation represent the starting point of an attempt to determine the mechanical properties of the smooth muscle cell. These studies may provide further insight into the mechanism underlying the generation of force in smooth muscle.

ACKNOWLEDGMENTS

This work was supported by a grant from the National Institutes of Health (HL-14523) to Fredric S. Fay and grants from the Sedgwick County Chapter of the Kansas Heart Association (KR-726) and Muscular Dystrophy

Association of America to Peter H. Cooke. We gratefully acknowledge the skilled technical assistance of Ms. Betsy Moody, Ms. Linda Hassinger, and Ms. Lorraine Hammer in these studies.

REFERENCES

Ashford, T., Palmerio, C., and Fine, J. (1966): *Ann. Surg.,* 164:575–585.
Bagby, R. M., and Fisher, B. A. (1973): *Am. Zoologist,* 12:xl–xli.
Bagby, R. M., and Fisher, B. A. (1973): *Am. J. Physiol.,* 225:105–109.
Bagby, R. M., Young, A. M., Dotson, R. S., Fisher, B. A., and McKinnon, K. (1971): *Nature,* 234:351–352.
Burnstock, G. (1970): In: *Smooth Muscle,* edited by E. Bülbring, A. F. Brading, A. W. Jones, and T. Tomita, pp. 1–69. Williams and Wilkins, Baltimore, Md.
Casteels, R. (1970): In: *Smooth Muscle,* edited by E. Bülbring, A. F. Brading, A. W. Jones, and T. Tomita, pp. 70–99. Williams and Wilkins, Baltimore, Md.
Fay, F. S., and Delise, C. M. (1972): *Physiologist,* 15:132.
Fay, F. S., and Delise, C. M. (1973): *Proc. Natl. Acad. Sci., US,* 70:641–645.
Fay, F. S., and Singer, J. J. (1974): *Fed. Proc.,* 33(3):435.
Fisher, B. A., and Bagby, R. M. (1972): *Am. Zoologist,* 12:xl–xli.
Fisher, B. A., and Bagby, R. M. (1974): *Fed. Proc.,* 33(3):435.
Gordon, A. R., and Siegman, M. J. (1971): *Am. J. Physiol.,* 221:1243–1249.
Kelly, R. E., and Rice, R. V. (1969): *J. Cell Biol.,* 42:683–694.
Lane, B. P. (1965): *J. Cell Biol.,* 27:199–213.
Lee, C. Y. (1970): In: *Smooth Muscle,* edited by E. Bülbring, A. F. Brading, A. W. Jones, and T. Tomita, pp. 549–557. Williams and Wilkins, Baltimore, Md.
Merrillees, N. C. R. (1968): *J. Cell Biol.,* 37:794–817.
Nagasawa, J., and Suzuki, T. (1967): *Tohoku J. Exp. Med.,* 91:299–313.
Pease, D. C., and Molinari, S. (1960): *J. Ultrastruct. Res.,* 3:447–468.
Purves, R. D., Hill, C. E., Chamley, J. H., Mark, G. E., Fry, D. M., and Burnstock, G. (1974): *Pflügers Arch.,* 350:1–7.
Purves, R. D., Mark, G. E., and Burnstock, G. (1973): *Pflügers Arch.,* 341:325–330.
Rosenbluth, J. (1965): *Science,* 148:1337–1339.
Singer, J. J., and Fay, F. S. (1974): *Fed. Proc.,* 33(3):435.
Small, J. V. (1974): *Nature,* 249:324–327.

Physiology of Smooth Muscle, edited by
E. Bülbring and M. F. Shuba.
Raven Press, New York © 1976.

The Structure of Smooth Muscles of the Eye and the Intestine

Giorgio Gabella

University College London, London WC1E 6BT

During the past 15 years studies in electron microscopy have greatly contributed to our understanding of how smooth muscle works. Some of the problems which have received particular attention are sites of electrical coupling between the cells, the extent of the extracellular space, structural specializations of the cell membrane, distribution of the sarcoplasmic reticulum, structure and arrangement of the myofilaments, and innervation of the tissue. Some of these topics will be discussed briefly in the light of recent investigations on smooth muscles of the intestine and the eye—two examples of smooth muscle whose individual characteristics are very different.

SITES OF ELECTRICAL COUPLING

Since it was first observed by Dewey and Barr (1), the *nexus* has been considered the morphological correlate of the electrical coupling site. The evidence for this suggestion is indirect, based on the closeness of the junction (a gap of only about 2 nm occurs between the adjacent membranes) and on the permeability and coupling characteristics of similar junctions between epithelial cells (gap junctions). However, several observations have cast doubts on the general validity of this hypothesis. Although nexuses are described in a number of smooth muscles, there is not a good correlation between the presence of electrical coupling and the presence of nexuses. For example, nexuses are numerous in the circular muscle of the dog duodenum (2) and the guinea pig ileum (3), but none were found in the corresponding longitudinal muscle. In a recent investigation on the taenia coli of the guinea pig nexuses were not observed. These results (*unpublished data*) on the taenia coli were obtained by using a variety of fixation (with aldehydes or osmium, or both) and staining (with uranyl acetate and lead) procedures; by the same procedures nexuses were consistently observed in the iris and the circular musculature of the ileum. The results call for a reassessment of earlier conclusions that nexuses are present in the guinea pig taenia coli (1,4,5, and others); their presence has been claimed also on the evidence of freeze-etching preparations (6). However, judging from the published material, I feel that there is not yet convincing evidence showing nexuses in normal and adequately fixed taenia coli. A more critical approach to the image interpretation and to the problem of structure preservation and staining will be needed to clarify this question.

On the other hand, the muscle of the sphincter pupillae of guinea pig display a very large number of nexuses, greater than in any other muscle studied (7) (Fig. 1A). In transverse sections of the muscle, about one-half of the muscle cells appear to be connected to each other by a nexus, sometimes by two or three. Taking into account the length of the muscle cells, each of them must have scores of nexuses; in longitudinal sections rows of nexuses are commonly seen, connecting at several points two adjacent muscle cells. Some nexuses connect muscle cells which are closely related to the same nerve ending (neuromuscular cleft of about 20 nm), others are found between two processes of the same muscle cell (Fig. 1B). An occasional nexus between parts of the same cell has been observed in other muscles (3,8,9), but in the sphincter pupillae they are a common occurrence; they usually enclose a bundle of collagen fibrils.

If it is confirmed that nexuses do not occur in the taenia coli, other structures should be examined as sites of electrical coupling. Cytoplasmic processes protruding from one cell into another are sites of intimate contact between muscle cells. However, it has not yet been ascertained how consistent or frequent these structures are, and in some preparations of taenia none were found. Another structure which could play a role in the electrical coupling has recently been observed in the taenia coli. It consists of a wide, rather flat, area of apposition between two muscle cells (*paired cells*), involving about a third of their surface (Fig. 2A,B). The cleft varies between 10 and 150 nm or more, and contains material of medium electron density similar to the basal lamina and usually no collagen fibrils; in transversely sectioned muscles, the cleft is narrower at both ends of the pairing area (though never as close as to form a gap junction or nexus), producing an appearance suggestive of a chamber partly sealed from the bulk of the extracellular space (Fig. 2). Caveolae open into this cleft, and the cell membranes also show dense patches facing each other (Fig. 2A). These structures are difficult to preserve with the current histologic procedures, and are more numerous in taeniae fixed with osmium only. They have not yet been studied in serial sections nor in longitudinal sections. Series of "paired" cells can also be found, formed by three or four cells in a row. Finally, it should be mentioned that some authors have argued that it is unlikely that coupling occurs through membrane areas other than those where the extracellular space virtually disappears (10,11).

SURFACE OF MUSCLE CELLS

All muscle cells show characteristic inpocketings of the cell membrane, called *caveolae*. They occur in rows parallel to the major axis of the cell; in

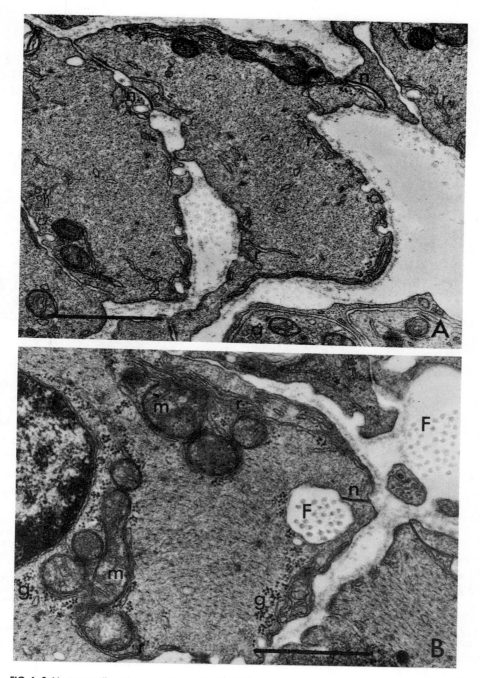

FIG. 1. Sphincter pupillae of guinea pig. A: Muscle cells in transverse section, with numerous sacs of sarco-plasmic reticulum (r). n, nexus; a, axon. Scale: 1 μm. B: A nexus (n) between two processes of the same muscle cell. F, collagen fibrils; m, mitochondria; r, sarcoplasmic reticulum; g, glycogen granules. Scale 1 μm. (reproduced from Ref. 7 by kind permission of the Royal Society.)

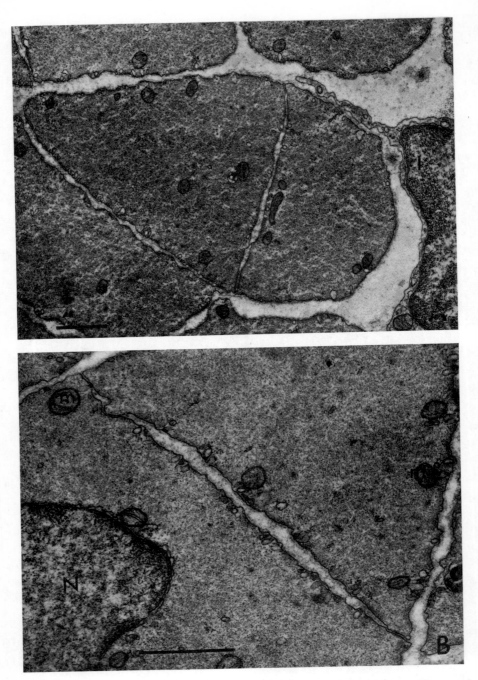

FIG. 2. Taenia coli of guinea pig, in transverse section. A: Two *paired* muscle cells, with a wide area of apposition. I, interstitial cell. Scale: 1 μm. B: A similar structure at higher magnification. N, nucleus; m, mitochondria. Scale: 1 μm.

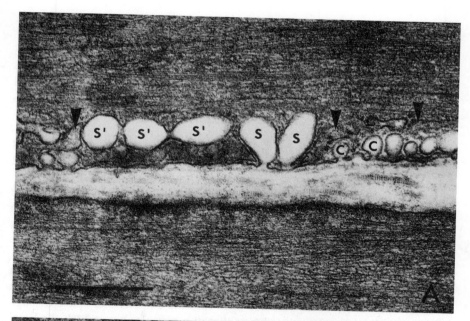

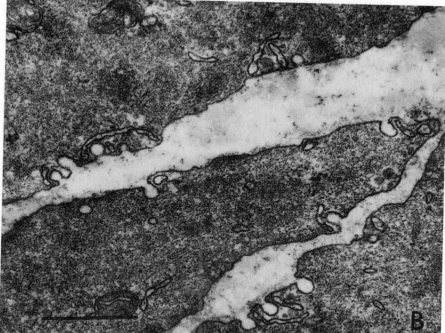

FIG. 3. Taenia coli of guinea pig. A: Relaxed muscle cut in longitudinal section. The cell membrane of one of the muscle cells shows caveolae (c) and larger invaginations (saccules) (s), which open into the bulk the extracellular space by a narrow neck. Other saccules (s') open into each other and communicate only indirectly with the extracellular space (the continuity of all the cavities is not visible in this section). Sacs of sarcoplasmic reticulum (*arrowheads*) lie close to the caveolae and the saccules. Scale: 0.5 μm. B: Muscle isometrically contracted, in transverse section. Numerous sacs of smooth sarcoplasmic reticulum are visible, particularly close to the caveolae. Scale: 0.5 μm.

the taenia it has been calculated that they increase the cell surface by more than 70% (12). There is no evidence that their number changes during contraction and relaxation, although it is possible that the intense mechanical reformation of contracted cells affects the structure or the properties of the caveolae. At the level of the caveolae the cell membrane is obviously very pliable, since thin and long evaginations, bearing caveolae, appear in contracted cells. It has been suggested that in addition to increasing the cell surface, the caveolae isolate a fraction of the extracellular space in a special compartment closely related to (and controlled by) the muscle cell itself (3). Since the caveolae are arranged to form a network, with very elongated meshes, on the muscle cell surface, this special compartment, although divided in many microunits, enwraps all parts of the muscle cell. It remains to be seen whether this compartment is important for conduction along the cell surface.

Besides the caveolae at the surface of muscle cells of the guinea pig taenia coli there are large invaginations, round or oval-shaped, up to 0.25 μm in diameter (Fig. 3A). These *saccules* open into the extracellular space through a narrow neck, about the same size as that of a caveola, over which the basal lamina passes without penetrating. The saccules occur singly or in rows, and they are much fewer in number than the caveolae; they often open into each other, so that many of them communicate with the extracellular space only through other saccules. Their content is totally electron-lucent, a characteristic which makes them easy to distinguish from cisternae of sarcoplasmic reticulum, which have a content of medium electron density. Caveolae can open into the saccules. Similar invaginations bearing caveolae on them have been described in the longitudinal muscle of the guinea pig ileum (13,14), but are not observed in the circular muscle.

SARCOPLASMIC RETICULUM

Sacs of sarcoplasmic reticulum are present in smooth muscle cells, particularly beneath the cell membrane and around the caveolae (Fig. 3B). Their function and their possible role in calcium transport and accumulation remain hypothetic. Quantitative studies have shown that in the rabbit the amount of sarcoplasmic reticulum is greater in the aorta and pulmonary artery (about 5% of the cell volume) than in the taenia coli and portal vein (about 2%), and this correlates with the different ability of smooth muscles to contract in the absence of extracellular calcium (15). Recently, calcium oxalate precipitates have been localized in the sarcoplasmic reticulum of the guinea pig taenia coli by a histochemical method; calcium deposits were also found in the perinuclear cisterna, mitochondria, caveolae, and cell membrane (16).

One of the characteristics of the sphincter pupillae of the guinea pig is the large amount of sarcoplasmic reticulum of its muscle cells (see Fig. 1B), probably even larger than in blood vessels. Cisternae or arrays of cisternae of smooth sarcoplasmic reticulum lie under large areas of the cell surface. Each nexus is usually accompanied by sarcoplasmic reticulum on either side. Some reticulum lies immediately underneath the neuromuscular junctions, and in such cases periodic densities between the cell membrane and the outer surface of these sacs of reticulum have been observed (17).

CONTRACTION AND RELAXATION

It is well known that great structural changes occur in smooth muscle cells with contraction and relaxation. In recent studies on the guinea pig taenia coli, lengths of muscle were elongated (with a load of 1 g) or shortened (with a K⁺-rich solution) and prepared for light and electron microscopy (18) (Fig. 4A). The transverse-sectional area of the taenia increases with shortening up to four times, when there is a decrease in length of similar magnitude. A similar increase in transverse sectional area is shown by the individual muscle cells (Fig. 4B,C). In longitudinal sections the stretched taenia shows a regular array of parallel muscle cells (Fig. 4D). On the other hand, in the shortened taenia the muscle cells run at an angle to each other and show a marked twisting (Fig. 4E), which is related to the degree of shortening.

In stretched (relaxed) muscle cells, longitudinally sectioned, the cell surface is smooth (Fig. 5A), whereas shortened (contracted) muscle cells display a very irregular surface with numerous evaginations of various sizes and shapes (Fig. 5B). Many evaginations are finger-shaped, up to 1 μm in length, and interlock with digitations from a neighboring muscle cell; they have numerous caveolae and cisternae of sarcoplasmic reticulum. Other evaginations are less prominent and with a larger base, and have mainly dense patches; the latter are found also on most of the nonevaginated parts of the cells. Several authors have observed that contracted muscle cells have an irregular surface and interdigitating processes. Single muscle cells isolated from the toad stomach develop upon contraction bulbous and moundlike evaginations (19), which are much larger than those observed here in the taenia coli, where the restraint imposed by the collagen network and the high packing density of the muscle cells probably play an important role. In this material evaginations are finer and more numerous. Although the caveolae predominate, dense patches can also be present in the evaginations (Fig. 5B). The cisternae of sarcoplasmic reticulum can perhaps also facilitate the sliding of the superficial parts of the cell on the bulk of myofilaments.

In the taenia coli the collagen fibrils are mostly parallel or nearly parallel to the muscle cells in the relaxed muscle (Fig. 5A), while in the shortened muscle most of them are nearly orthogonal to the major axis of muscle cells (Fig. 5B).

The presence of a thick network of collagen plays an essential role in the

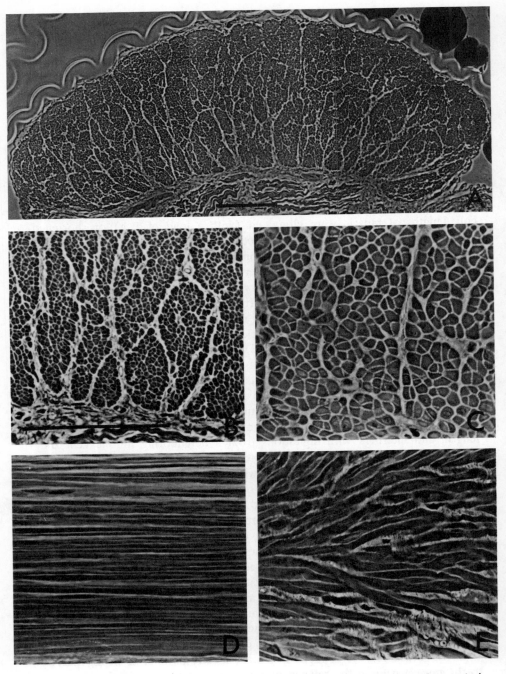

FIG. 4. Teania coli of guinea pig, photographed in phase contrast. A: Transverse section of a stretched taenia (1 g load). Photographic montage. Scale: 100 μm. B: Transverse section of a stretched taenia (1 g load). C: Transverse section of a taenia coli (from the same guinea pig as B) shortened (1 g load) with 117 mM K$^+$. D: The same taenia as in B, in longitudinal section. E: The same taenia as in C, in longitudinal section. B–E: same magnification. Scale: 100 μm. (Reproduced from Ref. 18 by kind permission of the Physiological Society.)

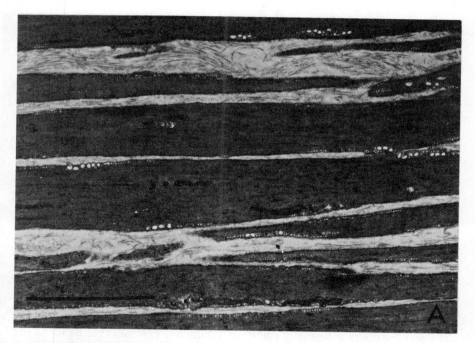

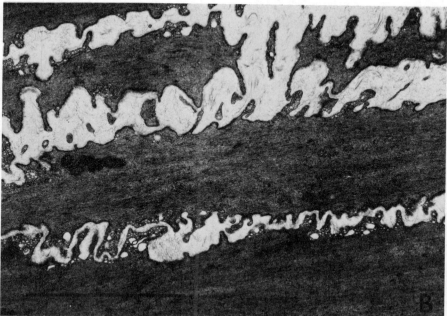

FIG. 5. Taenia coli of guinea pig in longitudinal section. A: Muscle stretched with 1 g load. Scale: 5 μm B: Muscle shortened with a K⁺-rich solution (1 g load). Numerous evaginations of the cell membrane are apparent. Scale: 5 μm.

transmission of the tension along a smooth muscle. The arrangement of the collagen fibrils varies in different muscles and is a determinant factor of the mechanical characteristics of muscle contraction. An interesting example is provided by two muscles circularly arranged around a lumen—the media of arteries and the circular muscle of the small intestine. The latter has a collagen network predominantly parallel to the muscle cells, the former a network predominantly orthogonal. This partly explains why upon contraction arteries change little their length, whereas with maximal contraction of the circular layer the small intestine of the guinea pig can reach a length almost twice the resting length.

INNERVATION

Studies in electron microscopy have shown that neuromuscular relations vary enormously from one smooth muscle to another (20). In their terminal length within a smooth muscle nerve fibers are varicose, and the varicosities contain large numbers of synaptic vesicles. In the intestinal musculature the distance between nerve endings and muscle cells is rarely less than 100 nm (21) and the fibers are usually gathered in bundles. On the other hand, in the vas deferens many nerve endings are closely apposed to the muscle cell membrane, with a gap of only 20 nm, and some lie embedded in deep invaginations of the cell membrane (22–24).

The sphincter pupillae of the guinea pig is supplied by a very large number of nerve fibers (7). In one transverse section of the sphincter pupillae more than 800 nerve fibers (and about 2,800 muscle cell profiles) were counted, half of which contained clusters of synaptic vesicles. The axons run singly or in very small bundles. Just under half of the vesicle-containing endings lay within 20 nm from a muscle cell. In some nerve endings dense projections and clustering of vesicles near the prejunctional membrane were visible, often with some specialization of the postjunctional membrane of the smooth muscle cell (Figs. 6A,B). The significance of the high density of innervation and of these junctional specializations is still unknown; they are similar to those found in the synapses of the central nervous system and are very rarely seen in other smooth muscles.

The innervation of the alimentary tract is extremely complicated owing to the presence of intramural ganglia. Examining quantitatively the innervation of the circular muscle layer of the guinea pig ileum, it appears that the nerve fibers are not evenly distributed throughout its thickness (25). Nearly two-thirds of the vesicle-containing axons are situated in the deep part of the circular muscle layer, the rest is scattered throughout its thickness. Other vesicle-containing axons are situated in the gap between circular and longitudinal layer, mainly at the surface of ganglia and connecting strands of the myenteric plexus. The longitudinal layer is usually devoid of nerves (21). In the innermost part of the circular layer, where there is the richest in-

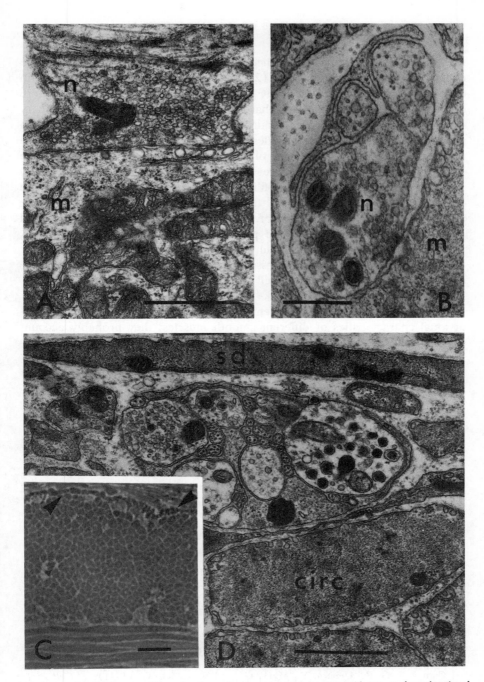

FIG. 6 A and B: Sphincter pupillae of guinea pig. Neuromuscular junctions with pre- and postjunctional specializations. n, nerve endings; m, muscle cells. Scale: A, 0.5 μm; B, 0.25 μm. C: Ileum of rabbit in transverse section, photographed in phase contrast miscoscopy, showing a unicellular layer of small and dark muscle cells (*arrows*) in the innermost part of the circular layer. Scale: 10 μm. D: Ileum of guinea pig, showing the small and dense muscle cells (sd) and ordinary muscle cells of the circular layer (circ). A nerve bundles runs between the two components of the circular layer. Scale: 1 μm. [Courtesy of the Physiological Society (27).]

nervation, there is a unicellular sheet of special muscle cells, which can be distinguished in light microscopy (26) and have recently been described in electron microscopy (27) (Fig. 6C,D). They are smaller in size and more electron-dense (and darker in phase contrast microscopy) than ordinary muscle cells, and are readily identified in the ileum of several mammalian species. They show all the organelles of smooth muscle cells and are in close relationship with a great number of axons issuing from the neurons of the intramural ganglia and of the prevertebral ganglia. It has been tentatively suggested that some of the nerve fibers in this part of the circular layer are motor, controlling the length of the special muscle cells, and some are afferent, detecting radial stretch of the intestinal wall or increased luminal pressure. It is known that in the presence of an intact myenteric plexus, radial stretch is an effective stimulus to elicit peristaltic contraction of the muscle layers (28). It may be possible to show experimentally whether the special small-and-dense muscle cells are indeed part of a stretch receptor.

The structures which have been briefly described are some of the individual characteristics of different smooth muscles; they probably account for the special properties of their diverse activities. Although structure–function correlations remain to a great extent obscure, morphologic observations are important because they suggest that the functional characteristics of a smooth muscle are specified and controlled at many structural levels.

ACKNOWLEDGMENTS

I thank Prof. Edith Bülbring, F.R.S., for valuable comments on the manuscript, and the Physiological Society for a travel grant. The work reported is supported by grants from the Wellcome Trust and the Medical Research Council.

REFERENCES

1. Dewey, M. M., and Barr, L. (1962): *Science,* 137:670–672.
2. Henderson, R. M., Duchon, G., and Daniel, E. E. (1971): *Am. J. Physiol.,* 221: 564–574.
3. Gabella, G. (1973): *Philos. Trans. R. Soc. Lond. [Biol. Sci.],* 265:7–16.
4. Bennett, M. R., and Rogers, D. C. (1967): *J. Cell Biol.,* 33:573–596.
5. Cobb, J. L. S., and Bennett, T. (1969): *J. Cell Biol.,* 41:287–297.
6. Geisweind, G., and Wermbter, G. (1974): *Cytobiologie,* 9:121–130.
7. Gabella, G. (1974): *Proc. R. Soc. Lond. [Biol. Sci.],* 186:369–386.
8. Iwayama, T. (1971): *J. Cell Biol.,* 49:521–525.
9. Campbell, G. R., Uehara, Y., Mark, G., and Burnstock, G. (1971): *J. Cell Biol.,* 49:21–34.
10. Bennett, M. R. (1972): *Autonomic Neuromuscular Transmission.* University Press, Cambridge.
11. Bennett, M. V. L. (1973): In: *Intracellular Staining in Neurobiology,* edited by S. D. Kater and C. Nicholson, pp. 115–133. Elsevier, New York.

12. Goodford, P. J., Johnson, F. R., Krasucki, Z., and Daniel, V. (1967): *J. Physiol.* (*Lond.*), 194:77–78P.
13. Gabella, G. (1971): *J. Cell Sci.*, 8:601–609.
14. Godfraind, T., Sturbois, X., and Verbeke, N. (1973): *Arch. Int. Physiol. Biochim.*, 81:786–791.
15. Devine, C. E., Somlyo, A. V., and Somlyo, A. P. (1972): *J. Cell Biol.*, 52:690–718.
16. Popescu, L. M., Diculescu, I., Zelck, U., and Ionescu, N. (1974): *Cell Tissue Res.*, 154:357–378.
17. Uehara, Y., and Burnstock, G. (1972): *J. Cell Biol.*, 53:849–853.
18. Gabella, G. (1974): *J. Physiol.* (*Lond.*), 242:36–38P.
19. Fay, F. S., and Delise, C. M. (1973): *Proc. Natl. Acad. Sci. US*, 70:641–645.
20. Campbell, G. (1970): In: *Smooth Muscle*, edited by E. Bülbring, A. F. Brading, A. W. Jones, and T. Tomita, pp. 451–495. Arnold, London.
21. Taxi, J. (1965): *Ann. Sci. Nat. Zool.*, 7:413–674.
22. Richardson, K. C. (1962): *J. Anat.*, 96:427–442.
23. Merrillees, N. C. R. (1968): *J. Cell Biol.*, 37:794–817.
24. Furness, J. B., and Iwayama, T. (1972): *J. Anat.*, 113:179–196.
25. Gabella, G. (1972): *J. Neurocytol.*, 1:341–362.
26. Li, P.-L. (1937): *Z. Anat. Entwicklungsgesch*, 107:212–222.
27. Gabella, G. (1974) *J. Physiol.* (*Lond.*), 240:1–3P.
28. Kosterlitz, H. W., and Robinson, J. A. (1956): *J. Physiol.* (*Lond.*), 146:369–379.

Physiology of Smooth Muscle, edited by
E. Bülbring and M. F. Shuba.
Raven Press, New York © 1976.

Spatial Neuromuscular Relations in Rabbit Ear Arteries

V. A. Govyrin

Laboratory of the Adaptive-Trophic Function of the Nervous System, Institute of Evolutionary
Physiology and Biochemistry Academy of Sciences of the USSR, Leningrad

The development of new histochemical and electron-microscopic techniques during the last 10–15 years has enabled us to demonstrate (1–4) that various forms of adrenergic innervation of the main blood vessels in mammals and birds may be classified into the four following patterns.

1. The adrenergic plexus of vasomotor nerves is confined to the adventitia-medial border (the majority of blood vessels in rat, rabbit, cat, pigeon, hen).

2. Single adrenergic fibers pass into the media from the plexus in the adventitia-medial border (pulmonary artery in rabbit, femoral vein in cat, splenic artery in pigeon).

3. In addition to the main plexus, nerve bundles penetrate into the media. With decreasing size of the blood vessel and decreasing thickness of the media, deep adrenergic fibers disappear (e.g., in smaller branches of large coronary arteries in pigs).

4. Adrenergic fibers form two plexuses, each related to the longitudinal and circular muscle layers (portal vein in rat, mesenteric arteries, and some of the veins in birds).

Adrenergic innervation of mammalian veins is poorer than that of arteries. Venous innervation is significantly richer in birds than in mammals.

It is generally accepted now that motor transmission to the vascular wall occurs via the terminal effector plexus close to the outer surface of the media (5,6). This plexus includes nerve bundles formed by one to three adrenergic axons with varicosities which lack Schwann cell processes (5). It is also assumed that the mediator is released from axonal varicosities located at constant distances from the surface of smooth muscle cells and that the nerve-muscle clefts (different for different vessels) remain constant (5,7). Based on electron-microscopic measurements of the width of nerve–muscle clefts and assuming that they remain constant, some authors have estimated the effective concentration of the sympathetic mediator in these clefts (6,8). It should be noted that in these calculations the functional condition of the vascular wall was not taken into account.

In view of these considerations, we decided to investigate spatial nerve-muscular relations in the wall of constricted and dilated blood vessels. Experiments were carried out on the auricular artery of rabbit, since this vessel

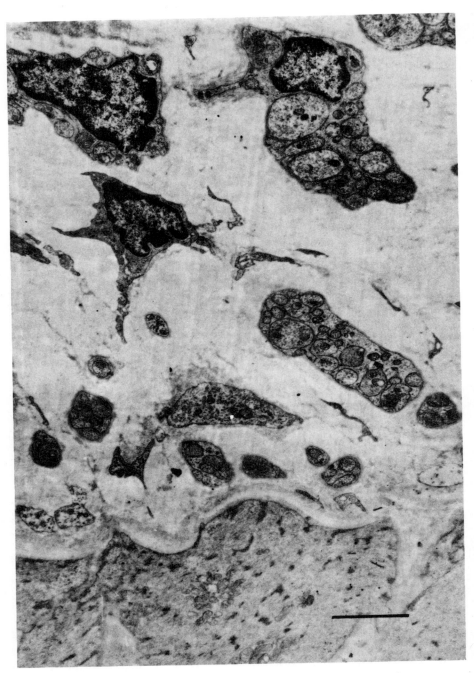

FIG. 1. Constricted auricular artery, transverse section. Bundles of unmyelinated fibers are distributed within a wide adventitial zone at large distance from smooth muscle cells.

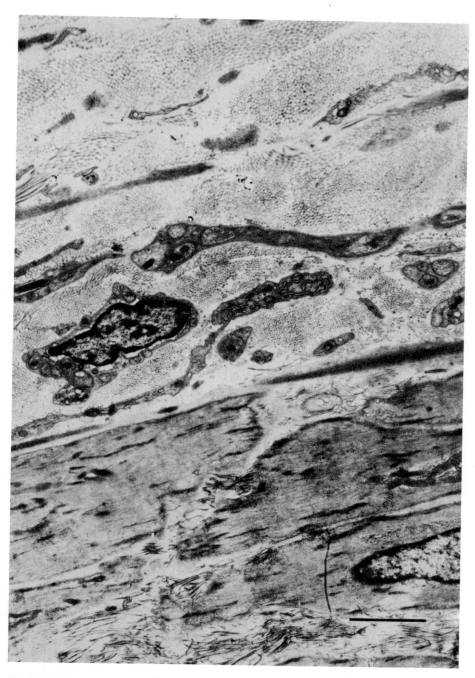

FIG. 2. Dilated auricular artery. All bundles are shifted to the outer surface of the muscle coat. Calibration: 1 μm.

has a rich adrenergic innervation. After a perfusion of the isolated ears with Krebs solution, one of the arteries was fixed in a 5% glutaraldehyde solution in the constricted state (eight specimens at low perfusion pressure), whereas the other was fixed in the dilated state (six specimens) at the 80–100 mmHg. The final fixation was achieved by OsO_4.

Under these conditions, it was rather difficult to reveal vesicles with a dense core in varicosities. However, the electron micrographs showed differences in spatial distribution of the nerve fibers within constricted and dilated arteries (Figs. 1 and 2; Table 1).

In the constricted vessels, single axons and bundles of axons were distributed within a wide zone of the adventitia over a thickness of about 7.0 μm. Whereas in the constricted vessel half of the nerve fibers were found in a 4.6 μm thick adventitial layer, nerve fibers occupy a zone only 1.8 μm thick in the dilated vessel. Usually, in bundles of axons and also in single axons, the varicosities were partially covered by Schwann cells. Often in the dilated vessels, the nerve bundles formed rather close contacts with muscle cells. In Fig. 3 this distance is about 2,000 Å. Single axons with varicosities might be observed at different distances from smooth muscle cells (Table 1). Only

TABLE 1. *Spatial distribution of the nerve fibers in constricted and dilated auricular arteries*

Distance from muscle coat (μm)	Constricted fibers (%)		Dilated fibers (%)	
	Bundles of axons	Single axons	Bundles of axons	Single axons
0–1	8.9	2.2	23.2	7.0
1–2	11.1	1.1	15.1	8.1
2–3	15.6	—	12.8	3.5
3–4	7.8	—	10.5	—
4–5	5.6	—	8.1	—
5–6	6.7	—	8.1	—
6–7	3.3	1.1	3.5	—
7–8	4.4	—	—	—
8–9	5.6	—	—	—
9–10	—	—	—	—
10–11	3.3	—	—	—
11–12	2.2	1.1	—	—
12–13	3.3	—	—	—
13–14	2.2	—	—	—
14–15	2.2	—	—	—
15–16	3.3	—	—	—
16–17	2.2	1.1	—	—
17–18	1.1	—	—	—
18–19	—	—	—	—
19–20	2.2	—	—	—
20–21	2.2	—	—	—
Total:	90		86	

Mean data for three constricted and four dilated vessels.

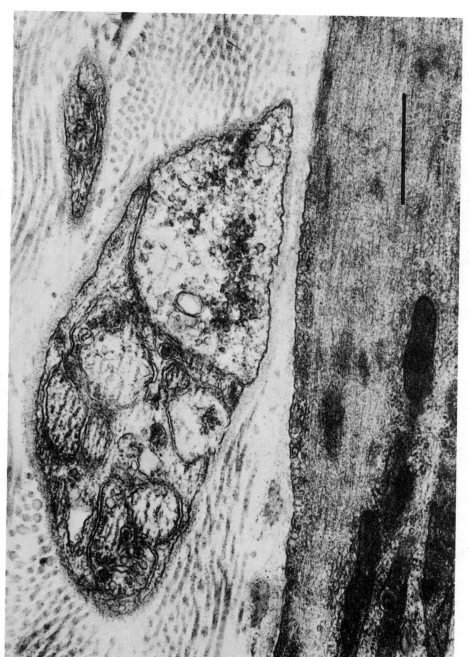

FIG. 3. Dilated auricular artery. Contact zone between a varicosity of fiber bundle and a muscle smooth cell.

in two cases, were they found in the constricted vessel at a distance of 150–400 Å from a muscle cell (Fig. 4).

Analysis of the distribution of nerve fibers in the vascular wall does not allow identification of a terminal effector plexus (5), although the relative number of single fibers somewhat increases in the proximity of the muscle coat. The presence of synaptic vesicles in the single fibers as well as in fiber

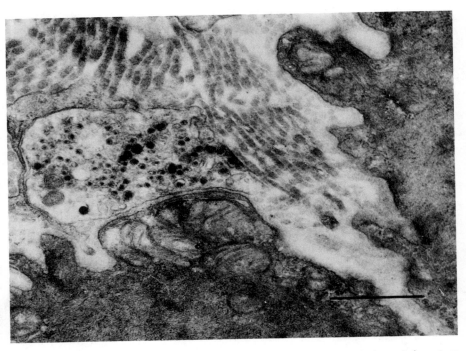

FIG. 4. Occasional close contact between a varicosity and muscle cell surface in the auricular artery. Calibration: 1 μm.

bundles suggests that the sympathetic mediator in the vascular wall is released by the whole net of adrenergic fibers.

Therefore, these observations indicate that, due to mechanical factors, spatial nerve–muscular relations in vascular walls are not constant, but change with constriction or dilatation of the vessel. Probably, blood vessels lack synaptic connections between varicosities of the adrenergic axons and smooth muscle cells, the width of a neuromuscular cleft varying over a wide range, so that mediator concentrations may become effective or ineffective. In constricted ear arteries the neuromuscular distances were significantly greater than in dilated arteries. At the same time, the muscle coat in the dilated vessel was thinner. These findings led to the conclusion that adrenergic nerves may have a greater influence on dilated than on constricted arteries. This mechanism may be involved in regulation of vascular tone, especially in those vessels which undergo significant changes of their lumen (e.g., veins and small arteries).

REFERENCES

1. Govyrin, V. A. (1965): *Dokl. Akad. Nauk SSSR,* 160:1179–1181.
2. Govyrin, V. A., and Leont'eva, G. R. (1971): *Zh. Evol. Biokhim. Fiziol.,* 7:145–149.
3. Bukinich, A. D. (1973): *Zh. Evol. Biokhim. Fiziol.,* 9:211–214.
4. Govyrin, V. A., and Bukinich, A. D. (1974): *Arkh. Anat.,* 67:30–37.
5. Verity, M. A. (1971): In: *Physiology and Pharmacology of Vascular Neuroeffector Systems,* edited by J. A. Bevan, R. F. Furchgott, R. A. Maxwell, and A. R. Somlyo, pp. 2–12. S. Karger, Basel.
6. Bevan, J. A. (1973), *Annu. Rev. Pharmacol.,* 13:269–285.
7. Su, Ch., and Bevan, J. A. (1971): In: *Physiology and Pharmacology of Vascular Neuroeffector Systems,* edited by J. A. Bevan, R. F. Furchgott, R. A. Maxwell, and A. P. Somlyo, pp. 13–21. S. Karger, Basel.
8. Ljung, B., and Wennergren, G. (1972): *Acta Physiol. Scand.,* 85:289–296.
9. Govyrin, V. A., and Khorkov, A. D. (1975), *Zh. Evol. Biokhim. Fiziol. (In press).*

Physiology of Smooth Muscle, edited by
E. Bülbring and M. F. Shuba.
Raven Press, New York © 1976.

Electrophysiological and Pharmacological Properties of Synaptic Processes in Smooth Muscle Cells of Stomach

I. A. Vladimirova

Department of Nerve—Muscle Physiology, The A. A. Bogomoletz Institute of Physiology, Kiev, USSR

Many experimental data concerning the synaptic processes in intestinal smooth muscle cells were obtained in recent years (1–6). As to the synaptic transmission of excitation and inhibition in smooth muscle cells of the stomach, there exist only a few studies (7,8). The aim of the present work was to investigate the postsynaptic potentials of smooth muscle cells in the minor curvature of the stomach. It was supposed that intramural stimulation combined with various blocking agents would make it possible to study in detail the excitatory and inhibitory influences on the smooth muscle cells in the stomach wall.

METHODS

Guinea pigs weighing 200 to 300 g were used. Muscle strips were excised from the minor curvature of the stomach. One of the stimulating electrodes was placed in Krebs solution; the other was attached to the end of the muscle strip, isolated from the Krebs solution by oil. Intramural stimulation of the nerve structures was produced by 0.2 to 0.5 msec pulses. The potentials were recorded intracellularly by means of glass electrodes with a resistance of 30 MΩ or more. Simultaneously, the contractile activity of the strip was recorded by a mechanotransducer.

RESULTS

Our experiments have shown that the smooth muscle cells in different parts of the minor curvature are subjected to various synaptic influences. In muscle strips excised from the dorsal part of the minor curvature, single stimuli evoked excitatory postsynaptic potentials (EPSPs) in the majority of cells, whereas in the ventral part they evoked inhibitory postsynaptic potentials (IPSPs). Combined responses like EPSP–IPSP were also observed. The amplitude of both EPSP and IPSP was enhanced when the intensity of stimulation was increased. The mean latency for EPSP and IPSP was 150 msec at the maximum stimulation, the duration of EPSP was 500 to 600 msec and that of IPSP was 900 to 1500 msec. The amplitude of the postsynaptic potentials was 5 to 15 mV. EPSPs reached their maximum after 150 to 200 msec and IPSPs after 200 to 300 msec.

Atropine abolished the EPSP; and the same cell which had responded with EPSP now began to produce only IPSP in response to stimulation. In young animals (1.5 to 2 months old) the EPSP of stomach muscle cells usually fail to initiate an action potential even with maximal intramural stimulation (6).

Paired stimulation with short interstimulus intervals caused some increase in the amplitude of the total EPSP. (Fig. 1A, 2–4). The second EPSP was

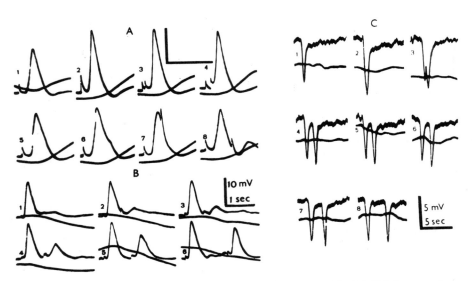

FIG. 1. EPSP and IPSP evoked by paired stimuli. A and B: EPSP, C: IPSP. A(1), B(1), C(1): Postsynaptic potentials evoked by single stimuli. A (2–8): EPSP evoked by paired stimuli with interstimulus intervals of 10, 60, 120, 200, 230, 350, and 560 msec. B (2–6): Interstimulus intervals of 600, 700, 800, 1,000, and 1,500 msec. C (2–8): IPSP evoked by paired stimuli with intervals of 200, 600, 1,200, 1,500, 1,600, 2,200, and 3,200 msec.

fully suppressed when the interval between two stimuli was 300 to 400 msec (Fig. 1A, 7). Further increase of the interval between stimuli (500 msec and more) (Fig. 1A, 8; 1B, 2–6) resulted in a gradual increase in amplitude of the second EPSP which had a long latency and a smaller amplitude than the first (Fig. 1B). The amplitude and latency of the second EPSP was restored when the interstimulus interval was 1,000 to 1,500 msec or more. As one can see in Fig. 1B, there could be a discrepancy between the electrical and contractile responses of the muscle strip: the strip relaxed despite the development of EPSP. It shows perhaps that IPSPs developed in neighboring smooth muscle cells.

As distinct from the EPSP, the IPSP showed facilitation in response to paired stimuli: Summation of IPSPs was observed at short intervals (Fig.

1C, 2–3) and potentiation of the second IPSP was seen if the interval after the preceding stimulus was long (Fig. 1C, 4–8).

Repetitive stimulation with a frequency of 1 stim/sec evoked an EPSP to each stimulus (Fig. 2,A,1). The amplitude of the second and subsequent EPSPs decreased considerably with a frequency of 2 stim/sec. Stimulation with a frequency of 5 to 10 stim/sec evoked no response at all by the postsynaptic membrane. The cell generated a single EPSP but with increased amplitude (Fig. 2A, 4–6).

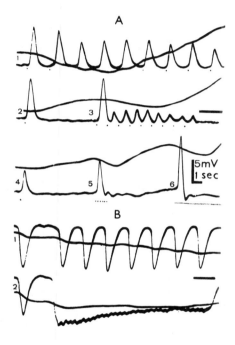

FIG. 2. EPSP and IPSP dependence upon the stimulation frequency. A (2, 4): EPSP evoked by single stimuli. A (1, 3, 5, 6): Stimulation with frequencies of 1, 2, 5, and 10 stim/sec, respectively. B (1): IPSP evoked by stimulation with 1 stim/sec; B (2): frequency changed to 5 stim/sec.

The amplitude of successive IPSPs clearly increased at frequencies of 1 to 2 stim/sec, and at 5 stim/sec IPSPs merged into a continuous hyperpolarization which, having attained a certain level, only slightly declined during stimulation (Fig. 2B).

Adrenergic blocking agents (phentolamine and propranolol) in a concentration of 1×10^{-6} g/ml had no effect on either EPSP or IPSP and they did not reduce the EPSP depression during rhythmic stimulation. Larger concentrations of these substances (5×10^{-6} to 5×10^{-5} g/ml) also had no effect on this depression but gradually reduced the EPSP. In the presence of these substances a large portion of cells generated IPSP in response to single stimuli. If a cell generated initially an IPSP, then, under the influence of adrenergic blocking agents in high concentration, the amplitude of the IPSP increased to twice its initial size or more (Fig. 3A, 1–5). IPSP summation in

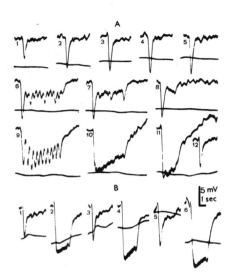

FIG. 3. Propranolol effect on IPSP of smooth muscle cells. A (1): IPSP in Krebs solution; A (2–5): IPSP recorded 2, 4, 6, and 15 min after propranolol (2.5 × 10⁻⁵ g/ml) addition. A (6–8): IPSP at stimulation 2, 5, and 10 stim/sec 20 min after the addition of propranolol. A (9–11): Stimulation with frequencies of 2, 5, and 10 stim/sec during washing the preparation by Krebs solution. A (12): IPSP in response to single stimulus. B (1, 3, 5): IPSP recorded in response to single stimulus; B (2, 4, 6): stimulation with frequency of 10 stim/sec; B (1, 2): synaptic potentials in Krebs solution; B (3, 4): at the 5th, B (5, 6): at the 15th minute after addition of propranolol (2.5 × 10⁻⁵ g/ml).

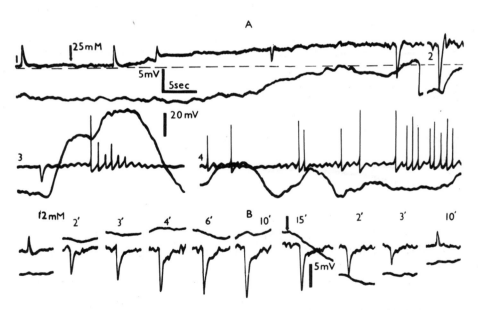

FIG. 4. The influence of imidazole. A (1): EPSP in normal Krebs solution and the changes after addition of imidazole to the solution (25 mM) at arrow. A (2, 3): Synaptic potentials at the 2nd and 10th minute of imidazole action. A (4): Appearance of spontaneous activity at the 10th minute of imidazole action. B: Changes of the synaptic potentials under imidazole influence in a concentration of 12 mM, at 2, 3, 4, 6, 10, and 15 min of its action. B (at arrow): Washing out with Krebs solution.

response to rhythmic stimulation with frequencies of 2, 5, and 10 stim/sec decreased after 20 min exposure to propranolol (Fig. 3A, 6–8), although the amplitude of initial IPSP was considerably larger than that of single IPSP in Krebs solution. As the preparation was washed with normal solution, the IPSP summation recovered and the amplitude of single IPSP decreased (Fig. 3A, 9–12). During rhythmic stimulation, the increase in the amplitude both of single IPSP and of the hyperpolarization due to summation was always noted with phentolamine, but only sometimes with propranolol (Fig. 3B, 1–6). High concentrations of these substances (5×10^{-5} to 5×10^{-4} g/ml) led to reduction and complete disappearance of IPSP. Propranolol did not cause the noticeable changes in the resting membrane potential and phentolamine evoked a slight depolarization of the muscle cell membrane.

Imidazole, the suspected blocker of the inhibitory action of ATP, had no facilitatory effect on the transmission of excitation in concentrations of 50, 25, and 12 mM. Imidazole, 25 mM, caused a slight depolarization of the membrane and in this condition an EPSP turned into an EPSP (Fig. 4A).

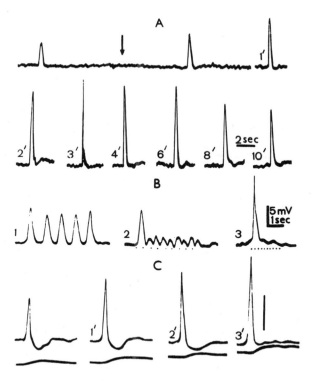

FIG. 5. Changes in EPSP under the influence of hexamethonium. A (arrow): Addition of hexamethonium (1×10^{-4} g/ml). EPSP at 1, 2, 3, 4, 6, 8, and 10 min of hexamethonium action. B (1–3): EPSP with stimulation frequency of 1, 2, 5 stim/sec, respectively, at the 25th minute of hexamethonium action. C: Mixed EPSP–IPSP response. IPSP decreasing under hexamethonium influence at 1, 2, 3 min of its action.

Then the membrane potential became stable but the IPSP amplitude usually continued to increase (Fig. 4A, 2). At the 10th minute of the imidazole action, the amplitude of the IPSP was very great, it was followed by rebound excitation and spontaneous activity was observed (Fig. 4A, 3–4). EPSPs turned into IPSPs also at lower concentrations of imidazole (Fig. 4B). All these changes were completely reversible.

To exclude the possible effects of stimulation of preganglionic nerve fibers and axons of sensitive cells we used hexamethonium in a concentration of 1×10^{-4} g/ml which selectively blocks N-cholinergic transmission in the ganglia. Hexamethonium brought about a gradual increase in the EPSP amplitude which reached the maximum after 3 to 5 min. (Fig. 5A), after which the EPSP amplitude gradually decreased and became stable after 10 to 15 min in spite of the continued presence of hexamethonium. A weakening depression of EPSPs during repeated stimulation was not observed under the influence of hexamethonium (Fig. 5B, 1–3). When a cell generated a complex response like EPSP–IPSP, it was clearly seen that the increase in the EPSP amplitude took place simultaneously with the decrease in the IPSP amplitude (Fig. 5C).

The amplitude of the inhibitory potential was reduced by hexamethonium even when atropine was present in the solution. Such a decrease reached its

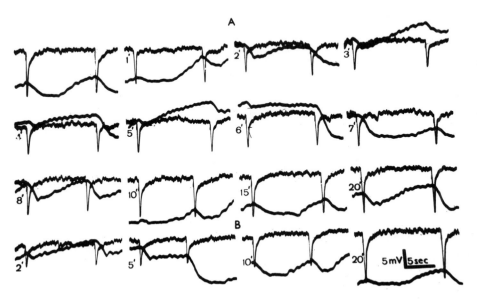

FIG. 6. Effect of hexamethonium on IPSP. A: Two IPSPs recorded in Krebs solution before and after adding atropine and hexamethonium (1×10^{-4} g/ml). The numbers indicate time of exposure. B: IPSP changes in the same cell at the 2, 5, 10, and 20 min during washing with Krebs solution.

maximum during 3 to 5 min and then the IPSP amplitude gradually recovered (Fig. 6A). When the preparation was washed with Krebs solution, a similar decrease in the IPSP amplitude with subsequent recovery to the initial size took place (Fig. 6B).

DISCUSSION

The experiments described have shown that the smooth muscle cells of stomach, as those of other viscera, are under the influence of excitatory and inhibitory nerve fibers (9–13). Interaction between excitatory and inhibitory (nonadrenergic) synaptic influences upon the smooth muscle cells is different in various parts of the minor curvature. One can believe that such irregular distribution of nerve fibers occurs also in other parts of the stomach. Whether depolarization or hyperpolarization is recorded depends on the predominance of excitatory or inhibitory innervation. The most convincing evidence for a mixed excitatory and inhibitory innervation of the same cell is the observation that, in the presence of atropine, EPSPs turn into IPSPs. Such simultaneous initiation of EPSP and IPSP caused by intramural stimulation appears to prevent EPSPs initiating action potentials in smooth muscle cells of stomach in young animals. This suggestion is supported by the fact that, when IPSPs are decreased by hexamethonium, EPSPs turn into action potentials.

The cause of EPSP depression evoked by frequent stimulation is a very complicated question. Such a depression could be evoked by simultaneous stimulation of excitatory and inhibitory nerve fibers (adrenergic and nonadrenergic). But EPSP depression is not apparently connected with adrenergic inhibition because adreno-blocking agents in a concentration of 1×10^{-6} g/ml did not abolish it.

Our experiments do not allow any conclusion about the possible role of the inhibitory nonadrenergic innervation in the EPSP depression. According to the hypothesis of Burnstock (14) that ATP may be the transmitter of the nonadrenergic IPSPs, we attempted to block them by imidazole. Since imidazole is known to abolish the inhibitory effect of low ATP concentrations and not to prevent its action in high concentrations we hoped to notice at least a slight decrease of the inhibitory influence on the smooth muscle cells (15–17). In our experiments imidazole not only did not decrease IPSPs but, on the contrary, contributed to their appearance. Further investigation will only be possible when a specific blocking agent of the nonadrenergic IPSPs will be found.

Analysis of the results obtained with hexamethonium shows that there are complex interneuronal connections within the intramural structures. The biphasic action of hexamethonium seems to indicate two sites of action, but this problem requires further investigation.

REFERENCES

1. Bennett, M. R. (1966): *J. Physiol. (Lond.)*, 185:124–131.
2. Bennett, M. R. (1966): *J. Physiol. (Lond.)*, 185:132–147.
3. Bülbring, E., and Tomita, T. (1967): *J. Physiol. (Lond.)*, 189:299–316.
4. Kuriyama, H., Osa, T., and Todia, N. (1967): *J. Physiol. (Lond.)*, 191:257–270.
5. Furness, J. B. (1969): *J. Physiol. (Lond.)*, 205:549–562.
6. Vladimirova, I. A., and Shuba, M. F. (1970): *Neurofiziologia*, 2:541–551.
7. Beani, L., Bianchi, C., and Crema, A. (1971): *J. Physiol.*, 217:259–279.
8. Atanasova, E., Vladimirova, I., and Shuba, N. (1972): *Neurofiziologia*, 4:216–226.
9. Orlov, R. (1962): *Fiziol. Zh. SSSR*, 48:342–348.
10. Burnstock, G., and Holman, M. (1961): *J. Physiol. (Lond.)*, 160:446–460.
11. Gillespie, J. (1962): *J. Physiol. (Lond.)*, 162:76–92.
12. Speden, R. (1964): *Nature*, 202:193–194.
13. Burnstock, G., Campbell, G., and Rand, M. (1966): *J. Physiol. (Lond.)*, 182:504–526.
14. Burnstock, G., Campbell, G., Satchell, D., and Smythe, A. (1970): *Br. J. Pharmacol.*, 40:668–688.
15. Bueding, E., Bülbring, E., Gercken, G., Hawkins, J., and Kuriyama, H. (1967): *J. Physiol. (Lond.)*, 193:187–212.
16. Rikimaru, A., Fukushi, Y., and Suzuki, T. (1971): *Tohoku J. Exp. Med.*, 105:199–200.
17. Tomita, T., and Watanabe, H. (1973): *J. Physiol. (Lond.)*, 231:167–177.

Physiology of Smooth Muscle, edited by
E. Bülbring and M. F. Shuba.
Raven Press, New York © 1976.

Pharmacological and Electrophysiological Properties of the Anococcygeus Muscles

Kate E. Creed and J. S. Gillespie

Department of Pharmacology, The University, Glasgow G12 8QQ, Scotland

Rat anococcygeus muscle has recently been introduced as a smooth muscle preparation (1). It is a narrow band of muscle, comparatively free of connective tissue and can be easily removed from the animal. The muscle is closely associated with the alimentary canal anatomically but differs from it in having no rhythmic activity and responding with a powerful contraction to norepinephrine. In view of the controversy over the role of adrenergic nerves in the vas deferens (2,3) it is of interest to study other muscles that respond to norepinephrine. We have investigated the pharmacology and electrophysiology of the anococcygeus of the rat and also of the rabbit.

THE RAT ANOCOCCYGEUS MUSCLE

The anococcygeus muscles are paired smooth muscles arising in the midline immediately behind the colon by a true tendinous origin from the coccygeal vertebrae. The muscles extend caudally round the colon to unite in front as a ventral bar. In the rat, the muscles consist of parallel bundles of two to eight smooth muscle cells which have an irregular outline because of numerous caveolae and projections. The extrinsic nerves lie in the perineal and genitofemoral nerves, and enter the muscle near the colon. Within the muscle, innervation consists of unmyelinated axons running in Schwann cell sheaths with usually one to five axons in each. These axons have intermittent varicosities containing small vesicles. Tissues preincubated with 5-hydroxy-dopamine and fixed with $KMnO_4$ for the electron microscope show that many of these have a dense core typical of adrenergic nerves. At intervals the Schwann cell withdraws from the varicosities to leave regions of presumed functional innervation of the muscle. The distance between these bare varicosities and the smooth muscle cells is large, averaging 2,600 Å, and is never less than 550 Å (4).

The anococcygeus muscle of the rat lacks tone or spontaneous activity *in vitro* but responds to field stimulation or to drugs. It is caused to contract by norepinephrine, by acetylcholine, and by 5-hydroxytryptamine (5-HT) (Fig. 1) and the first two effects are blocked by phentolamine and atropine, respectively, suggesting the presence of α-adrenergic and muscarinic receptors

(1). The muscle is unaffected by histamine, and isoprenaline only produces contraction at high concentrations. The effects of stimulation of extrinsic nerves or field stimulation are identical. Field stimulation at all frequencies causes only contraction, an effect abolished by phentolamine but unaffected by atropine or neostigmine, and therefore presumably due to stimulation of adrenergic motor nerves.

In addition to the excitatory innervation, the muscle has an inhibitory innervation. If the tone of the muscle is raised and particularly if the motor adrenergic nerves are blocked by guanethidine then a powerful inhibitory response can be obtained to field stimulation. The relaxation is caused by

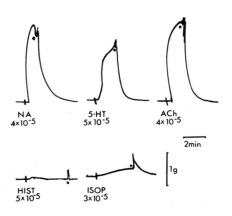

NA
4×10⁻⁵

5-HT
5×10⁻⁵

ACh
4×10⁻⁵

2min

HIST
5×10⁻⁵

ISOP
3×10⁻⁵

1 g

FIG. 1. Motor responses of the rat anococcygeus muscle to various drugs. Concentrations are molar. Norepinephrine (NA, noradrenaline) 5-hydroxytryptamine (5-HT) and acetylcholine (ACh) produced powerful contraction; histamine (HIST) was without effect and isoprenaline (ISOP) only produced contraction at high concentration. The drugs were washed out at the dots. Tension calibration for the first trace is 2 g; for the remaining traces, 1 g.

stimulation of inhibitory nerves, since it is abolished by tetrodotoxin and the origin of the nerves can be traced to the sacral region of the spinal cord (5). This response is mediated by an unknown transmitter since no drug so far used, including ATP, was found to cause relaxation or to specifically block the response to field stimulation (1).

THE ELECTRICAL BASIS OF CONTRACTION AND RELAXATION

Microelectrode recording showed that the smooth muscle cells have a mean resting potential of −60 mV and completely lack spontaneous electrical activity (6). Field stimulation with single pulses produces a small depolarization of up to 10 mV and, despite the absence of spikes, contraction occurs (Fig. 2). Repetitive stimulation produces summation of mechanical and electrical responses with the appearance of a second faster and larger component in both at about 3 Hz. The maximum electrical response can be produced with short bursts at 30 Hz when the transmembrane potential at the peak of the response is about −21 mV. Spike potentials are rare and never exceed 10 mV. Measurement of input resistance during the depolarization produced by stimulation at 30 Hz shows a fall which reaches a maximum during the rising phase, suggesting an increase in ionic conductance. When the membrane potential is displaced by passage of current between two extracellular electrodes (7), field stimulation produces responses, the peak of which remains at about −21 mV. This suggests that the ion or ions responsible have an equilibrium potential at this value (8).

The effect of stimulating inhibitory nerves can be demonstrated after blocking motor nerves with guanethidine at a concentration of 3×10^{-5} M. This drug causes depolarization to −20 to −30 mV and increase in tone. Field stimulation under these conditions produces relaxation of the muscle but preliminary experiments suggested that this was not accompanied by a

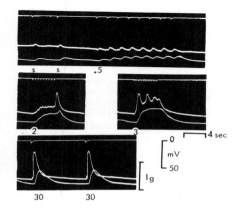

FIG. 2. The mechanical responses of the rat anococcygeus muscle (*bottom trace*) and the electrical response of a single cell (*middle trace*) to single pulses and stimulation at frequencies of 0.5, 2, 3, and 30 Hz. (*Top trace*) Zero potential; shows the stimulus. At all frequencies a depolarization occurs which is associated with contraction and is maximal with short bursts of 30 Hz.

consistent change in membrane potential, although any oscillation in membrane potential was damped (6). Measurement of input resistance failed to show any change in membrane conductance during inhibition.

THE RABBIT ANOCOCCYGEUS MUSCLE

Recently we have turned our attention to the rabbit and have found some important pharmacological and electrophysiological differences between the anococcygeus of this species and that of the rat. As well as norepinephrine and 5-HT, histamine also produces a powerful contraction (Fig. 3) that is antagonized by mepyramine. Acetylcholine has no excitatory action but in some preparations produces relaxation. Isoprenaline always produces a powerful relaxation which is antagonized by propranolol indicating the existance of β-receptors.

Histologic examination of the anococcygeus with the Falck technique indicated that the density of adrenergic innervation is considerably less in the rabbit. It also differs from the rat in the rhythmic appearance of spontaneous tone. Correlated with the scarcity of adrenergic nerves and the existence of spontaneous tone, field stimulation produces varied responses. In general,

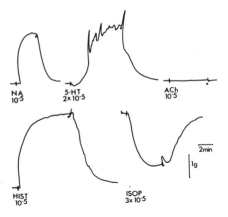

FIG. 3. Motor responses of the rabbit anococcygeus muscle to various drugs (molar concentrations). Before the application of isoprenaline the tone of the muscle had been raised with histamine (2×10^{-6} M). No spontaneous activity was present in these preparations. In this species, histamine, but not acetylcholine, produced contraction; isoprenaline produced relaxation.

however, stimulation at low frequencies (up to 8 Hz) results in relaxation, with higher frequencies producing contraction followed by a mixture of contraction and relaxation. At the end of stimulation, irrespective of frequency, a rebound contraction is normally observed.

The rabbit anococcygeus muscle has a lower resting potential than that of the rat even in the absence of spontaneous tone (-48 mV compared with -60 mV). There is often intermittent spontaneous electrical activity consisting of brief depolarizations with spike potentials or prolonged steady depolarizations. Field stimulation with single pulses (Fig. 4a) normally causes a hyperpolarization of up to 14 mV, which is followed by a rebound depolarization with spike and contraction. Rarely does a single stimulus directly cause depolarization and contraction. Low-frequency repetitive stimulation of 1 to 5 Hz results in summation of these individual hyperpolarizations (Fig. 4b,c) producing a maintained hyperpolarization of 5 to 23 mV. In the absence of tone this is not accompanied by a mechanical response. At higher frequencies, direct depolarization and contraction nor-

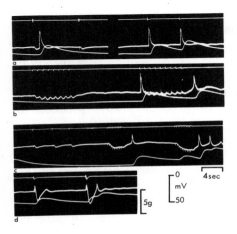

FIG. 4. Mechanical (*bottom*) and electrical responses (*middle*) of the rabbit anococcygeus to field stimulation. (a) Single pulses; (b) stimulation at 1 Hz; (c) stimulation with single pulses and at 3 Hz; (d) short bursts at 30 Hz. The *top* trace is zero potential and shows the stimulus. The upper two records (a, b) are from a different preparation from the lower two (c, d). In these muscles there was no resting tone and the only mechanical response was contraction following either the directly evoked or rebound depolarization. With field stimulation at low frequencies hyperpolarization predominated.

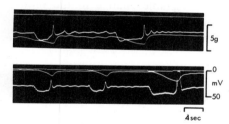

FIG. 5. Mechanical and electrical responses of the rabbit anococcygeus to field stimulation when the tone was raised. In the upper record the increase in tone was spontaneous and in the lower record was induced by guanethidine (4×10^{-5} M). The hyperpolarizing response to field stimulation was now associated with relaxation.

mally occur with or without subsequent hyperpolarization. The hyperpolarization does not accelerate the relaxation following contraction (Fig. 4d) but prevents further contraction. On ending stimulation there is usually a rebound depolarization at all frequencies usually proceeding to spike formation followed by oscillation in membrane potential which declines over several seconds. In muscles in which the tone is increased, either spontaneously or in the presence of guanethidine, the membrane is depolarized to as little as -20 mV, and a slight oscillation of the membrane potential is sometimes seen. Field stimulation in these circumstances again produces hyperpolarization but this is now associated with relaxation (Fig. 5). Guanethidine abolishes the direct depolarizing response to stimulation, but the hyperpolarization is still followed by rebound depolarization with restoration of tone.

Because of the clear evidence of hyperpolarization in response to field stimulation in the rabbit muscle, we have recently reexamined the response to stimulation of inhibitory nerves in the rat. Although in some cells inhibition was not accompanied by hyperpolarization, in others there was a small hyperpolarization (Fig. 6). Maximum relaxation occurred with frequencies of 5 to 8 Hz, but even at these frequencies the hyperpolarization, if present, was never more than 15 mV. At the end of stimulation large rhythmic fluctuations were sometimes seen which were associated with an increase in tone. In

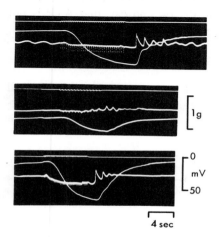

FIG. 6. Mechanical and electrical responses of the rat anococcygeus muscle in the presence of guanethidine (3×10^{-5} M). Field stimulation produced relaxation, which may be associated with abolition of oscillations of the membrane potential (*top record*), no change in membrane potential (*middle*) or hyperpolarization (*bottom*). Stimulation was followed by rebound depolarization.

neither the rat nor the rabbit was there a detectable change in the input resistance during the hyperpolarization. More direct methods are now being used to analyze the nature of the inhibitory response in the two species.

CONCLUSION

The results suggest that in the rat anococcygeus muscle, which has a dense adrenergic innervation, field stimulation produces contraction associated with graded depolarization of the muscle cell membrane but usually not with action potentials. The depolarization results from an increase in membrane permeability so that the membrane potential is displaced toward an equilibrium potential of about −21 mV (8). In the rabbit, adrenergic innervation is more sparse, and field stimulation only rarely produces contraction, which may be associated with spike potentials. The usual response is large hyperpolarizations and inhibition of any spontaneous tone. In the rat relaxation can only be seen if the tone is increased with drugs and it is accompanied by small or no changes in membrane potential.

Another interesting feature is the different responses to drugs. Although norepinephrine and 5-HT cause contraction in both species, acetylcholine is excitatory only in the rat, and histamine only in the rabbit. Isoprenaline relaxes the muscle of the rabbit but not that of the rat. In the cat both acetylcholine and isoprenaline relax the muscle (9). It is difficult to interpret these differences in response to field stimulation and to drugs in a muscle, the function of which is not known. Clearly, however, in all species the excitatory nerves are adrenergic.

The putative inhibitory transmitter is still unknown. In view of the anatomical association between the anococcygeus and the colon, and the similarity in the rabbit between the electrical and mechanical responses of the ileum (10) and the anococcygeus to stimulation of nonadrenergic inhibitory nerves, it is tempting to believe that the nerves in the two tissues are the same. However, there is clear evidence at least in the rat and cat, that ATP is not the inhibitory transmitter in the anococcygeus (1,9).

ACKNOWLEDGMENTS

This work was carried out with financial assistance from the Medical Research Council and from the Medical Research Funds of Glasgow University.

REFERENCES

1. Gillespie, J. S. (1972): *Br. J. Pharmacol.*, 45:404–416.
2. Ambache, N., Dunk, L. P., Verney, J., and Zar, M. A. (1972): *J. Physiol.*, 227:433–456.
3. Furness, J. B. (1974): *Br. J. Pharmacol.*, 50:63–68.
4. Gillespie, J. S., and Lüllmann-Rauch, R. (1974): *Cell Tiss. Res.*, 149:91–104.

5. Gillespie, J. S., and McGrath, J. C. (1973): *J. Physiol.*, 230:659–672.
6. Creed, K. E., Gillespie, J. S., and Muir, T. C. (1975): *J. Physiol.*, 245:33–47.
7. Abe, Y., and Tomita, T. (1968): *J. Physiol.*, 196:87–100.
8. Creed, K. E. (1975): *J. Physiol.*, 245:49–62.
9. Gillespie, J. S., and McGrath, J. C. (1974): *Br. J. Pharmacol.*, 50:109–118.
10. Furness, J. B. (1969): *J. Physiol.*, 205:549–562.

Physiology of Smooth Muscle, edited by
E. Bülbring and M. F. Shuba.
Raven Press, New York © 1976.

The Soma Spike in Myenteric Plexus Neurons with a Calcium-Dependent After-Hyperpolarization

R. A. North and S. Nishi

*Department of Pharmacology, University Medical Buildings, Foresterhill, Aberdeen, AB9 2ZD,
Scotland and Department of Pharmacology, Loyola University Medical Center, Maywood, Illinois
60153*

Intracellular recording from the myenteric plexus of the guinea pig intestine has revealed at least three cell types (1,2). Between two-thirds and three-quarters of the cells are inexcitable and are presumed to be glial cells. More than half the neurons have properties characteristic of autonomic ganglion cells and can be excited synaptically and "antidromically" by close focal stimulation. Less than half the neurons are of a second type, with somewhat different membrane properties and showing a slow after-hyperpolarization following a single action potential (1–5). This paper reports and reviews some properties of this second type of neuron.

METHODS

Intracellular recordings were made from single cells of the isolated myenteric plexus–longitudinal muscle preparation of the guinea pig ileum. Full details of the methods of making the preparation, composition of the bathing solutions, immobilization of single ganglia, intracellular recording, focal stimulation, and acetylcholine (ACh) iontophoresis have been described (1). During penetration, anodal pulses at 10 Hz were passed which enable the increased RC load to be seen at once.

RESULTS

General Properties

Stable recordings have been made from almost 500 myenteric plexus neurons. Approximately 40% of these were type 2 cells (1); their defining property was a marked accommodation to depolarizing currents. Some of these cells could not be excited by passage of a depolarizing current through the recording microelectrode although at the time of impalement they gave an "off-spike" in response to repetitive anodal pulses (anodal break excitation) and they could be excited by focal stimulation of one of the cell processes which then conducted to the soma.

At the time of impalement, type 2 cells seldom gave more than one "off-spike" and they achieved their final steady resting membrane potential within a minute. By contrast, type 1 cells often gave a short burst of "off-spikes" at the time of penetration and took up to 15 min to acquire their steady resting membrane potential. Differences in the passive electrical properties of the two cell types have been described (1). Successive recordings from neighboring cells showed no anatomic organization of cell types within a ganglion; both type 1 and type 2 cells were sometimes found immediately adjacent to a cell of the same type.

Cell Geometry

Experiments in which the position of the stimulating microelectrode was progressively moved across the surface of the ganglion enabled the location of the cell processes to be determined up to a distance of 150 μm from the soma. These experiments demonstrated that type 2 cells are often bipolar, the soma giving rise to at least one process running in each direction along the long axis of the ganglion.

Slow After-Hyperpolarization

Almost all the type 2 cells showed a long-lasting after-hyperpolarization following a single soma action potential. This occurred whether the action potential was elicited by depolarization of the soma membrane by passing current through the recording microelectrode, or by focal excitation of a cell process which then conducted to the cell soma. Those cells which did not show the slow after-hyperpolarization tended to have relatively high resting membrane potentials and in these cells a slow after-hyperpolarization could be evoked by changing to a K^+-free solution. The slow after-hyperpolarization was associated with a decreased membrane resistance; it was increased in amplitude and duration by K^+-free solutions and much reduced by solutions containing 16 mM K^+. It was not affected by Cl^--free solutions. The amplitude of the slow after-hyperpolarization was increased by membrane depolarization and decreased by membrane hyperpolarization; progressive membrane hyperpolarization nullified the slow after-hyperpolarization and eventually reversed it (1,5). The membrane potential at which the reversal occurred was approximately -90 mV.

The slow after-hyperpolarization was reversibly abolished by Ca^{2+}-free solutions and by the addition of $LaCl_3$ (1 mM) or $MnCl_2$ (1 mM) to the bathing solution: the action potential in the soma was not affected by these procedures other than those changes attributable to a fall in threshold and resting potential. When the slow after-hyperpolarization had been abolished by a Ca^{2+}-free solution, it could be restored to its original amplitude by a solution in which the $CaCl_2$ of the control solution had been replaced by a equimolar amount of $SrCl_2$. When Ca^{2+} was completely replaced by Sr^{2+}, the maximum rate of fall of the action potential was reduced and the slow after-hyperpolarization took longer to reach its maximum value. Mg^{2+} did not substitute for Ca^{2+} in mediating the slow after-hyperpolarization. Caffeine (1 and 10 mM) did not change the membrane potential or resistance of type 2 neurons.

Soma Action Potential

Tetrodotoxin (TTX) (200 nM) rapidly abolished the action potential in type 1 cells (3,4). In type 2 cells, the soma action potential evoked by conduction from a stimulated cell process was rapidly and reversibly abolished by TTX, although an action potential could still be elicited by passing a depolarizing current through the intracellular electrode. The action potential which persisted in the presence of TTX was reversibly abolished by Ca^{2+}-free solutions or by the addition of $CaCl_3$ (1mM) or $MnCl_2$ (1 mM); it was increased in amplitude by a solution containing 10 mM Ca^{2+}. The TTX-resistant action potential was followed by a slow after-hyperpolarization.

Exposure for more than 30 min to a Na^+-free solution [tris(hydroxymethyl)aminomethane substitution] did not affect the slow after-hyperpolarization. The soma action potential evoked by conduction from a focally stimulated cell process was quickly abolished by Na^+-free solutions. The action potential evoked by passing a depolarizing current through the intracellular electrode persisted in Na^+-free solutions in 35% of type 2 cells: in the 65% of cells in which it disappeared, it could be restored by changing to a Na^+-free solution, which contained 5 or 10 mM Ca^{2+} (control solution 2.5 mM Ca^{2+}). Further increase in Ca^{2+} concentration in Na^+-free solutions caused an increase in spike amplitude, but this proved difficult to quantify.

ACh Iontophoresis

Although ACh potentials could be recorded from type 1 cells (1), close iontophoretic application of ACh to type 2 cells never evoked any response. This finding was unlikely to be due to any technical reason; the same iontophoresis microelectrode was sometimes employed on both types of cell and its tip was placed upon the cell soma under visual control.

Synaptic Input

As has been described (1,2), synaptic responses could not be evoked in type 2 cells, although they were frequently observed in type 1 cells.

DISCUSSION

It seems likely that the type 2 cells as defined here are identical with the AH cells reported by Hirst and co-workers (2), although that group did not observe the different electrical properties of the two cell types. The absence of a synaptic response is an unsatisfactory criterion for classification when close focal stimulation is used; in some type 1 cells a synaptic response could only be obtained by trying many different positions of the stimulating micro-electrode. The failure to evoke any response by iontophoretic application of ACh is further evidence that the type 2 cells do not receive cholinergic synapses as do the type 1 cells.

One type 2 cell from which a long electrophysiologic recording was obtained was subsequently identified in the electron microscope (6). In more than 200 serial sections of the cell, no synaptic boutons were observed which made synaptic contact with the cell; numerous typical synaptic contacts could be seen on neighboring cells in other parts of the sections. The electrophysiologic finding that many of the type 2 cells are bipolar was confirmed by the intracellular injection of procion yellow (1,5,7) and this adds further weight to the suggestion that the type 2 cells may function as afferent neurons.

All the experimental evidence is compatible with the hypothesis (1,4) that, in normal solution, Na^+ carries almost all the current in both soma and processes. However, in the soma, the Ca^{2+} which does enter is essential to trigger the slow after-hyperpolarization. This slow after-hyperpolarization which follows a single soma spike is mediated by an increase in K^+ conductance brought about by the influx of Ca^{2+} during the action potential. The observation that the action potential is essentially unchanged by Ca^{2+} removal or by addition of Mn^{2+} or La^{3+} (3) suggests that in normal conditions only a small proportion of the inward current is carried by Ca^{2+}. However, in the absence of Na^+ or in the presence of TTX, the Ca^{2+} entry may be sufficient to produce an action potential in response to direct depolarization of the soma membrane by intracellular current injection.

Two observations suggest that the Ca^{2+} entry during the action potential is a property of the soma but not the processes of the type 2 neurons. First, when a cell process is stimulated at a distance from the soma by a just-threshold stimulus, an action potential is initiated which propagates to the soma. Throughout the course of the ensuing slow after-hyperpolarization, further stimuli continue to cause propagated spikes, although the soma membrane hyperpolarization may cause them to be fractionated (that is, they fail to invade the soma and are recorded as all-or-nothing potential changes from the proximal part of the process). If the slow after-hyperpolarization were also affecting the processes (other than by electrotonic spread), then a just-threshold stimulus would fail to excite the process, or else the action potential would be blocked before its arrival at the soma. Second, the observation that the soma action potential evoked by focal stimulation of a cell process was rapidly abolished by TTX or Na^+-free solutions (although the action potential could still be evoked by passing a depolarizing current through the recording microelectrode) suggests that the neuronal process does not become permeable to Ca^{2+} during depolarization. If it does become permeable, then the rate of depolarization by the Ca^{2+} current must be insufficient to cause a propagated spike.

The slow after-hyperpolarization provides a long-lasting depression of excitability of a substantial proportion of myenteric plexus neurons. Much evidence is suggestive that these neurons may be afferent; in that case, the input to the cell soma will sooner or later be cut off by the membrane hyperpolarization even though the sensory stimulus to the distal part of the cell process is continuing. Such a mechanism might contribute to the smooth gradation of inhibition and excitation which occurs during peristalsis.

ACKNOWLEDGMENTS

Supported in part by a M.R.C. (U.K.) Junior Research Fellowship and Wellcome Research Travel Grants (to R. A. N.) and by N.I.H. Research Grant NS06672 and N.S.F. Research Grant GB30360 (to S. N.).

REFERENCES

1. Nishi, S., and North, R. A. (1973): *J. Physiol. (Lond.)*, 231:471–491.
2. Hirst, G. D. S., Holman, M. E., and Spence, I. (1974): *J. Physiol. (Lond.)*, 236:303–326.
3. North, R. A. (1973): *Br. J. Pharmacol.*, 49:709–711.
4. Hirst, G. D. S., and Spence, I. (1973): *Nature (New Biol.)*, 243:54–56.
5. North, R. A., and Nishi, S. (1974): *Proc. Int. Symp. Gastrointestinal Motility, 4th*, pp. 667–676. Mitchell, Vancouver.
6. Gabella, G., and North, R. A. 1974): *J. Physiol. (Lond.)*, 240:28P–30P.
7. North, R. A. (1973): Ph.D. Thesis, University of Aberdeen, Aberdeen, Scotland.

Physiology of Smooth Muscle, edited by
E. Bülbring and M. F. Shuba.
Raven Press, New York © 1976.

Nervous Pathways Excited During Peristalsis

G. D. S. Hirst, Mollie E. Holman, and H. C. McKirdy

Department of Physiology, Monash University, Clayton, Victoria, Australia

Isolated segments of small intestine show an integrated response to appropriate stimuli (for example, radial distension), which involves temporally organized changes in the mechanical activity of both the longitudinal and circular layers of smooth muscle (Bayliss and Starling, 1899). In the guinea pig small intestine, radial distension gives rise to inhibitory junction potentials (IJPs) in the circular muscle which have been recorded up to 5 cm below the region of distension (descending inhibition). Excitatory junction potentials (EJPs) occur in both layers of smooth muscle below the region of distension, but only after a longer delay of 2 to 10 sec after the onset of distension (Hirst, Holman, and McKirdy, 1974).

In order to explain these observations and to clarify the neuronal basis of peristalsis in the small intestine of guinea pigs, intracellular recordings were made from the neurons of the myenteric and submucous plexuses. Two classes of neurons could be distinguished in the myenteric plexus; about one-third of these neurons did not appear to receive any synaptic input, whereas the remaining two-thirds responded to transmural stimulation of the myenteric plexus with excitatory synaptic potentials (ESPs) which were readily blocked by *d*-tubocurarine. Action potentials generated in the soma of those neurons that did not appear to have a synaptic input were characterized by a prolonged after-hyperpolarization which restricted the rate at which they could be made to fire repetitively to one or a short burst of action potentials separated by an interval of several seconds. These neurons have been referred to as "after-hyperpolarizing" cells (Hirst, Holman, and Spence, 1974) and are probably the same population of neurons studied in detail by North and Nishi (*this volume*). Both "after-hyperpolarizing" cells and those with synaptic input could show a response to transmural stimulation which was not blocked by curare and resembled that ascribed by previous workers to antidromic activation. We cannot rule out the possibility that some of these responses may have resulted from electrical coupling between cells.

In subsequent experiments segments of myenteric plexus were maintained in continuity with intact segments of small intestine which could be stimulated by distension or electrically, by transmural electrodes (Hirst and McKirdy, 1974). It was possible to record from myenteric neurons situated

2 cm or more from the site of stimulation (Hirst and McKirdy, 1974). The only responses observed were ESPs which were blocked by curare. Responses were rarely observed to occur above the site of distension or electrical stimulation. However, nearly all of the neurons situated up to 2.5 cm below the stimulated region (the longest distance studied so far) gave ESPs in response to either stimulus.

Two patterns of response were distinguishable. About half the neurons which gave ESPs in response to close transmural stimulation were excited by distension after a delay of less than 1.2 sec. This "short latency" response was transient and only occurred at the onset of distension. It appeared to be correlated with the initiation of IJPs in the circular smooth muscle (Hirst and McKirdy, 1974). The second group of neurons that responded to radial distension did so after a longer latency (up to 11 sec). These neurons continued to show ESPs throughout and after the period of distension. The EJPs recorded from both layers of smooth muscle in response to distension showed a similar long latency; occasionally more than one EJP was initiated by a maintained distension. These experiments also indicated that both the long latency discharge of ESPs in myenteric neurons and the long latency EJPs recorded for the two smooth muscle layers could be evoked by transient distension (1-sec duration) of the intestine. In this case the discharge of ESPs occurred several seconds after the stimulus had been removed. In contrast with the descending inhibition, descending excitation could not be recorded after removal of the submucosa.

Therefore we suggest that there may be two separate descending nervous pathways in the small intestine of guinea pigs which mediate descending inhibition and excitation of the smooth muscle; the descending excitatory pathway may involve both the myenteric and submucous plexuses.

The characteristics of the neurons of the myenteric plexus and their synaptic input did not seem to be adequate to explain their response to distension of a neighboring segment of gut or that recorded from the smooth muscle. For example, observations on myenteric neurons did not provide any explanation for the delayed response of some myenteric neurons to distension. We have therefore begun to study isolated segments of the submucous plexus. In contrast to the myenteric plexus, responses to transmural stimulation could only be recorded from neurons which were close to the transmural stimulating electrodes (within 4 mm). This finding suggests that the submucous plexus may not be involved in "long" ascending or descending pathways in the wall of the gut but that the role of this plexus may be confined to a more local integrative action. We have failed to record so far from any neuron in the submucous plexus whose properties resembled those of the "after-hyperpolarizing" cells of the myenteric plexus. Low-intensity transmural stimulation caused ESPs in many neurons which were blocked by curare. Stronger stimuli caused inhibitory synaptic potentials (ISPs) in some neurons which also received an excitatory synaptic input. These ISPs were of exceedingly long duration (up to 5 sec), they were associated with an increase in membrane conductance, probably for potassium ions, and were blocked by BOL and by guanethidine. We would suggest that these inhibitory synapses may lie on the descending excitatory pathway and that their activation may in part be responsible for the delay before the onset of descending excitation.

REFERENCES

Baylis, W. M., and Starling, E. H. (1899): The movements and innervation of the small intestine. *J. Physiol.* (*Lond.*), 24:99–143.

Hirst, G. D. S., Holman, Mollie E., and McKirdy, H. C. (1974): Two descending nerve pathways activated by distension of guinea-pig small intestine. *J. Physiol.* (*Lond.*), 244:113–127.

Hirst, G. D. S., and McKirdy, H. C. (1974): A nervous mechanism for descending inhibition in guinea-pig small intestine. *J. Physiol.* (*Lond.*), 238:129–144.

Physiology of Smooth Muscle, edited by
E. Bülbring and M. F. Shuba.
Raven Press, New York © 1976.

Control of Gastrointestinal Motility by Prevertebral Ganglia

J. H. Szurszewski and William A. Weems

Department of Physiology and Biophysics, Mayo Medical School, Rochester, Minnesota 55900

It is generally held that inhibition of gastrointestinal motility by post-ganglionic, noradrenergic neurons is under the control of preganglionic input from the central nervous system. In this short review, based on our observations and those of others, we propose that the inhibitory commands of these noradrenergic neurons are the integrated result of neural inputs not only from the central nervous system but also from sensory receptors located in various regions of the gastrointestinal system.

The cell bodies of the noradrenergic neurons of the sympathetic nervous system which control gastrointestinal motility are located in prevertebral ganglia (celiac, superior mesenteric, and inferior mesenteric) (1–3). Nor-epinephrine released from these neurons may produce inhibition of gastrointestinal motility by reducing the release of acetylcholine from cholinergic neurons in the enteric nervous system (4–6), by acting directly on the smooth muscle (3), or possibly by decreasing the blood flow through the system (7). Regardless of the mechanism underlying noradrenergic inhibition, it is caused by impulses in sympathetic, postganglionic neurons.

Since the classic studies of Langley (8) in 1921, these postganglionic neurons have often been considered to receive synaptic input only from preganglionic fibers which carry neural information from the central nervous system. Based on this view, sympathetic prevertebral ganglia are considered to function only as relay stations between the central nervous system and peripheral organs.

Since Langley's observations, however, considerable evidence has accumulated which suggests that neurons of prevertebral ganglia also receive synaptic input from neurons in the gastrointestinal tract which they innervate. Kuntz (9,10) found that decentralization of celiac and inferior mesenteric ganglia in the cat did not result in degeneration of all axons terminating in these ganglia, and sectioning the nerves between these ganglia and visceral organs did not cause degeneration of all fibers distal to the section. From these observations, he concluded that neurons in prevertebral ganglia receive synaptic input from neurons located in the enteric nervous system and that these neural connections may function as pathways for a peripheral reflex. McLennan and Pascoe (11) came to a similar conclusion when they found that decentralization, splanchnichotomy, bilateral sympathectomy, and bilateral vagotomy did not abolish C-fiber input to the rabbit inferior mesenteric ganglion. They concluded that slow conducting C fibers originate in the colon possibly as axons of cells lying in the enteric plexus.

Several physiologic studies have indicated that these peripheral pathways with afferent to efferent synapses in prevertebral ganglia are functional components of reflexes such as the intestinointestinal inhibitory reflex. Kuntz and co-workers (10,12,13) published several studies in which they found that distension of one segment of intestine reflexively inhibited another segment of intestine even after the associated prevertebral ganglia had been decentralized and, presumably, all the associated visceral afferent fibers issuing from the dorsal root ganglia had been cut. These observations were confirmed by Semba (14). Kock (15) found that cutting the splanchnic nerves did not abolish the intestinointestinal reflex, but that nerve ramifications along the mesenteric vascular branches must be left intact for the reflex to occur.

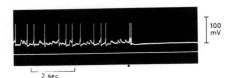

FIG. 1. Intracellular recording from a ganglion cell in the inferior mesenteric ganglion before, during, and after cutting the lumbar colonic nerve. The moment of cutting is indicated by the dot. (From ref. 17, with permission.)

Crowcroft and Szurszewski (16) recorded electrical activity of neurons in the inferior mesenteric ganglia of the guinea pig using intracellular techniques. In these *in vitro* preparations, they found that stimulation of inferior splanchnic nerves and colonic nerves—hitherto considered to be a postganglionic nerve trunk—produced excitatory postsynaptic potentials (EPSPs) in approximately 90% of the postganglionic neurons from which they recorded. Crowcroft et al. (17) recorded spontaneous EPSPs in these neurons when the inferior mesenteric ganglion remained attached to a segment of distal colon. This spontaneous input was immediately abolished when the colonic nerve connecting the colon segment to the inferior mesenteric ganglion was cut (Fig. 1). Synaptic input was increased by colonic distension. We recently found that this "spontaneous" synaptic input received by postganglionic neurons of the inferior mesenteric ganglion results from the activity of colonic mechanoreceptors which monitor the mechanical state of the colon (18,19). Relaxation of colonic smooth muscle by superfusion of the colon with smooth muscle relaxants decreased this input, while increased smooth muscle activity during superfusion with muscle stimulants increased the input. It is clear then from our work as well as from the work of others that prevertebral ganglia are involved in peripheral reflexes the afferent limb of which consists of axons of cholinergic neurons within the wall of the intestine. The postganglionic fibers of these ganglia form the efferent limb.

In addition to synaptic inputs from the central nervous system and from the segment of the gastrointestinal tract which is associated with a prevertebral ganglion, there is considerable neurophysiological evidence indicating that prevertebral, postganglionic neurons are innervated by peripheral nerves arising from other viscera or other segments of the gastrointestinal tract. For example, Job and Lundberg (2) found that electrical stimulation of the cat hypogastric nerve produced action potentials, recorded extracellularly, in the colonic nerve after section and degeneration of preganglionic nerves. Several others, see review by Skok (21), have demonstrated the existence of many such pathways in prevertebral ganglia which connect peripheral nerves with peripheral nerves. Some of these pathways are synaptically interrupted within these ganglia and are activated by stimulation of attached viscera. For example, we have obtained evidence (22) that indicates that postganglionic neurons in the inferior mesenteric ganglion receive sensory

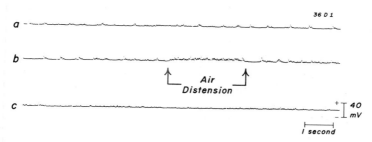

FIG. 2. Synaptic response of a neuron in the superior mesenteric ganglion of the guinea pig to distension of the distal colon. Record is continuous from a through c.

input from proximal regions of the gastrointestinal system and that the neurons in the superior mesenteric ganglion receive input from mechanoreceptors in the distal colon (Fig. 2). This interganglionic connection occurs via the intermesenteric nerve.

These briefly reviewed studies indicate that many postganglionic neurons of prevertebral ganglia receive neural input from both the central nervous system via preganglionic pathways and from viscera via afferent fibers. This raises the question: What is the physiologic significance of this peripheral input to noradrenergic control of visceral function? We claim no originality in posing this question, for others have already asked it, beginning with the work described by Kuntz and his co-workers. They found that the intestino-intestinal reflex was mediated by decentralized ganglia (Fig. 3). Their findings, however, are often considered to represent a pseudoreflex because the intraluminal pressure they needed to produce this reflex through decentralized ganglia was considerably greater than the physiological pressures required to elicit the reflex when spinal connections to the ganglia remained intact. For example, when intestinal segments are distended with physiological pressures, the intestinointestinal reflex is abolished by cutting dorsal roots D-7 to L-1 (23) or by producing spinal anesthesia (4,24). These observations are often

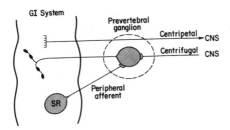

FIG. 3. Summary diagram of the neural circuitry which peripherally mediates the intestinointestinal reflex through prevertebral ganglia when centripetal (sensory) and centrifugal (preganglionic) fibers to the central nervous system have been cut. SR, sensory receptor of enteric neurons; CNS, central nervous system.

cited as proof that the intestinointestinal reflex is spinally mediated (Fig. 4). However, a major difficulty with the spinal reflex hypothesis has been the continued failure to unequivocally demonstrate that sympathetic preganglionic nerves (splanchnic nerves) constitute the efferent limb of the intestinointestinal reflex (7,15). Stimulation frequencies required to elicit prompt intestinal inhibition by direct stimulation of splanchnic nerves (8 to 10 impulses/sec) are above the observed physiological range of discharge rates for these nerve fibers, whereas stimulation of visceral afferents at frequencies of 1 to 4 impulses/sec produce prompt inhibition of intestinal muscle. The intestinointestinal reflex, therefore, is not adequately explained by either the hypothesis proposing spinal reflex pathways or by the hypothesis proposing reflex pathways in prevertebral ganglia.

Our recent studies (employing intracellular recording techniques) of synaptic input to postganglionic neurons in the inferior mesenteric ganglion of the guinea pig have led us to conclude that the intestinointestinal reflex is mediated simultaneously by both spinal and ganglia reflex pathways. We have found that approximately 90% of the neurons in the inferior mesenteric ganglion we have recorded from received synaptic input from both splanchnic nerves (classic preganglionic input) and from mechanoreceptors in the distal colon via afferent fibers in the colonic nerve (colonic afferent input). In these neurons, the resting membrane potential averaged −54 mV ± 10.5 ($N =$ 324) and the absolute voltage threshold for the production of a single action potential ranged from −40 to −35 mV. Therefore, synaptic input to these

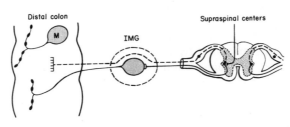

FIG. 4. Summary diagram of the neural circuitry for spinal mediation of the intestinointestinal reflex. This circuit does not include the neural connections known to occur between peripheral afferent fibers and postganglionic neurons of prevertebral ganglia. M, Cholinergic motor neuron of enteric nervous system; IMG, inferior mesenteric ganglia.

neurons must produce a minimum depolarization of 14 to 10 mV to trigger action potentials. The level of spontaneous input from colonic afferents was usually subthreshold and only reached threshold after the colon had been distended to supraphysiologic volumes; thus, it is likely that the level of colonic afferent input to postganglionic neurons under physiologic conditions is insufficient by itself to trigger action potential discharge. These neurophysiological observations explain why Kuntz and co-workers were able to elicit the intestinointestinal reflex through decentralized ganglia only when intraluminal pressures exceeded physiological values.

In an intact animal, however, the postganglionic neurons receive input from not just the colonic afferents but also from preganglionic neurons of the central nervous system. Because this dual input to postganglionic neurons can result in spatial summation (16), inhibition of intestinal motility may

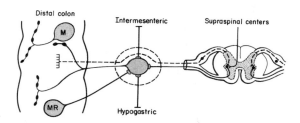

FIG. 5. Diagram illustrating the circuit integration known to occur at prevertebral ganglia between peripheral and spinal reflex pathways. MR, mechanoreceptor of enteric nervous system; M, cholinergic motor neuron of enteric nervous system.

occur as a result of ganglionic summation of subthreshold inputs from the central nervous system and from peripheral organs. We propose that this summation is an integral part of the intestinointestinal reflex. Its inclusion in the neural system associated with the reflex clarifies the discrepancies which have plagued previous attempts to describe it adequately. Figure 5 is a general representation of neural pathways known to be associated with noradrenergic inhibition of colonic smooth muscle. According to this diagram, spinal and peripheral reflex pathways do not exist as separate neural circuits but must be considered to be functionally integrated. As discussed above, it was difficult to explain how the spinal pathways produced intestinal inhibition, because the natural discharge frequency of preganglionic fibers is below the frequency required to induce inhibition by electrically stimulating the same fibers. On the other hand, elimination of preganglionic input to postganglionic neurons prevents the intestinointestinal reflex, because the peripheral input by itself is insufficient to trigger the reflex. It appears, therefore, that the intestinointestinal reflex results from spatial summation by postganglionic neurons of subthreshold inputs from both preganglionic and peripheral afferent fibers.

Presently, the functional importance of peripheral input to sympathetic noradrenergic neurons to gastrointestinal control cannot be fully assessed. It appears likely, however, that gastrointestinal motility patterns and reflexes, such as the enterogastric reflex (25), are modulated by neural commands of noradrenergic neurons which result from integration in prevertebral ganglia of neural information received directly from the viscera with neural information received from the central nervous system.

ACKNOWLEDGMENTS

This investigation was supported by Research Grant AM17632 from the National Institutes of Health, Public Health Service. This work was performed by Dr. Szurszewski during the tenure of an Established Investigatorship of the American Heart Association. Dr. Weems is a fellow of the Minnesota Heart Association.

REFERENCES

1. Youmans, W. B. (1968): Innervation of the gastrointestinal tract. In: *Alimentary Canal*, Vol. 4: *Motility*, edited by C. F. Code, pp. 1655–1663. American Physiological Society, Washington, D.C.
2. Campbell, G. (1970): Autonomic nervous supply to effector tissues. In: *Smooth Muscle*, edited by E. Bülbring, A. F. Brading, A. W. Jones, and T. Tomita, pp. 451–495. Edward Arnold, London.
3. Furness, J. B., and Costa, M. (1974): The adrenergic innervation of the gastrointestinal tract. *Ergeb. Physiol.*, 69:1–51.
4. Jansson, G., and Martinson, J. (1966): Studies on the ganglionic site of action of sympathetic outflow to the stomach. *Acta Physiol. Scand.*, 68:184–192.
5. Paton, W. D. M., and Vizi, E. S. (1969): The inhibitory action of noradrenaline and adrenaline on acetylcholine output by guinea-pig ileum longitudinal muscle strip. *Br. J. Pharmacol.*, 35:10–28.
6. Beani, L., Bianchi, C., and Crema, A. (1969): The effect of catecholamine and sympathetic stimulation on the release of acetylcholine from the guinea-pig colon. *Br. J. Pharmacol.*, 36:1–17.
7. Celander, O. (1959): Are there any centrally controlled sympathetic inhibitory fibers to the musculature of the intestine? *Acta Physiol. Scand.*, 47:299–309.
8. Langley, J. N. (1921): *The Autonomic Nervous System*, Part I. Heffer, Cambridge.
9. Kuntz, A. (1938): The structural organization of the celiac ganglia. *J. Comp. Neurol.*, 69:1–12.
10. Kuntz, A. (1940): The structural organization of the inferior mesenteric ganglia. *J. Comp. Neurol.*, 72:371–382.
11. McLennan, H., and Pascoe, J. E. (1954): The origin of certain nonmedullated nerve fibers which form synapses in the inferior mesenteric ganglion of the rabbit. *J. Physiol. (Lond.)*, 124:145–156.
12. Kuntz, A., and VanBuskirk, C. (1941): Reflex inhibition of bile flow and intestinal motility mediated through decentralized celiac plexus. *Proc. Soc. Exp. Biol.*, 46: 519–523.
13. Kuntz, A., and Saccomanno, G. (1944): Reflex inhibition of intestinal motility mediated through decentralized prevertebral ganglia. *J. Neurophysiol.*, 7:163–170.
14. Semba, T. (1954): Intestino-intestinal inhibitory reflexes. *Jap. J. Physiol.*, 4:241–245.
15. Kock, N. G. (1959): An experimental analysis of mechanisms engaged in reflex inhibition of intestinal motility. *Acta Physiol. Scand.: Suppl.* 47, 164, 1–54.

16. Crowcroft, P. J., and Szurszewski, J. H. (1971): A study of the inferior mesenteric and pelvic ganglia of guinea-pigs with intracellular electrodes. *J. Physiol. (Lond.)*, 219:421–441.
17. Crowcroft, P. J., Holman, M. E., and Szurszewski, J. H. (1971): Excitatory input from the distal colon to the inferior mesenteric ganglion in the guinea-pig. *J. Physiol. (Lond.)*, 219:443–461.
18. Weems, W. A., and Szurszewski, J. H. (1974): Input to the inferior mesenteric ganglion from mechano-receptors in the colon. *Fed. Proc.*, 33:392.
19. Weems, W. A., and Szurszewski, J. H. (1974): Physiological significance of colonic mechano-receptor input to neurons in the inferior mesenteric ganglion. *Gastroenterology*, 66:870.
20. Job, C., and Lundberg, A. (1952): Reflex excitation of cells in the inferior mesenteric ganglion on stimulation of the hypogastric nerve. *Acta Physiol. Scand.*, 26:366–382.
21. Skok, V. I. (1973): *Physiology of Autonomic Ganglia*. Igaku Shoin, Tokyo.
22. Weems, W. A. (1974): An electrophysiological investigation of neural pathways in the intermesenteric nerve. *Physiologist*, 17:355.
23. Chang, P.-Y., and Hsu, F.-Y. (1942): The localization of the intestinal inhibitory reflex arc. *Quart. J. Exp. Physiol.*, 31:311–318.
24. Johansson, B., and Langston, J. B. (1964): Reflex influence of mesenteric afferents on renal, intestinal and muscle blood flow and on intestinal motility. *Acta Physiol. Scand.*, 61:400–412.
25. Schapiro, H., and Woodward, E. R. (1959): Pathway of enterogastric reflex. *Proc. Soc. Exp. Biol.*, 101:407–409.

Physiology of Smooth Muscle, edited by
E. Bülbring and M. F. Shuba.
Raven Press, New York © 1976.

Neuronal Interactions Within Ganglia of Auerbach's Plexus of the Small Intestine

J. D. Wood

Department of Physiology, University of Kansas Medical Center, Kansas City, Kansas 66103

Neural influence on movement of the gastrointestinal musculature originates at three principal levels of organization of the nervous system. First, gastrointestinal motility is altered by activation of central nervous centers both spinal and suprasegmental (1); second, peripheral reflexes with synaptic connections within prevertebral ganglia, such as the inferior mesenteric ganglion, apparently are involved (2,25); and third, it is evident that much of the programming and integration of the various motility patterns of the bowel are performed by intrinsic neuronal mechanisms within the enteric plexuses (3).

Included in Auerbach's and Meissner's plexuses are small ensembles of ganglion cells which contribute fine nerve fibers to a synaptic neuropil within each ganglion. It is reasonable to assume that these neural connections compose an integrative circuitry which processes sensory information and generates nervous output that is immediately appropriate for control of a localized section of the gastrointestinal musculature. This implies that a requisite for understanding nervous control of gastrointestinal motor function is information on the nature of the interactions between the neurons present in enteric ganglia. For the most part, it seems that these interactions involve synaptic connections. Results obtained from light and electron microscopy (4–8), histochemistry (9,10), pharmacological studies (11,12), and electrophysiological investigations (13–15) suggest that extensive synaptic connections do exist within enteric ganglia and support the concept of synaptic involvement in the integrative function of the neural network within each ganglion.

One advantage of multiunit recording with relatively large electrode tips in Auerbach's plexus is that it is possible to record simultaneously, with a single electrode, the activity of two or more units within one ganglion and to investigate relationships between the discharge patterns of different neurons (16). In this chapter the various kinds of interaction between single neuronal units in Auerbach's plexus are described and discussed as observed with techniques of extracellular recording.

METHODS OF MULTIUNIT RECORDING

The results presented here were obtained by extracellular recording with metal microelectrodes within ganglia of cat and guinea pig jejunum viewed at 35× magnification with a stereomicroscope after vital staining with methylene blue. Details of the method appear in Wood (17) and Wood and Mayer (18). Action potentials of single neurons were distinguished by spike amplitude, the pattern of discharge, the waveform of the spikes as viewed on a storage oscilloscope, and by the first time derivative of the action potential voltage.

INTERACTIONS BETWEEN BURST UNITS

Burst-type units (Fig. 2) are continuously active neurons within the myenteric plexus which discharge periodic bursts of action potentials with silent intervals separating the bursts (3,16–20).

Multiunit recording from the myenteric plexus of the small intestine reveals the following kinds of neuronal interaction involving burst-type units: (a) sequential temporal coupling between two neurons both of which discharge burst-type patterns of spikes; (b) sequential temporal coupling between single spikes of one neuron and bursts of spikes of a second neuron; (c) synchronous discharge of two burst-type units which appear to be activated by presynaptic fibers common to both cells; (d) inhibitory interactions in which the ongoing discharge of one burst-type neuron is halted during the discharge of a second neuron; (e) interactions of a diffuse nature which involve modulation of the frequency of discharge of bursts of spikes over relatively long spans of time.

Sequential Coupling Between Single Spikes and Bursts

When sequential temporal coupling is observed (Fig. 1A), the second unit of the paired bursts always discharges with a relatively constant latency, greater than 20 msec, following the discharge of the first unit. This is illustrated by the paired bursts of Fig. 1A for which the mean and standard deviation of the time intervals separating the last spike of the first burst unit and the first spike of the second burst unit were 33 ± 3 msec for 59 paired bursts. The time intervals between the paired bursts are not significantly different from the interburst intervals of burst-type units which on the recordings do not show temporal coupling to the discharge of a second burst neuron (16,18). The mean and standard deviation of interburst intervals (time between the first spike of the first unit of consecutive paired bursts) for the coupled bursts of Fig. 1A were 9.7 ± 1.7 sec for 117 successive bursts. When paired coupling is observed, it seems to be obligatory only for the second unit of the pair. For the units of Fig. 1A, discharge of the first unit was not accompanied by discharge of the second unit in 58 of 117 bursts. The second unit never discharged in the absence of the first.

Interspike-interval analysis for the bursts of spikes of each member of a coupled pair shows distinct differences between the discharge characteristics of the first and second neurons (16,18). The first of a coupled pair usually discharges fewer spikes per burst than the second, and the mean interspike interval is less for the first than the second member of a coupled pair. Spike-interval histograms also show a smaller coefficient of variation and less positive skewness for the first unit of coupled pairs (18).

The neuronal connections which account for temporal coupling of pairs of burst units are not understood. The time delay between the first and second bursts is comparable to synaptic delays in other autonomic ganglia (25) and might be attributable to synaptic delay between a "driver" and directly driven "follower" cell. Recently I have observed that elevation of the concentration of magnesium in the organ bath from 1.2 to 12 mM selectively stopped the discharge of the second neuron of two different coupled pairs (*unpublished observations*). It also is possible that the burst discharge of the second member of a coupled pair could represent an event such as post-inhibitory rebound excitation of the second cell (21) rather than a direct excitatory input.

Sequential Coupling Between Single Spikes and Bursts

This kind of interaction in which a single spike of one unit precedes the burst-type discharge of a second unit (Fig. 1A–E) is often observed. In some ganglia, the single-spike unit discharges only in conjunction with the burst-type unit, whereas in others, the single-spike discharge may be observed both in conjunction with and independent of the burst discharge. When the single spike occurs in conjunction with the burst discharge, the latency between the single spike and the first spike of the burst is relatively constant and greater than 10 msec. Figure 1A is an example of a coupled pair of burst units, the first of which was preceded by a single spike from a different cell. In this case, the mean and standard deviation of the latency between the single spike and the initial spike of the first burst was 15 ± 2 msec for 59 occurrences.

Action potentials of single-spike units may also be coupled by fixed latencies to the discharge of cells with firing patterns that consist of doublets and single spikes (Fig. 1C–E). The mean and standard deviation of the latency between the single spike and first spike of the discharge of the second unit of Fig. 1C–E was 47 ± 9 msec for 11 occurrences. The small single spike (Fig. 1C–E) was present in 11 of 39 bursts.

In some experiments, the single spikes that precede the burst-type discharge have been identified as neurons with behavior similar to that of slowly adapting mechanoreceptors (3). Figure 1B is an example of one such case that was reported for the small intestine of guinea pig (20). The neuron that

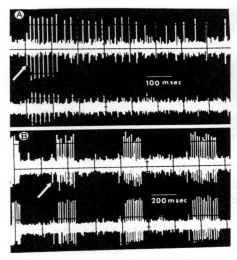

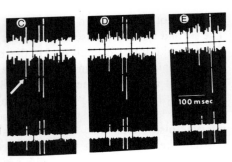

FIG. 1. Examples of neuronal interactions involving burst-type units within Auerbach's plexus. A: Sequential temporal coupling between bursts of spikes from two different units recorded from cat duodenum. A single spike from a third unit (*arrow*) was coupled to the paired bursts and appears to precede the initial burst of spikes. B: Sequential temporal coupling between a single-spike unit (*arrow*) and bursts of spikes in guinea pig jejunum. C–E: Successive impulses from a single-spike unit (*arrow*) that was coupled to discharge of a second neuron which consisted of either doublets (C and D) or single spikes (E) in cat jejunum. Action potentials appear in upper trace of each record; lower trace is first time derivative of action potential voltage.

generated the small amplitude single spikes which preceded the burst discharge of Fig. 1B also discharged in the absence of any activity of the burst unit. When the burst unit was active (Fig. 1B), the single spikes occurred in conjunction with each burst and the latency between the single spike and first spike of the burst was 42 ± 3 msec; $N = 49$.

Synchronous Discharge of Two Burst Units

Burst-type patterns of discharge in which some of the spike bursts are comprised of action potentials from two different neurons are often recorded within the myenteric plexus (3,18). Figure 2 is an example of burst-type discharge in which 31 of 67 bursts of the control contained spikes from two different units (Fig. 2C,D). It was the large amplitude spikes (Fig. 2B) which occurred either in pure bursts or mixed with smaller amplitude spikes (Fig. 2C,D); the smaller spikes did not occur in the absence of the large amplitude spikes.

Effects of norepinephrine and nicotine were tested on the preparation of Fig. 2. After norepinephrine (1×10^{-5} g/ml), 18 of 24 bursts contained spikes from the two different units. Following washout of norepinephrine, 19 of 27 bursts were mixed. In the presence of nicotine (1×10^{-5} g/ml), 42 of 44 bursts were mixed bursts, and after removal of nicotine from the bath, 35 of 66 bursts contained spikes from both cells.

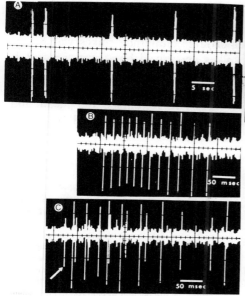

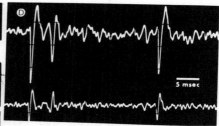

FIG. 2. Synchronous discharge of two burst-type units within Auerbach's plexus of cat duodenum. A: Pattern of burst discharge recorded with a slow time base. B: Single bursts of spikes consisting only of large amplitude spikes from first unit. C: Single burst of spikes showing synchronous discharge of spikes by the same units as in B and by a second unit firing smaller amplitude spikes (arrow). D: Waveforms of the spikes comprising the mixed bursts. Same burst as in C. Action potentials appear on upper trace; lower trace is first time derivative of action potential voltage.

Figure 3 is a frequency distribution histogram of combined interburst intervals for mixed and pure bursts of the units of Fig. 2 during the initial control period. This histogram is multimodal with peaks that tend to be multiples of a 5-sec interval, and in this respect, the discharge of these cells resembles burst units reported by Wood and Mayer (16) which discharged bursts every 6 sec or multiples of a 6-sec interval.

A sequential interspike-interval histogram for the bursts which consisted only of large amplitude spikes (pure bursts; Fig. 2B) is compared in Fig. 4 with the same kind of histogram for the large amplitude spikes when they were accompanied by the smaller spikes from the second unit (mixed bursts; Fig. 2C,D). This demonstrates that in the mixed bursts the duration of the second through sixth sequential interspike intervals of the large amplitude spikes were significantly greater ($p < 0.05$) than the same intervals of the pure bursts of large amplitude spikes (Fig. 4).

The number of large amplitude spikes of the mixed and pure bursts of

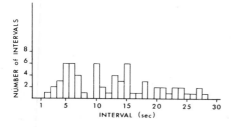

FIG. 3. Histogram of the interval distribution of consecutive bursts of spikes of the same units as in Fig. 2. An interburst interval is the time between the first spike of consecutive bursts. Ordinate represents the number of intervals for the duration indicated on the abscissa. Histogram is constructed with 1-sec interval classes. Given are number of intervals ($N = 67$), mean interval ($\bar{X} = 14$), and SD = 9.8.

Fig. 2 were not significantly different. The mean and standard deviation of the number of large amplitude spikes for 36 pure bursts was 8.9 ± 2.5 and for 31 mixed bursts the mean number of large amplitude spikes was 8.1 ± 2.4, and the mean number of small spikes per mixed burst was 5.3 ± 2.2.

With the exception of one burst, the first small amplitude spike of the mixed bursts always preceded the first large amplitude spike of the burst by a relatively constant latent period (Fig. 2C). In 30 bursts, the mean and standard deviation of the time between the first small spike and first large spike of the mixed bursts was 17.5 ± 1.1 msec.

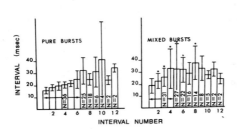

FIG. 4. Histograms of the interval distribution of consecutive spikes of the burst unit responsible for the large amplitude spikes of Fig. 2. Histograms labeled *pure bursts* correspond to bursts containing only large amplitude spikes. *Mixed bursts* refers to bursts showing synchronous discharge of both units (Fig. 2C). Ordinate represents the mean duration of the sequential intervals indicated on the abscissa. Given for each histogram are number of intervals (N) and standard deviation (*vertical bars*). Asterisk at top of histogram bars for mixed bursts indicates that the duration of the indicated spike interval is significantly longer than the corresponding interval for pure bursts ($p < 0.05$, Student's *t*-test).

The mechanisms that produce bursts containing spikes from two different units are not clear. It is possible that presynaptic fibers common to both neurons produce the relatively synchronous discharge of the two different units. I have observed on two different preparations which showed this kind of interaction that elevation of the concentration of magnesium in the bathing solution to 12 mM abolished the discharge of both neurons, and this was reversed when the magnesium concentration was returned to 1.2 mM (*unpublished observations*).

Inhibitory Interactions

A fourth kind of interaction involving burst units appears to be inhibitory in nature (Fig. 5). As shown in Fig. 5, 35% of 112 bursts of large amplitude spikes were accompanied by smaller amplitude action potentials of a second neuron (Fig. 5B), the discharge of which appeared to inhibit firing of the first burst-type unit. When this kind of interaction occurred, the spikes of the inhibitory unit were not coupled to a particular phase of the burst discharge, but instead they occurred at any point within the burst as well as preceding the burst (18). Measurements of latent periods between the first inhibitory spike

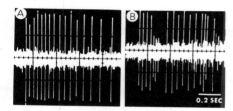

FIG. 5. Inhibition of burst-type discharge by activity of a second neuron in Auerbach's plexus of cat jejunum. A: Single burst of spikes showing no inhibitory discharge of second neuron. B: Same burst unit, discharge of second unit occurred following fourth spike of burst and inhibited discharge of burst unit.

and each of the preceding spikes of the burst revealed no consistent temporal relationships between the discharge of the two units. This would seem to rule out the possibility that the unit which appeared to be inhibitory was activated by discharge of the burst-type neuron, and that the inhibition of the burst unit was recurrent inhibition analogous to the negative feedback produced by the Renshaw system of the spinal cord.

Diffuse Interactions of Burst Units

Wood and Mayer (16) recorded concurrently the discharge of four different burst-type units within a single myenteric ganglion. We analyzed simultaneous changes in frequency of discharge of spike bursts for each of the four units and found that a decline in burst frequency of one of the units corresponded in time to an increase in burst frequency of two of the three other units. Cross-interval analysis for each possible paired combination of these three units indicated that there was no close temporal coupling between any of the units. This suggested that this kind of interaction was of a diffuse nature and could have resulted from any one of the following: (a) local interactions among the three individual units; (b) an influence of other neurons in the same or a different ganglion; (c) from a common effect of some other factor on each of the three units.

INTERACTIONS BETWEEN MECHANOSENSITIVE NEURONS

Three different kinds of units in ganglia of Auerbach's plexus respond with increased frequency of discharge to mechanical distortion of the ganglion and can be classified as mechanosensitive neurons (3). One kind of mechanosensitive unit behaves like a typical slowly adapting mechanoreceptor and another like a fast-adapting mechanoreceptor. The third kind of mechanosensitive unit is activated by mechanical stimulation to discharge prolonged trains of spikes (Fig. 6) and is called a tonic-type neuron (3). The spike trains from these cells behave as all-or-nothing events. Once the cell is activated to fire by mechanical distortion of the ganglion, it discharges a complete train of spikes with a characteristic pattern even if the original stimulus is withdrawn.

Action potentials of units with properties of slowly adapting mechanoreceptors are often observed preceding the discharge of tonic-type neurons on multiunit records (17,18), and it has been suggested that tonic-type neurons may be triggered by input derived from mechanoreceptors (3,18).

On some records of spike activity of tonic-type neurons, two to three units are observed to be activated simultaneously (Fig. 6). Analyses of these records reveal no fixed temporal relationships between individual spikes of the different units which make up the trains of spikes. This implies that each of the units which contribute to the spike train is activated simultaneously by the same factor related to the mechanical stimulation, rather than the mixed

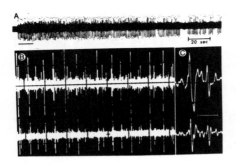

FIG. 6. Tonic-type discharge of multiunits in Auerbach's plexus of cat jejunum. A: Tonic-type discharge elicited by transient forward–reverse movement of a glass-Pt electrode. Horizontal bar indicates stimulus. B: Expanded trace of same tonic-type discharge as in A shows spikes from two or more different units. C: Waveform of spikes appears on upper trace. Lower trace is first derivative of action potential voltage. *Time calibration* in C is 5 msec and same calibration for B is 100 msec.

spikes of the train resulting from interactions between the individual tonic-type units.

DISCUSSION

There is little doubt that much of the nervous control involved in the various motility patterns of the intestinal musculature originates from local integrative networks within ganglia of the enteric nervous system. The intestinal musculature behaves physiologically as an extensive functional syncytium with myogenic pacemaker mechanisms. It may be that the integrated activity of the enteric ganglia provides nervous modulation of the myogenic system rather than direct nervous excitation. In recent years, considerable evidence has accrued which indicates that nervous inhibition of the musculature is a significant aspect of intrinsic neuronal control (3,15,22–24). It has been proposed that myogenic excitation of the musculature and the extent of spread of excitation within the muscular syncytium are regulated directly by a nonadrenergic inhibitory transmitter substance that is released from continuously active ganglion cells of Auerbach's plexus, and that one mechanism of neural control is integration of the activity of this neurogenic inhibitory system (3,24). Because discharge of burst-type neurons is the predominant neural activity when inhibition of the circular muscle layer is produced *in vitro,* it has been assumed that these units are involved in the continuously active inhibitory pathways to the muscle. If this is the case, then some of the different kinds of neuronal interactions involving burst-type units may represent correlates of neural connections involved in inhibitory control of the musculature.

ACKNOWLEDGMENT

Supported by National Science Foundation Grant GB-31292.

REFERENCES

1. Pick, J. (1970): *The Autonomic Nervous System,* p. 61–91. Lippincott, Philadelphia, Pa.
2. Crowcroft, P. J., Holman, M. E., and Szurszewski, J. H. (1971): Excitatory input from the distal colon to the inferior mesenteric ganglion in the guinea-pig. *J. Physiol. (London),* 219:443–461.
3. Wood, J. D. (1974): Neurophysiology of ganglia of Auerbach's plexus. *Am. Zool.,* 14:973–989.
4. Shul'pin, G. V. (1968): Morphology of autonomic ganglia in ontogenesis of the rabbit. *Neurosci. Trans.* 5:514–522.
5. Richardson, K. C. (1958). Electron microscopic observations on Auerbach's plexus in the rabbit with special reference to the problem of smooth muscle innervation. *Am. J. Anat.,* 103:99–136.
6. Taxi, J. (1965). Contribution a l'etude des connections des neurones motoeur du systeme nerveux autonome. *Ann. Sci. Nat. Zool., Paris,* Ser. 12, 7:413–674.
7. Baumgarten, H. G., Holstein, A.-F., and Owman, C. H. (1970): Auerbach's plexus of mammals and man: electron miscroscopic identification of three different types of neuronal processes in myenteric ganglia of the large intestine from rhesus monkeys, guinea pigs and man. *Z. Zellforsch. Mikrosk. Anat.,* 106:376–397.
8. Gabella, G. (1972): Fine structure of the myenteric plexus in the guinea-pig ileum. *J. Anat.,* 111:69–97.
9. Gershon, M. D., Drakontides, A. B., and Ross, L. L. (1965): Serotonin: Synthesis and release from the myenteric plexus of mouse intestine. *Science,* 149:197–200.
10. Furness, J. B., and Costa, M. (1971): Morphology and distribution of intrinsic adrenergic neurons in the proximal colon of the guinea-pig. *Z. Zellforsch. Mikrosk. Anat.,* 120:346–363.
11. Daniel, E. E. (1968): Pharmacology of the gastrointestinal tract. *Handbook of Physiology: Alimentary Canal,* Sect. 6, Vol. 4, pp. 2267–2324. Williams and Wilkins, Baltimore, Md.
12. Kosterlitz, H. W. (1968): Intrinsic and extrinsic nervous control of motility of the stomach and the intestines. *Handbook of Physiology: Alimentary Canal,* Sect. 6, Vol. 4, pp. 2147–2171. Williams and Wilkins, Baltimore, Md.
13. Hirst, G. D. S., Holman, M. E., and Spence, I. (1974): Two types of neurons in the myenteric plexus of duodenum in the guinea pig. *J. Physiol. (London),* 236: 303–326.
14. Nishi, S., and North, R. A. (1973): Intracellular recording from the myenteric plexus of the guinea-pig ileum. *J. Physiol. (London),* 231:471–491.
15. Hirst, G. D. S., and McKirdy, H. C. (1974): A nervous mechanism for descending inhibition in guinea-pig small intestine. *J. Physiol. (London),* 238:129–143.
16. Wood, J. D., and Mayer, C. J. (1973): Patterned discharge of six different neurons in a single enteric ganglion. *Pflügers Arch.,* 338:247–256.
17. Wood, J. D. (1970): Electrical activity from single neurons in Auerbach's Plexus. *Am. J. Physiol.,* 219:159–169.
18. Wood, J. D., and Mayer, C. J. (1974): Discharge patterns of single enteric neurons

of the small intestine of the cat, dog and guinea pig. *Proc. Int. Symp. Gastrointestinal Motility, 4th,* p. 387–408. *Banff, Alberta, Canada.* Mitchell Press, Vancouver, Canada.

19. Ohkawa, H., and Prosser, C. L. (1972): Electrical activity in myenteric and submucous plexuses of cat intestine. *Am. J. Physiol.,* 222:1420–1426.

20. Wood, J. D. (1973): Electrical discharge of single enteric neurons of guinea-pig small intestine. *Am. J. Physiol.,* 225:1107–1113.

21. Perkel, D. H., and Mulloney, B. (1974): Motor pattern production in reciprocally inhibitory neurons exhibiting postinhibitory rebound. *Science,* 185:181–182.

22. Burnstock, A. (1972): Purinergic neurons. *Pharmacol. Rev.,* 34:509–581.

23. Furness, J. B., and Costa, M. (1973): The nervous release and the action of substances which affect intestinal muscle through neither adrenoreceptors nor cholinoreceptors. *Philos. Trans. R. Soc. Lond. [Biol. Sci.],* 265:123–133.

24. Wood, J. D (1975): Neurophysiology of Auerbach's plexus and control of intestinal motility. *Physiol. Rev.,* 55:307–324.

25. Nozdrachev, A. D., Besenkina, G. I., and Efimova, N. I. (1970): Conduction pathways of the cat's caudal mesenteric ganglia. *Fiziol. Zh. SSSR,* 4:543–551.

Physiology of Smooth Muscle, edited by
E. Bülbring and M. F. Shuba.
Raven Press, New York © 1976.

Voltage Clamp of Taenia Smooth Muscle Using Intracellular Recording of Membrane Potential: Action of Carbachol

T. B. Bolton

University Department of Pharmacology, Oxford, England

The aim of a voltage-clamp technique is to hold the whole of the membrane through which current is passing at a known and constant potential so that its conductance can be studied. So far, the voltage-clamp methods applied to smooth muscle have utilized extracellular methods of potential recording. These suffer from the disadvantage that it is impossible to be certain during voltage clamp that the membrane of all cells everywhere is held at the same potential, either because of a decrease in the space constant or because of a significant resistance in series with some part of the membrane (7).

It is important, therefore, that an independent check be made of the constancy of potential during voltage clamp; this can be done only by inserting a microelectrode into cells at various points within the voltage-clamped region to record the potential. This chapter describes some preliminary results (5) obtained in this way and also reports on the action of carbachol studied with this technique. My main purpose has not been to evaluate the efficiency of the extracellular method of potential recording as used by others, but to attempt to define conditions under which reasonable constancy of potential can be achieved, while, at the same time, the electrophysiological properties of the muscle are not changed too drastically, because, as we shall see, as the region of active muscle is reduced in an attempt to achieve spatial constancy of potential, the activity of the muscle is altered. I have chosen to record the membrane potential intracellularly at a point within the voltage-clamped region, rather than use the extracellular method of potential recording, since the microelectrode method would seem to record a larger fraction of the true membrane potential.

METHOD

Strips of smooth muscle 0.3 to 0.5 mm wide and about 0.1 mm thick were cut from the taenia of the guinea pig caecum. These were introduced into a double sucrose-gap apparatus which allowed one or more microelectrodes (resistance 30 to 50 MΩ) to be inserted into the cells in the voltage-clamped region (node). The width of the node was varied from 0.1 to 1 mm or more in different experiments by altering the positions of two chambers, about 0.8 mm wide, through which deionized sucrose solution (containing 10^{-5} M $CaCl_2$) flowed. The ends of the muscle were depolarized in isotonic potassium chloride solution. The total length of muscle between high potassium solutions was therefore about 2 mm when the node width was 0.5 mm.

The nodal solution (Krebs, composition given in Ref. 3) was maintained at earth potential by means of one voltage-clamp circuit. A second voltage-clamp circuit was used to clamp the potential recorded between the tip of a microelectrode and an indifferent silver–silver chloride electrode in the nodal solution placed to almost touch the muscle (10). Alternatively, this circuit was used to pass rectangular current pulses.

Current was passed between either silver–silver chloride or calomel electrodes connected to the nodal solution (downstream) or high potassium solution by 3 M KCl agar bridges. In experiments where the effects of varying node width were studied, current was passed across only one sucrose gap. In the experiments where two independent microelectrodes were inserted into the node, or where the action of carbachol was studied, the isotonic Krebs solution bathing the node was changed to one which had been made hypertonic by the addition of sucrose (11). This latter solution was used in order to reduce artifacts due to contraction and to prevent the resulting dislodgement of electrodes. In these experiments current was passed across both sucrose gaps.

RESULTS

Effects of Varying Node Width

Using node widths of 1 mm or more it was found that the smooth muscle was often spontaneously active. Hyperpolarizing currents elicited electrotonic potentials of the usual form. Depolarizing currents produced repetitive spiking. The membrane potential was about 60 mV and spikes showed overshoot. The responses were very similar to those recorded from larger pieces of taenia (12), indicating that these narrow strips were capable of essentially normal electrical activity.

In nodes 0.3 to 0.5 mm wide, the muscle was generally not spontaneously active. Inward current hyperpolarized the membrane, while outward current produced regenerative (all-or-none) spikes. Often only a single spike was discharged during a depolarization of more than 1 sec.

In nodes 0.1 to 0.3 mm wide, the muscle was never spontaneously active; nor could regenerative spikes be elicited. The time course of the hyperpolarizing electrotonic potential was noticeably faster than in wider nodes. Relatively strong depolarizing currents were required to elicit spike-like responses, and spikes were graded. In good preparations, 10 μA passed into the node gave extracellular potential steps of about 1 mV. Assuming all the current passes into the muscle (almost certainly not true) this gives a resistance in series with the node of 100 Ω ($V = 10^{-3}$ V, $I = 10^{-5}$ A, and $R = V/I = 10^2$ Ω). Clearly, the electrical properties of the muscle were changed by restricting the volume of active muscle by reducing node width. The main changes were a loss of spontaneous electrical activity and a loss of the ability to produce regenerative spikes. Probably these changes result from an increase in the proportion of current flowing which is a leakage current. Whatever the reason, a similar detrimental action of small node widths has been observed in cardiac muscle (8,10).

Voltage Clamp

When responses to current pulses had been obtained, the membrane potential was clamped at the resting membrane potential, and in the same cell rectangular voltage-clamp pulses were imposed. In wider nodes (0.3 to 0.5 mm) the initial current was inward after the capacitive transient. Current then became outward and an early, and often a late, peak in outward current occurred. It seems unlikely that the potential was uniform across the node in these wider nodes. Additionally, contraction of the muscle which generally dislodged the microelectrode eventually, may have introduced some artifact.

In narrow nodes (0.1 to 0.3 mm) a net inward current was not seen. Outward current reached an early peak and then declined. Measured at 1.5 sec the current–voltage relationship showed outward rectification which was increased if the current was measured at the time of peak outward current.

Uniformity of Potential Across the Node

Because the electrical properties of the muscle were considerably changed when narrow nodes were used, it was decided to use wider nodes (0.4 to 0.5 mm) and to pass current into the node across both sucrose gaps instead of across only one sucrose gap as in the experiments described so far. This ought to improve the uniformity of potential across the node. In previous voltage-clamp studies on smooth muscle using extracellular potential recording, current has been passed into the node across only one sucrose gap. The uniformity of potential achieved using *one-sided current injection* into the node, should be about the same as that obtained when *two-sided current injection* is made into nodes *twice as wide,* i.e., the node width in the present experiments is equivalent to a node width of 0.2 to 0.25 mm in sucrose-gap experiments where extracellular recordings of potential are made. Two microelectrodes were first inserted into smooth muscle cells at different points within the node. Responses to rectangular current pulses were obtained (Fig. 1) and then the potential recorded by one microelectrode was clamped at the resting membrane potential. Rectangular voltage-clamp steps were imposed.

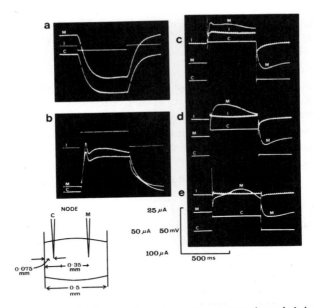

FIG. 1. Variation in nodal potential during voltage clamp of taenia smooth muscle in hypertonic solution. The potential was recorded intracellularly at two points within a node (0.5 mm wide) at the positions shown in the diagram. Responses to rectangular current pulses (a, b) and to rectangular command potentials under voltage clamp (c, d, e) were obtained. Current is labeled I and potentials recorded by micro- electrodes C and M, labeled accordingly. In c, d, e potential recorded by electrode C was clamped at the resting membrane potential and subjected to voltage-clamp steps while potential was recorded inde- pendently by electrode M. Notice the difference in potential between M and C following either a step- depolarization or a step-repolarization to the resting membrane potential. Vertical calibration: 25 μA in a, b; 100 μA in c; and 50 μA in d, e.

In most preparations, the potential recorded by the independent monitor- ing electrode was within a few millivolts of the clamp potential at times greater than 200 msec following a depolarizing step; in other experiments differences existed between the two potentials for up to 500 msec (Fig. 1). Following repolarization, potential again varied across the node for periods of up to 200 msec or more. Figure 1 shows the result obtained in the experi- ment where voltage variation across the node persisted longest following a step change in potential. The results are similar to those obtained by McGuigan (8) in cardiac muscle. In these experiments it would clearly have not been justified to analyze the current flowing at early times following a step depolarization or repolarization on the assumption of a uniform nodal potential.

Action of Carbachol

The action of carbachol has been studied in a few preliminary experiments. The current–voltage relationships were obtained under voltage clamp by using a ramp command potential. The rate of ramp rise was about 25 mV/sec. It was shown in several preparations that the potential recorded at another point within the node by an independent microelectrode followed clamp potential quite closely at this low rate of ramp-rise.

Carbachol (10^{-6} to 10^{-5} g/ml) depolarized the membrane. The current–voltage relationship was linear in a hyperpolarizing direction and remained so in the presence of carbachol, although conductance was increased. In a

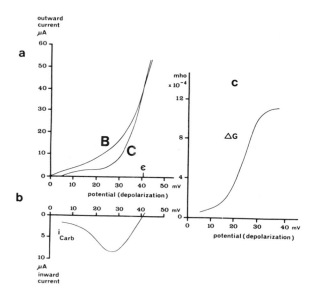

FIG. 2. (a) Action of carbachol on the current–voltage relationships of taenia. Depolarizing ramp command potentials (rate of rise 30 mV/sec) were applied under voltage clamp in sucrose hypertonic solution before (B) and in the presence of carbachol (C, 5×10^{-6} g/ml) to obtain the current–voltage relationships. Zero potential is the resting membrane potential before application of carbachol. The holding potential in each case was the same as the existing potential, i.e., it was 5 mV depolarized in carbachol. ϵ is the equilibrium potential. (b) The variation with potential of the additional current flowing in the presence of carbachol (i_{carb}). This line was obtained by subtracting line B in (a) from line C. (c) The variation with potential of the conductance opened by the action of carbachol, $\Delta G[= i_{carb}/(V - \epsilon)]$.

depolarizing direction, the current–voltage relationship was nonlinear in the absence of carbachol. Carbachol had the effect of increasing the inward current flowing at all potentials negative to an equilibrium potential (ϵ) at which the current flowing was unaffected by the presence of carbachol (Fig. 2a). At potentials positive to ϵ, carbachol caused additional outward current to flow. In some experiments, the additional current flowing at any potential in the presence of carbachol (i_{carb}) was roughly linearly related to potential as described previously in experiments using extracellular recording of potential (4). In others, where leakage current was small, additional inward current initially increased at potentials positive to the resting membrane potential before declining (Fig. 2b). If we define the additional conductance appearing in the presence of carbachol at any potential V (volts) as

$$\Delta G = \frac{i_{\text{carb}}}{(V - \epsilon)}$$

then the conductance ΔG so defined apparently increased substantially in the region of potential between the resting membrane potential and ϵ (Fig. 2c). If this result can be confirmed, it might provide an explanation for the enhancement of the slow potential brought about by muscarinic stimulants in smooth muscle (2).

DISCUSSION

Although it is probably more difficult to insert and maintain a micro-electrode within a cell during voltage-clamp than to record the membrane potential extracellularly, this voltage-clamp method involving intracellular recording would seem to offer several advantages. With a microelectrode it is possible to explore the potential at any point within the node during voltage-clamp either of the potential recorded by a second microelectrode, or of the potential recorded by extracellular electrodes. The method also has the advantage of current injection across both sucrose gaps, thus effectively allowing a doubling of node width. This was useful since taenia smooth muscle at least showed considerable loss of normal electrophysiologic properties as the node width was reduced to a fraction of the normal space constant. This seems to be a general property of multifiber preparations (8). The resistance in series with the nodal membrane was also apparently more than an order of magnitude smaller than the 10 KΩ encountered when using extracellular potential recording (6). Additionally, the microelectrode method would seem to be capable of recording a greater fraction of the true membrane potential.

The most significant result obtained was the evidence that potential was not constant across the node for 200 msec or more following a step depolarization or step repolarization. Analysis of early inward currents, or tail currents upon repolarization, assuming constant potential, would therefore have been an error in these experiments. The narrowest node width used in the double microelectrode experiments was equivalent, where current injection can be made across only one sucrose gap, to a node width of 0.2 mm. Most workers using extracellular potential recording have claimed gap widths of 0.1 mm or less. It is possible therefore that better potential uniformity was present in their experiments than in mine. However, some facts must be borne in mind. First, the early inward current system in the taenia strips used by me was rather weak, incapable in hypertonic solution of producing regeneratve spikes, whereas in the experiments of others, generally on uterine muscle, the inward current system may well have been stronger, even capable of producing regenerative spikes. (e.g., see refs. 1 and 9). These would probably be more difficult to clamp, particularly if the series resistance was an order of magnitude greater, and if a smaller fraction of the true membrane potential was recorded due to the use of extracellular electrodes.

Besides being a method of investigating nodal potential variation under voltage clamp and the actions of drugs as I have described, the technique would seem to have some usefulness for studying the ionic basis of the slower potential changes of smooth muscle. During these the conductance of the membrane is increased only to a moderate extent so that it should be possible to use relatively wider nodes in which the electrophysiologic properties of the smooth muscle are not changed too drastically.

ACKNOWLEDGMENT

This work was done during the tenure of The Royal Society Locke Fellowship and was supported by the Medical Research Council.

REFERENCES

1. Anderson, N. C. (1969): *J. Gen. Physiol.*, 53:145–165.
2. Bolton, T. B. (1971): *J. Physiol. (London)*, 216:403–418.
3. Bolton, T. B. (1972): *J. Physiol. (London)*, 220:647–671.
4. Bolton, T. B. (1974): *Br. J. Pharmacol.*, 51:129–130P.
5. Bolton, T. B. (1975): *J. Physiol. (London)*, 250:175–202.
6. Connor, J. A., Prosser, C. L., and Weems, W. A. (1974): *J. Physiol. (London)*, 240:671–701.
7. Johnson, E. A., and Lieberman, M. (1971): *Ann. Rev. Physiol.*, 33:479–532.
8. McGuigan, J. A. S. (1974): *J. Physiol. (London)*, 240:775–806.
9. Mironneau, J., and Lenfant, J. (1972): *C. R. Acad. Sci. Paris*, 274:3269–3272.
10. New, W., and Trautwein, W. (1972): *Pflügers Arch.*, 334:1–23.
11. Tomita, T. (1966): *J. Physiol. (London)*, 183:450–468.
12. Tomita, T. (1967): *J. Physiol. (London)*, 191:517–527.

Physiology of Smooth Muscle, edited by
E. Bülbring and M. F. Shuba.
Raven Press, New York © 1976.

Changes in Ionic Fluxes in Uterine Smooth Muscle Induced by Carbachol

M. Worcel and G. Hamon

Physiologie et Pharmacologie, INSERM Unite 7, Hopital Necker, Paris 15, France

The excitatory and inhibitory actions of mediators on the cell membranes of skeletal, smooth, and cardiac muscles as well as on synapses, is generally the consequence of an increase in the membrane permeability to certain ions (1). We have shown previously that angiotensin II depolarizes rat uterus smooth muscle through a primary increase of membrane permeability to Na^+ (2). In order to compare this action with the effect of another spasmogen, we examined the influence of carbachol (CCh) on the ionic movements of the same smooth muscle. Durbin and Jenkinson (3), using isotope techniques, have shown that carbachol increases the outward and inward fluxes of K^+, Br^- and Cl^-, and Na^+ and Ca^{2+} efflux in depolarized preparations. In the same muscle, Bülbring and Kuriyama (4), using the intracellular electrical recording technique, have shown that acetylcholine depolarizes the guinea pig taenia coli cells through an increase in Na^+ and K^+ permeabilities. In another smooth muscle, the guinea pig uterus, Szurszewski and Bülbring (5) have found that acetylcholine depolarization may be due to the increase in Na^+, K^+, and Ca^{2+} permeabilities. The action of CCh on Na^+, K^+, and Cl^- movements of rat myometrium was examined on strips obtained from estrogen-treated rats.

METHODS

Uterine horns were obtained from female Wistar rats 180/200 g wt injected the day before the experiment with 0.5 mg/kg of diethylstilbestrol dipropionate. At 24 hr later, the animals were killed by a blow on the head. Both uterine horns were excised, cleaned, and then opened in the midline. The endometrium was carefully removed with forceps and scissors. The strips, consisting of the peritoneum, subserosal connective tissue, longitudinal smooth muscle, and part of the circular smooth muscle cut transversely, had 10 mm length, 1 mm width, and 0.1 to 0.15 mm depth. They were mounted isometrically at *in vivo* length on Pyrex holders and equilibrated in Ringer solution for at least 90 min before use. The Ringer solution used had the following composition (mM); Na^+ 136.9, K^+ 5.9, Ca^{2+} 2.5, Mg^{2+} 1.2, Cl^- 133.5, HCO_3^- 15.5, $H_2PO_4^-$ 1.2, and glucose 11.5, and was aerated with 95% O_2–5% CO_2; pH 7.30 (35°C).

Measurement of Ionic Fluxes

The tissues were loaded with either ^{24}Na or ^{42}K obtained from the Commissariat de l'Energie Atomique (91 Saclay, France), or ^{36}Cl obtained from the Radiochemical Center (Amersham, Great Britain). Sodium and potassium were supplied as the solution of the chloride salt, and the radioactive chloride as the sodium salt. In all cases the final concentrations of the saline solution were adjusted to compensate for the addition of the radioactive salts. The loading time for efflux experiments was 60 min for ^{36}Cl and ^{24}Na and more than 2 hr for ^{42}K. Equilibration was not attained with this latter isotope. After loading, the tissues were transferred through a series of tubes containing the inactive Ringer solution, bubbled with the O_2–CO_2 mixture. The amount of radioactivity in the tissues at the end of the washout was measured. The strips were cut between ligatures, blotted, weighed, dissolved in concentrated HNO_3, and counted in a well counter, as well as the radioactivity in the washout tubes. The efflux curve was calculated by adding in a reverse order the washout curve to the remaining radioactivity in the tissue at the end of the experiment. The uptakes of ^{36}Cl, ^{24}Na, and ^{42}K were measured by loading the tissues during fixed times at 35°C, according to the technique used by Casteels (6). Effluxes were performed at 4°C, and uptake was calculated by extrapolating to zero time the slow component of the efflux curve (7). Potassium-42 uptakes were also performed by measuring the radioactivity left in the tissues following a 10-min loading time and after washing the strips for 1 min in Ringer solution.

RESULTS

The action of CCh on ^{24}Na, ^{42}K, and ^{36}Cl efflux was studied by adding a certain amount of the drug to some of the washout tubes at times that corresponded to efflux of the ions from the cellular compartment. These parts of the efflux are the last exponentials of the curves for ^{42}K at 35°C, and ^{24}Na and ^{36}Cl at 20°C (7). The actions of the drug are expressed as a function of the rate coefficient for isotope loss

$$r = \frac{C_1 - C_2}{\Delta[(C_1 - C_2)/2]}$$

where C_2 is the radioactivity left in the tissue after a given time Δt has elapsed from the previous level C_1. CCh increases markedly ^{42}K and ^{36}Cl efflux, without modifying ^{24}Na efflux (Fig. 1). At the CCh concentration of 1.6×10^{-4} M, that produces a maximum contractile response of uterine smooth muscle, ^{42}K efflux is increased by 160%, as compared with the increase of ^{36}Cl efflux by 60%.

The CCh effect on ^{42}K and ^{36}Cl efflux is dose-dependent. We can see the effect of different concentrations of CCh on ^{42}K efflux in Fig. 2. A dose–ionic

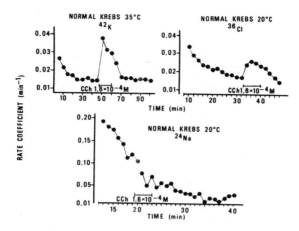

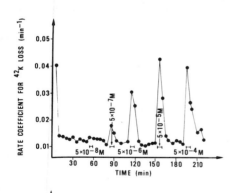

FIG. 1. Action of CCh on rate coefficients of ^{42}K efflux at 35°C and ^{36}Cl, and ^{24}Na effluxes at 20°C. The curves are the average of at least six experiments.

effect curve may be drawn that is perfectly superimposable on the dose–contractile response curve of the muscle. Our findings contrast with the results obtained by Burgen and Spero (8) that indicate that contraction and Rb efflux curves of CCh in guinea pig ileum are different. This is not the case in rat uterine smooth muscle.

Ionic uptake was measured using Casteels' technique (6) of extrapolation to zero time of the slow exponential component of efflux curves, performed

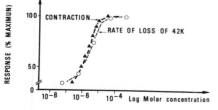

FIG. 2. Upper panel: Single experiment showing the dose–response relationship for the CCh action on ^{42}K efflux coefficient. Lower panel: Average curves of at least six experiments showing dose–contraction and dose–^{42}K rate of efflux curve.

TABLE 1. *Uptake in polarized muscles* (mM/kg)

| | Extrapolation technique | | |
	Controls	Carbachol	P
Na+ 5 min	7.76 ± 0.52 (13)	9.57 ± 0.51 (6)	<0.05
K+ 10 min	9.66 ± 0.64 (24)	12.42 ± 1.42 (8)	<0.05
Cl− 5 min	4.18 ± 0.41 (6)	3.45 ± 0.57 (5)	NS
Isotope Content in Whole Tissues			
K+ 3 min	4.76 ± 0.20 (8)	5.31 ± 0.22 (8)	NS
K+ 5 min	6.57 ± 0.46 (8)	9.10 ± 0.36 (8)	<0.001

at low temperature, after 5 min loading time for ^{24}Na and ^{36}Cl and 10 min for ^{42}K. The use of this technique in uterine smooth muscle has been evaluated in a previous paper (7). As shown in Table 1, CCh 1.6×10^{-4} M increases ^{24}Na and ^{42}K uptakes in polarized muscles significantly. We have also estimated ^{42}K uptake by a direct method. The tissues were loaded during 3 and 5 min in the radioactive solutions. Afterward the strips were rinsed in a nonradioactive Krebs solution to wash out extracellular ^{42}K, and then blotted, weighed, dissolved, and counted. The ^{42}K uptake values per minute with both extrapolation and direct methods are similar, as would be expected and, of course, CCh 1.6×10^{-4} M increases ^{42}K uptake.

CCh produced an increase of ^{42}K and ^{36}Cl effluxes and an increase in

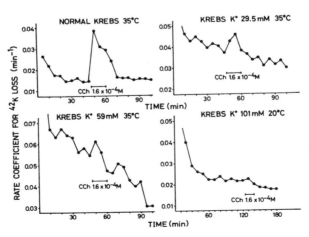

FIG. 3. Average curves of at least five experiments showing the action of CCh (1.6×10^{-4} M) on ^{42}K rate of efflux at different [K+$_0$].

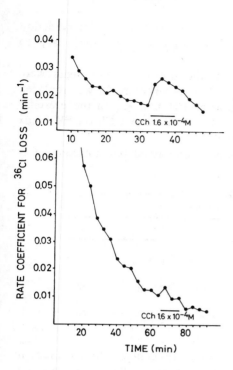

FIG. 4. Average curves of at least five experiments showing the action of CCh (1.6×10^{-4} M) on ^{36}Cl efflux-rate coefficient at 20°C in polarized preparations (*upper panel*) and preparations depolarized by K^+ 101 mM (*lower panel*)

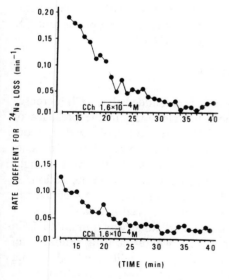

FIG. 5. Average curves of at least six experiments showing the action of CCh 1.6×10^{-4} M on ^{24}Na efflux-rate coefficient at 20°C in polarized preparations (*upper panel*) and preparations depolarized by K^+ 101 mM (*lower panel*).

^{24}Na and ^{42}K uptakes. These effects may be the result of a primary action of the drug on membrane permeability, or they may be secondary to depolarization and spike production.

In order to evaluate these possibilities, the previous experiments were repeated in the presence of depolarizing solutions where NaCl was replaced by KCl in order to obtain a concentration of 101 mM K$^+$. In the presence of increasing concentrations of K$^+$ (Fig. 3), the effect of CCh on ^{42}K efflux was progressively reduced but not abolished.

The action of CCh on ^{36}Cl efflux was also diminished but not completely

TABLE 2. Uptake in depolarized muscles (K$^+$ 101 mM) (mM/kg) using extrapolation technique

	Controls	Carbachol	P
Na$^+$ 5 min	1.59 ± 0.026 (7)	1.78 ± 0.068 (8)	<0.05
K$^+$ 10 min	19.93 ± 1.79 (12)	18.13 ± 1.35 (6)	NS
Cl$^-$ 5 min	10.82 ± 1.00 (6)	11.70 ± 1.33 (6)	NS

abolished by depolarizing solutions (Fig. 4). In contrast to its effect in polarized muscles CCh 1.6×10^{-4} M increased ^{24}Na efflux of depolarized muscles significantly (Fig. 5).

Ionic uptakes by depolarized muscles were measured with the extrapolation technique (Table 2). CCh 1.6×10^{-4} M increased ^{24}Na uptake, but ^{42}K and ^{36}Cl uptakes were not changed in the presence of CCh.

DISCUSSION

From the experimental results described, it is concluded that CCh increases the Na$^+$ permeability of rat uterine smooth muscle since, in the presence of the drug, there is an increased uptake of ^{24}Na in both polarized and depolarized preparations, and ^{24}Na efflux is increased in the depolarized muscle. CCh may also increase K$^+$ and Cl$^-$ permeabilities, as suggested by the slight increment of ^{36}Cl and ^{42}K efflux of depolarized muscles under its action. It is obviously difficult to explain why the drug does not modify ^{24}Na effluxes in polarized muscles, nor ^{42}K and ^{36}Cl influxes in depolarized muscles.

Our results in rat uterine smooth muscle are in some way similar to CCh effects in depolarized guinea pig taenia coli (3), where the drug produces an increase in K$^+$, Cl$^-$, Na$^+$, and probably in Ca^{2+} permeabilities. On the other hand, the analysis of the dose–^{42}K-efflux curve and dose–contraction curve shows a different behavior of rat uterine smooth muscle from what has been observed by Burgen and Spero (8) in intestinal smooth muscle. This suggests that in rat uterine smooth muscle the receptor sites for CCh are uniform, and the limitation of the response at the level of the excitation mechanism is no evidence for the absence of a CCh receptor reserve (9), if it exists in rat myometrium.

ACKNOWLEDGMENT

This work was supported by grants from the I.N.S.E.R.M. and D.G.R.S.T. (France).

REFERENCES

1. Ginsborg, B. L. (1967): *Pharmacol. Rev.*, 19:289.
2. Hamon, G. and Worcel, M. (1973): *J. Physiol. (Lond.)*, 232:99p.
3. Durbin, R. P., and Jenkinson, D. H. (1961): *J. Physiol. (Lond.)*, 157:79.
4. Bülbring, E., and Kuriyama, H. (1963): *J. Physiol. (Lond.)*, 166:59.
5. Szurszewski, J. H., and Bülbring, E. (1973): *Philos. Trans. R. Soc. Lond. [Biol. Sci.]*, 265:149.
6. Casteels, R. (1969): *J. Physiol. (Lond.)*, 205:193.
7. Hamon, G., Papadimitriou, A., and Worcel, M. (1974): *In press*.
8. Burgen, A. S. V., and Spero, L. (1968): *Br. J. Pharmacol.*, 34:99.
9. Nickerson, R. (1957): *Pharmacol. Rev.*, 9:246.

Physiology of Smooth Muscle, edited by
E. Bülbring and M. F. Shuba.
Raven Press, New York © 1976.

Mechanism of the Excitatory and Inhibitory Actions of Catecholamines on the Membrane of Smooth Muscle Cells

M. F. Shuba, A. V. Gurkovskaya, M. J. Klevetz, N. G. Kochemasova, and V. M. Taranenko

Department of Nerve–Muscle Physiology, A. A. Bogomoletz Institute of Physiology, Kiev, USSR

Although, under special conditions, catecholamines can produce contraction and relaxation of smooth muscles without changes in membrane potential or conductance, it seems likely that, *in vivo,* membrane potential and conductance changes are of great importance in their mode of action.

Our reasons for believing this are based on the view that, especially in smooth muscles with a poor adrenergic innervation, some cells will not be directly influenced by neurally released norepinephrine (NE). However, the depolarization or hyperpolarization produced in the directly affected cells will spread electrotonically to the noninnervated cells. In addition, the increased membrane conductivity in directly affected cells, e.g., during adrenergic inhibition, will tend to shunt any excitatory membrane changes occurring in surrounding cells.

On the other hand, if NE acted directly on the contractile processes within the cell rather than on the electrical properties of the membrane, then the action of the transmitter would be much less effective because contraction or relaxation could not spread from one cell to another.

In this chapter, we describe some changes in membrane properties induced by catecholamines in several smooth muscle preparations.

THE INHIBITORY ACTION OF CATECHOLAMINES

In 1954, Bülbring (1), using microelectrodes, showed that the inhibitory action of epinephrine (E) on the guinea pig's taenia coli was associated with hyperpolarization and suppression of spontaneous spike discharge. Both the stimulation of an electrogenic sodium pump (2) and a decrease of the membrane Na conductance (3) were early suggestions as causes of this hyperpolarization.

In our own laboratory, however, we came to the conclusion that the hyperpolarization underlying the inhibitory action of E and NE on gut muscle is the result of an increased membrane permeability to potassium (4–6). Initial investigations carried out in 1961 showed that, in the smooth muscle of frog stomach electrotonic potentials evoked by constant current pulses decreased in the presence of E, indicating an increased membrane conductance (7). Substitution of the NaCl in the bathing solution by sucrose was without effect on this conductance increase (8). Further experiments on both the frog stomach and the guinea pig's taenia coli, using the double sucrose gap (9), again showed that the hyperpolarization evoked by E and NE, was associated with an increased membrane conductance, as measured by the change in amplitude of the electrotonic potentials (Figs. 1 and 2). Substitution of the NaCl in the bathing solution by TEA–Cl, choline–Cl, or sucrose did not alter the effects of the drugs (Figs. 1 and 2) (6,10–14). The ineffectiveness of Na^+ and Cl^- substitution indicated that the hyperpolarization produced by catecholamines was caused largely by an increase in the membrane permeability to potassium.

This suggestion was also confirmed in experiments in which the dependence of E or NE hyperpolarization on the level of the resting potential was studied. It was found that in K^+-free solution, or when the level of the membrane potential was decreased by passing constant currents, then the hyperpolarization produced by E or NE was increased (Fig. 2C). On the other hand, in K-rich solution, or when the membrane potential was hyper-

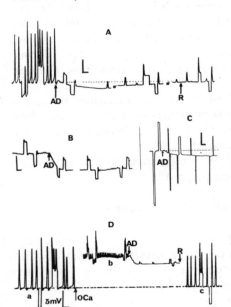

FIG. 1. Hyperpolarization, inhibition of spontaneous activity and of excitability, and decrease in amplitude of the electrotonic potentials in the muscle cells of frog stomach produced by epinephrine (E) (10^{-6} g/ml), in normal Ringer solution (A), Na-free choline solution, (B), sucrose–Ringer solution without NaCl (C), and in Ca-free solution (D). D shows the spontaneous and evoked activity and the electronic potentials in normal Ringer solution (a), on the minute in Ca-free solution with E (b), and again in normal Ringer solution (c).

Note: In this and the following figures, an arrow with the letters E, R (or NE, Kr, OCa) indicates the application of E (or NE), (AD, adrenaline; NA, noradrenaline of the normal Ringer solution (or Krebs). The dotted line indicates the relative zero potential. Numbers breaking these and other traces indicate gaps (in minutes) in the recording.

polarized by constant current, then the E or NE hyperpolarization was decreased. These effects are brought about presumably by changing E_K when $[K^+]_0$ is varied, or by varying the difference between the membrane potential and E_K in the case of constant currents.

Finally, the effect of Ca-free solutions was determined (Figs. 1 and 2). In such solutions, the muscle cells were markedly depolarized and the membrane resistance was decreased, suggesting a high Na^+ permeability, since these effects of Ca-free solution did not occur in Na-free sucrose solution (11,12,15). Nevertheless, in Ca-free solutions the hyperpolarization produced by E and NE persisted (Fig. 1D). Even in sucrose Krebs solution free of Na^+, Ca^{2+}, and Cl^- and containing EDTA the E and NE hyperpolarization was still observed (Fig. 2D). However, in Ca-free solution the catecholamine effect was reduced (6,11,12,14).

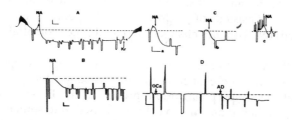

FIG. 2. Hyperpolarization, inhibition of spontaneous activity, and decrease of electrotonic potential in the guinea pig taenia coli produced by NE in normal Krebs solution (A), in a sucrose Krebs solution without NaCl (B), in a K-free solution (C, a), in normal Krebs solution (5.9 mM K^+) (C, b), and with 29.5 mM K^+ (C, c), and in a Ca-free sucrose Krebs solution with EDTA (D). In records A and B two different intensities of hyperpolarizing current were used. *Calibrations:* horizontal = 1 min; vertical = 1 mV (A–C), 5 mV (D).

In 1969 Bülbring and Tomita (16,17), using a double sucrose gap, confirmed the finding that an increased potassium permeability was the main cause of the hyperpolarization seen on application of NE or E to the guinea pig's taenia coli.

They suggested that the presence of Ca^{2+} was essential for the action of E, since the hyperpolarization and increase in membrane conductance by E was abolished by addition of Mn^{2+}, or Ca replacement with Ba^{2+} (42), or prolonged exposure to Ca-free solution in which either the Na^+ was also removed or 2 mM Mg^{2+} was added (43). Bülbring and Tomita also presented evidence supporting an increase in the chloride permeability of the membrane by E. No evidence of this was seen by Jenkinson and Morton (18,19) in their studies on the ion fluxes in the depolarized taenia coli. However, they reported an increased exchange of tracer K^+ on application of catecholamines. Both sets of authors showed the increase in K^+ permeability produced by

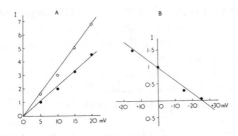

FIG. 3. A: The current–voltage characteristic of the membrane of the muscle cells of the guinea pig's taenia coli before (●) and during the action of NE (10^{-6} g/ml) (○). The holding potential, 55 mV, corresponds to the zero potential on the abscissa. B: The dependence of the intensity of the outward current evoked by NE(10^{-6} g/ml), on the level of the holding potential. Zero on the abscissa corresponds to the holding potential of 55 mV. The outward ionic current induced by NE is taken as 1 at this potential (see ordinate).

catecholamines to be an α effect. The β effect of catecholamines, produced by isoprenaline, consisted of suppression of spontaneous electrical and mechanical activity with no changes in membrane potential or resistance (17,19).

More recently, Inomata and Kao (20), using voltage-clamped taenia coli, found that E produced no change in membrane conductance. They therefore concluded that the hyperpolarization produced by E was caused by an increase in E_K rather than G_K. We have carried out similar experiments, and we found that NE produced not only an outward ionic current in clamped muscles, but also increased the membrane conductance (Figs. 3A,B) as has also been reported by Tomita (21). Furthermore, we found that decreasing the clamp potential increased the outward current evoked by NE, whereas increasing the clamp potential decreased this outward current. At a holding potential of 75 to 80 mV, application of NE produced no current flow. This potential presumably corresponds to the NE equilibrium potential (Fig. 3B).

TABLE 1. *The suggested ionic mechanisms of the inhibitory action of catecholamines*

Smooth muscles	Activation of receptors	Change in the resting potential	Change in the ion conductance of the membrane				Activation of ionic pump		Increase in E_K	References
			G_K	G_{Na}	G_{ca}	GC_l	Na	Ca		
GI tract	α	+	+	0	0	0	0	0	0	4–8, 10–12, 18, 19
			+	0	0	+	0	0	0	16, 17
			0	0	0	0	0	0	+	20
Myometrium	β	+	0	0	0	0	0	0	+	22
			0	0	0	0	+	0	0	23
			0	0	0	0	0	+	0	24
Ureter	β	+	+	0	0	0	0	0	0	25, 26
Blood vessels	β	+	0	0	0	0	+	0	0	27

Symbols: + = increase; 0 = no effect.

The suggested ionic mechanisms underlying the inhibitory actions of catecholamines on these and other smooth muscles are summarized in Table 1.

THE EXCITATORY ACTION OF CATECHOLAMINES

The probable permeability changes underlying the excitatory action of catecholamines on certain blood vessels (28–31), the ureter (25,26,32), vas deferens (33), and myometrium (34) suggested by different authors are shown in Table 2. Here experiments on the rat vena porta, rabbit pulmonary artery, and guinea pig ureter are described.

TABLE 2. The suggested ionic mechanism of the excitatory action of catecholamines

Smooth muscles	Activation of receptors	Change in the resting potential	Change in the ion conductance of the membrane				References
			G_K	G_{Na}	G_{ca}	GC_l	
Portal vein	α	−	0	+	+	+	28, 29
			0	0	0	+	36
Common carotid artery		−	+	+	0	+	31
Pulmonary artery	α	−	−	0	0	0	30
Other blood vessels	α	−	0	0	+	0	27
Vas deferens	α	−	+	+	+?	0	33
Ureter	α	−	−	0	0	0	25, 26, 32
Myometrium	α	−	+	0	+	+	34

Symbols: − = decrease; + = increase; 0 = no effect.

Using the double sucrose gap, it was found that both E and NE produced depolarization and increased spontaneous electrical activity in the rat vena porta. The size of the electrotonic potential to constant current pulses decreased, indicating that these changes were associated with a fall in membrane resistance (Fig. 4). Hence the depolarization produced was unlikely to be due to a fall in potassium permeability. In solutions where both the NaCl and NaHCO$_3$ were replaced by sucrose the spontaneous activity of the muscle ceased and the membrane resistance increased. Spikes, however, could still be evoked by direct electrical stimulation (35). In such solutions NE and E still produced depolarization, spontaneous activity, and lowered membrane resistance, the effect being larger than in normal solution. We next looked at the effects of Ca-free solution. In such a solution both drugs still caused depolarization even in the presence of EDTA. Thus, when either NaCl (sucrose substitution) or Ca were removed, the response was still observed. If NaCl and Ca were removed together, the preparation deteriorated and the response to catecholamines could not be tested satisfactorily. We conclude that in this tissue the excitatory action of catecholamines consists of increases in the ionic conductances of Ca^{2+}, Cl^-, and Na^+.

It is difficult to determine the relative contributions of these three ionic conductances. However, the depolarizations produced by drug application in Na-free sucrose solution were greater than those in Ca-free solution. This indicates that a change in G_{Ca} may be more important in the depolarizing

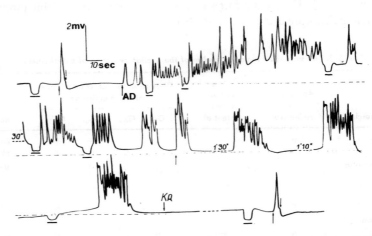

FIG. 4. Depolarization, appearance of spontaneous activity, and decrease of electronic potential in the muscle cells of rat vena porta under the influence of E (10^{-6} g/ml). Unlabeled arrows show the effect of rectangular depolarizing pulses. The bars show the anelectrotonic potential.

action of catecholamines in this tissue. However, the fact that in Ca-free solutions the cells are depolarized, whereas in sucrose solution they are hyperpolarized, may account for the differences seen.

Recently Wahlström (36) investigated the effects of Na on ion fluxes in this tissue. He found that the main effect was an increase of the chloride permeability.

Most recently we have studied the action of NE on the rabbit pulmonary artery, using intracellular microelectrodes. Under normal conditions depolarizing current pulses produced no spikes and no contraction in muscle cells of this artery (Fig. 5A). Nor did hyperpolarization lead to relaxation. Addition of NE led to a depolarization of 5 to 7 mV and contraction. The membrane resistance was increased (Fig. 5B) indicating that here the depolarization was caused by decreased G_K. In the presence of the drug, the hyperpolarizing pulses now relaxed and the depolarizing pulses now contracted the tissue (Fig. 5B). Whether this effect was due directly to NE or to the depolarization produced by NE is not yet known. Phentolamine pre-

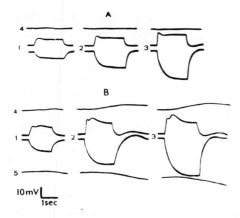

FIG. 5. The dependence of the amplitude of the electrotonic potentials and the contraction of muscle cells of arteria pulmonalis on the intensity of the rectangular current pulse. In traces 2 and 3 the current pulses were, respectively, 2 and 3 times greater than in trace 1. A in the absence, B in the presence of NE (10^{-7} g/ml). The traces labeled 4 show the contraction of the muscle strip during the depolarizing pulses; those labeled 5 show the relaxation during hyperpolarizing pulses.

vented the excitatory action of NE indicating that this effect is mediated by α receptors.

We have also studied the effects of catecholamines on guinea pig ureter. Here NE induced a small depolarization and some increase in membrane resistance (Fig. 6A). However, its most marked effect was a lengthening of the action potential's plateau and an increase in the amplitude and duration of the phasic contraction associated with the action potential. NE was without effect on the initial fast component of the action potential (Fig. 6B).

The depolarization produced by NE was associated with increased membrane resistance and was probably caused by a decreased K^+ permeability. It seems unlikely that the increased duration of the plateau is also associated with this depolarization, since application of constant depolarizing current did not lengthen the plateau. Thus the extension of the plateau by NE appears to be caused by a specific action on the mechanism of the action potential. This effect of NE could be due either to an increase in the inward NE^+ current, known to occur during the plateau (37–41), or a delay in the activation of a repolarizing outward K^+ current. To determine which ion

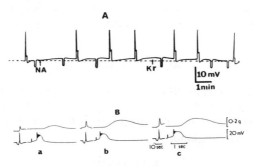

FIG. 6. The action of NE (10^{-5} g/ml) on the electrophysiologic properties of the muscle cells of guinea pig ureter. A: Depolarization and increase of electrotonic potentials. B: Increase of the duration of the plateau of AP and of the phasic contraction by NE. (a) Control; (b) during first minute; (c) during third minute of exposure to NE.

channels in the plateau NE was affecting, we looked at the effects of changing the ionic composition of the bathing solution and of applying specific blockers of certain ion channels.

The results of these experiments indicated that the prolongation of the action potential plateau caused by NE and E is due to a specific increase in the activation of the slow Na channels, the conductivity of which is potential-dependent (39–41). We feel that the increases seen in the amplitude and duration of the phasic contraction are secondary to these changes in the plateau of the action potential.

CONCLUSIONS

The ionic mechanisms underlying the inhibitory actions of catecholamines on smooth muscles are diverse (see Table 1). In gut muscle, inhibition seems to be caused by an increase in G_K leading to hyperpolarization and increased membrane conductance. Thus inhibition is achieved both by the hyperpolarization itself and by the electrical shunting created by the raised membrane conductance.

We consider that inhibitory hyperpolarization by electrogenic Na pumping is unlikely to be effective. First, the electrogenic pump is probably only capable of producing very small changes in membrane potential. Second, the shunting effect of a raised membrane conductance would be absent. Further, this type of inhibition may be metabolically too "expensive."

The membrane changes underlying the depolarization associated with the excitatory action of catecholamines also vary from tissue to tissue (see Table 2). Depolarization may be caused through increases in membrane conductances to K^+, Na^+, Cl^-, or Ca^{2+} in various combinations, or through a decrease in K^+ permeability.

Many of the ionic mechanisms, listed in Tables 1 and 2, which have been suggested as underlying the action of catecholamines, are rather tentative. In our view the most substantial evidence is for an increase in G_K during α-inhibition of gut muscle, and for an increased slow inward Na current during the action potential plateau during α excitation of the guinea pig's ureter.

ACKNOWLEDGMENT

The experimental results illustrated in Figs. 3 and 6b were obtained by M. F. Shuba during work, which was supported by the Medical Research Council, in the laboratory of Edith Bülbring in the Department of Physiology, University of Oxford, England.

REFERENCES

1. Bülbring, E. (1954): *J. Physiol. (Lond.)*, 125:302–315.
2. Bülbring, E. (1962): *Physiol. Rev.* 42, Suppl. 5:160–178.

3. Bülbring, E., and Kuriyama, H. (1963): *J. Physiol. (Lond.)*, 166:59–74.
4. Shuba, M. F. (1964): In: *Biophysics of Muscular Contraction*, edited by G. M. Franan, pp. 50–51. Moscow.
5. Shuba, M. F. (1965): In: *Role of Humoral Factors in Synaptic Transmission*, pp. 159–161. Kazan, USSR.
6. Shuba, M. F. (1966): In: *Biophysics of Muscular Contraction*, edited by G. M. Franu, pp. 126–131, Moscow, USSR.
7. Shuba, M. F. (1961): *Sechenov Physiol. J. USSR (English Transl.)*, 47;1068–1073.
8. Shuba, M. F. (1961): *Fiziol. Zh. Acad. Sci. Ukr. SSR*, 7:595–601.
9. Artemenko, D. P., and Shuba, M. F. (1964): *Fiziol. Zh. Acad. Sci. Ukr. SSR*, 10:403–407.
10. Klevets, M. Y. (1966): *Fiziol. Zh. Acad. Sci. Ukr. SSR.*, 12:763.
11. Shuba, M. F. (1967): Electrophysiological properties of smooth muscles. Synopsis of Doctor Diss. Kiev, USSR.
12. Shuba, M. F., and Klevets, M. Y. (1967): *Fiziol. Zh. Acad. Sci. Ukr. SSR.* 13:3–11.
13. Bogach, P. G., and Klevets, M. Y. (1967): *Biophysics (English Transl.)*, 12:1140–1146.
14. Klevets, M. Y., and Shuba, M. F. (1968): In: *Synaptic Processes*, pp. 92–108: Kiev, USSR.
15. Shuba, M. F. (1965): In: *Protoplasmatic Membranes, and Their Functional Role*, pp. 90–107. Kiev, USSR.
16. Bülbring, E., and Tomita, T. (1969): *Proc. R. Soc. Lond. B.* 172:89–102.
17. Bülbring, E., and Tomita, T. (1969): *Proc. R. Soc. Lond. [Biol. Sci.]*, 172:103–119.
18. Jenkinson, D. H., and Morton, I. K. M. (1967): *J. Physiol. (Lond.)*, 188:373–386.
19. Jenkinson, D. H., and Morton, I. K. M. (1967): *J Physiol. (Lond.)*, 188:387–402.
20. Inomata, H., and Kao, C. Y. (1972): *J. Physiol. (Lond.)*, 226:53–54P.
21. Tomita, T. (1974): *Nature*, 250:432–433.
22. Kao, C. Y., McCullough, J. R., and Davidson, H. L. (1971): *Fed. Proc.* 30:384.
23. Magaribuchi, Y., and Osa, T. (1971): *Jap. J. Physiol.*, 21:627–643.
24. Marshall, J. M., and Kroeger, E. A. (1973): *Philos. Trans. R. Soc. Lond. [Biol. Sci.]*, 265:135–148.
25. Kochemasova, N. G., and Shuba, M. F. (1972): *Sechenov Physiol. J. USSR* 58: 426–433.
26. Kochemasova, N. G., and Shuba, M. F. (1972): *Sechenov Physiol. J. USSR*, 58: 1287–1294.
27. Somlyo, A. P., and Somlyo A. V. (1970): *Pharmacol. Rev.*, 22:251–253.
28. Taranenko, V. M. (1970): *Bull. Exp. Biol. Med. (Engl. Transl.)*, 69:104–107.
29. Taranenko, V. M., and Shuba, M. F. (1970): *Neurophysiology (USSR)*, 2:643–653.
30. Gurkovskaya, A. V. (1974): In: *Physiology of Smooth Muscles*, pp. 43–44. Kiev.
31. Mekata, F., and Niu, H. (1972): *J. Gen. Physiol.*, 59:92–102.
32. Shuba, M. F. (1974): *J. Physiol. (London)*
33. Maganibuchi, T., Ito, Y. & Kuriyama, H. (1971): *Jap. J. Physiol.*, 21:691–708.
34. Bülbring, E., and Szurszewski, J. H. (1974): *Proc. R. Soc. Lond. [Biol. Sci.]*, 185:225–262.
35. Taranenko, V. M., and Shuba, M. F. (1970): *Sechenov Physiol. J. USSR*, 56:1149–1155.
36. Wahlström, B. (1972): *J. Physiol. (Lond.)*, 226:63P.
37. Kuriyama, H., and Tomita, T. (1970): *J. Gen. Physiol.*, 55:147–162.
38. Kochemasova, N. G. (1971): *Bull. Exp. Biol. Med. (Engl. Transl.)*, 72:983–987.
39. Shuba, M. F., and Bury, V. A. (1971): In: *Eur. Biophys. Congr., 1st*, edited by E. Broda et al., pp. 265–268, Baden, Austria.
40. Bury, V. A. (1973): *Schenov Physiol. J. USSR*, 59:1608–1616.
41. Bury, V. A., and Shuba, M. F. (1974): *Sechenov Physiol. J. USSR*, 60:1288–1297.
42. Bülbring, E., and Tomita, T. (1969): *Proc. R. Soc. Lond. [Biol. Sci.]*, 172:121–136.
43. Bülbring, E. (1970): *Rend. Gastroenterol.*, 2:197–207.

Physiology of Smooth Muscle, edited by
E. Bülbring and M. F. Shuba.
Raven Press, New York © 1976.

The Role of α-Adrenoceptors Situated in Auerbach's Plexus in the Inhibition of Gastrointestinal Motility

E. S. Vizi

Department of Pharmacology, Semmelweis University of Medicine, 1085 Budapest, Hungary

It is generally accepted that the parasympathetic cholinergic outflow to the intestine has a motor action on both longitudinal and circular muscle layers except in the region of the sphincters where it is inhibitory. It is now also established that stimulation of the thoracolumbar sympathetic outflow, of splanchnic, or of mesenteric nerves reduces or inhibits the motility of the gastrointestinal tract (19). The sympathetic outflow is commonly held to be antagonistic to the parasympathetic outflow. The inhibitory effect of the sympathetic transmitter (norepinephrine) has usually been considered in terms of a direct action on the smooth muscle (1,13,22). However, recent anatomic and physiologic findings have caused a reevaluation of the classical concepts of the inhibitory innervation of the intestine. It was shown by Paton and Vizi (44) and Vizi (50) that norepinephrine (NE) and epinephrine are capable of inhibiting the release of acetylcholine (ACh) from Auerbach's plexus and that this inhibitory action is mediated via α adrenoceptors. Sympathetic control of ACh output can be viewed as a kind of presynaptic inhibition and, when compared with an antagonism at the effector level, offers the physiologic advantage of economy in transmitter release. This type of interaction between the adrenergic and cholinergic nervous system exists mainly in the gastrointestinal tract (52). In addition, another inhibitory nonadrenergic, noncholinergic ("purinergic") innervation has recently been described (5,19), and it has been proposed that the transmitter may be adenosine triphosphate (ATP) or a closely related nucleotide (6).

INHIBITORY EFFECT OF NE ON THE RELEASE OF ACh FROM AUERBACH'S PLEXUS

If intestinal smooth muscle contracts spontaneously, or in response to drugs, or to stimulation of its cholinergic motor nerve, the increase in tension can be correlated with the frequency of firing of action potentials. ACh, the excitatory transmitter, depolarizes the membrane and increases spike activity, and any drug which decreases the release of ACh would be expected to reduce both the electrical and mechanical activity of the smooth muscle.

Stimulation of adrenergic nerves supplying the gastrointestinal tract reduces spontaneous intestinal motility (19,53) and inhibits contractions in response to stimulation of vagal cholinergic nerves (7,26,52,53). An interesting observation by McDougal and West (37) showed that, in the guinea pig ileum, the peristaltic reflect is blocked by concentrations of NE which are lower than those required to affect the sensitivity of smooth muscle to ACh.

Paton and Vizi (44) and Vizi (50) obtained direct evidence that NE and epinephrine inhibit the release of ACh from Auerbach's plexus in response to field stimulation. A similar finding has been described by others (3,28, 34,49). Del Tacca and co-workers (49) showed that NE is able to inhibit the release of ACh from human taenia coli. Indirectly acting sympathomimetics (30,31) or stimulation of the sympathetic nerves (3,53) can reduce the release of ACh. The inhibitory effect of NE was prevented by phentolamine and by other α adrenoceptor-blocking drugs (44,50,51,39) indicating that this action is mediated via α adrenoceptors.

Since the abolition of the sympathetic nervous action by reserpine or by guanethidine (3,28,44,50,53) and the prevention of the effect of NE on α adrenoceptors by α-blocking agents (44) enhances the release of ACh, it is suggested that there may be a permanent sympathetic control of ACh release. Reserpine pretreatment enhanced both the peristalsis and the release of ACh in whole ileum preparations of the guinea pig (28). These experimental data are in good agreement with the clinical observations that in patients treated with reserpine alkaloids, or with guanethidine, or with α-adrenoceptor blocking drugs, an increase of the gastrointestinal motility and diarrhea are very common side effects. The results appear to favor the role of NE as the transmitter predominantly responsible for the inhibition of the motor responses of the gut. It is also possible to explain the beneficial action of α-adrenoceptor blocking drugs (38,47) in paralytic ileus, which may be caused by an increased sympathetic outflow.

Evidence has been presented that gastrointestinal hormones (gastrin, cholecystokinin) may play a role as hormonal factors in regulating gastrointestinal motility via ACh release, and that their effect is continuously controlled by the tonic activity of the sympathetic nervous system which reduces ACh release (51,54,55,57). It has also been shown that gastrin and gastrin-like polypeptides stimulate gastrin-sensitive receptors of the ganglia in Auerbach's plexus leading to a release of ACh (51). The ACh release induced by gastrin-like polypeptides can be inhibited by NE (51,54), and α-adrenoceptor blocking agents are able to prevent the effect of NE (51) or of sympathetic nerve stimulation (54).

When the contractions of a longitudinal muscle strip were used as a measure of the ACh released by stimulation, β-adrenoceptor blocking agents were not able to prevent the inhibitory effect (31,34); however, α-adrenoceptor blocking agents prevented the inhibitory effect of NE or epinephrine.

These results are in good agreement with those from direct measurements of ACh release.

Although catecholamines have a direct α and β action on intestinal smooth muscle, the α adrenoreceptors situated in Auerbach's plexus appear to be of greater importance for the inhibition of intestinal movement. Ahlquist and Levy (1) and Furchgott (15) arrived at the conclusion that intestinal relaxation induced by sympathetic agents and by NE release is mediated through stimulation of α and/or β receptors of the smooth muscle. However, the fact that cold storage prevents the relaxation produced by stimulation of α receptors (36) while the sensitivity of the rabbit jejunum to β-adrenoceptor stimulation remains unchanged, indicates that the α adrenoceptors are located on the nervous tissue which is very sensitive to cold storage.

Prostaglandin E_1 has been shown to reduce the inhibitory effect of sympathetic nerve stimulation on intestinal motility (24a) and to enhance the responses to both nerve stimulation and exogenous ACh in the guinea pig longitudinal muscle strip (24a) and whole-ileum preparations (2,27). It was also shown that indomethacin, an inhibitor of endogenous prostaglandin synthesis, markedly reduced the contractions of whole ileum in response to nerve stimulation (27). It seems very likely that the ability of prostaglandin to remove noradrenergic restraint on ACh release (by inhibiting the release of NE) also plays some role in the enhancement of gastrointestinal motility by prostaglandin (4).

FREQUENCY DEPENDENCE OF THE INHIBITORY EFFECT OF NE ON ACh RELEASE

It has been shown that both exogenous and endogenous (released by stimulation of sympathetic nerves) NE can reduce the release of ACh from the myenteric plexus.

The inhibitory effect of α-adrenoceptor stimulation on ACh release is significant provided low frequency of stimulation (<3 Hz) or a short train of high-frequency stimulation is applied (30,31,52). The ACh release per

TABLE 1. *The rate of acetylcholine release from different tissues*

Rate of stimulation (Hz)	Acetylcholine release (pmol/g.min)($\pm$SEM)				
	0.1	1	5	10	Resting
Longitudinal muscle of guinea pig ileum	404.4 ± 31.7 (43.2)	846.8 ± 68.3 (11.1)	2117 ± 195 (6.0)	4081.2 ± 361.6 (6.45)	221.4 ± 11.8
Guinea pig taenia coli		58.2 ± 11.7 (0.39)	513.6 ± 144.2 (1.6)	1706.5 ± 210.5 (2.1)	29.5 ± 1.7

The ACh output per pulse (pmol/volley.g) is given in parentheses. For details of assay and calculation see Paton and Vizi (1969).

pulse is greater at low (11,30–32,50) than at high frequencies (Table 1). It is interesting that taenia coli behaves in a quite different way: the higher the frequency of stimulation, the higher the release per pulse (volley output). In the longitudinal muscle strip of the guinea pig ileum the ACh output per pulse can be reduced by NE to the level obtained by sustained stimulation with high frequency (10 Hz; Table 2). Knoll and Vizi (31) showed that the ACh output per pulse induced by the first shocks was reduced by NE even when a higher frequency of stimulation was used. Nishi and North (39) provided electrophysiologic evidence for this finding. It was shown that NE depresses the amplitude of the first EPSP but had little effect on subsequent EPSPs.

TABLE 2. Inhibitory effect of different conditions on the release of acetylcholine from Auerbach's plexus

	Acetylcholine release pmol/g.volley (±SEM)	P
0.1 Hz stimulation, control	49.5 ± 3.8 (10)	—
+Mg-excess (9.3 mM)	10.5 ± 1.2 (5)	<0.01
+(−)-norepinephrine (10⁻⁶ M)	7.6 ± 1.0	<0.01
+ATP, 10⁻⁵ M	21.0 ± 1.9 (4)	<0.01
+Morphine, 5 × 10⁻⁶ M	15.3 ± 1.1 (5)	<0.01
10 Hz stimulation, control	6.45 ± 0.34 (10)	<0.01

Number of experiments in parentheses.

In the isolated vagus–stomach preparation the contractions of the stomach in response to preganglionic stimulation (4 Hz) were completely inhibited either by morphine or by the stimulation of α adrenoceptors with clonidine (Catapresan). Figure 1 shows one of the tracings obtained. The responses of the stomach to ACh added to the bath was not affected. A similar observation was made (Szabolcsi et al., *in press*) in the isolated vagus–esophagus preparation.

The question now arises: why, during preganglionic stimulation, is NE effective at sustained low or high frequency of stimulation, whereas, when field, i.e., axonal, stimulation is applied NE is effective only at low frequency? (See Fig. 2.) One possible explanation is that, during preganglionic stimulation, the excitatory impulses are relayed at least once in the myenteric ganglia, where ACh, released from the preganglionic nerve terminals, is used to excite the intramural motor neuron. NE and α-adrenoceptor stimulation can completely inhibit the release of ACh by direct stimulation of ganglion

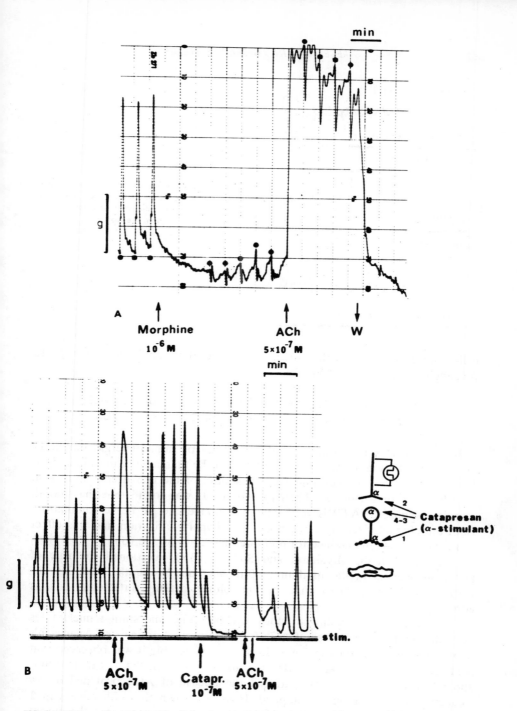

FIG. 1. Inhibitory effect of morphine (A) and α-adrenoceptor stimulation (Catapresan) (B), on the contractions of longitudinal muscle of rat stomach in response to electrical stimulation of the vagus nerve. Isolated vagus–stomach preparation (from Ref. 43). At dots, stimulation: 5 V, 0.2 msec, 4 Hz, 10 shocks every 30 sec. Isometric recording. Organ bath, 10 ml Krebs solution. W, Wash out. (A) Note, after acetylcholine (ACh) was added, there was a relaxation in response to vagus stimulation. (B) Catapresan (Catapr.) inhibited the contractions in response to vagal stimulation. Note that the response to acetylcholine (ACh) was not affected by Catapresan, indicating that the sensitivity of the effector cells of ACh was not impaired. The inset shows the possible sites of action of Catapresan (see Fig. 4). *Black bars* (stim): every 30 sec stimulation with above parameters.

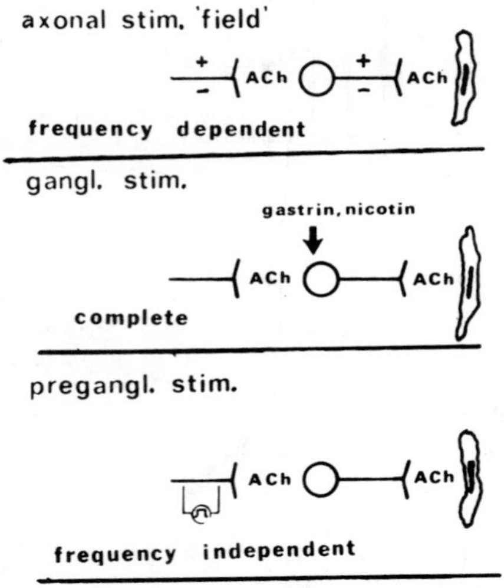

axonal stim. 'field'

frequency dependent

gangl. stim.

gastrin, nicotin

complete

FIG. 2. Inhibitory effect of α-adrenoceptor stimulation on acetylcholine release from myenteric plexus induced by different stimulation techniques. For description, see text.

pregangl. stim.

frequency independent

cells with nicotine or with gastrin (51), and this condition is similar to preganglionic stimulation (Fig. 2).

INHIBITORY EFFECT OF ATP AND RELATED NUCLEOTIDES ON THE RELEASE OF ACh FROM AUERBACH'S PLEXUS

Burnstock (5) has postulated the presence of nonadrenergic inhibitory neurons in addition to the adrenergic fibers in the mammalian intestinal tract. He proposed that the neurohumoral transmitter of the nonadrenergic neuron may be ATP or a closely related nucleotide. It was found that the smooth muscle response to ATP is very similar to nonadrenergic nerve stimulation. Both are characterized by a rapid onset of hyperpolarization and this effect is transient, being maintained for no more than 20 to 30 sec (5).

However, Kuchii, Miyahara, and Shibata (35) questioned the role of purinergic nerves in the inhibition of taenia coli. Weisenthal and his co-workers (57) postulated that the inhibitory activity in taenia coli is adrenergic.

It seemed very interesting to study whether ATP and similar nucleotides are able to affect the release of ACh from the nerve terminals of Auerbach's plexus. Using the longitudinal muscle strip of guinea pig it was observed that ATP (2×10^{-5} to 10^{-4} M), ADP (10^{-5} to 10^{-4} M), AMP (10^{-5} to 10^{-4} M), and adenosine (5×10^{-5} M) reduced the release of ACh provided a low frequency of stimulation was applied. However, when the ACh release in 1 min was measured in response to 10 Hz stimulation (600 pulses) the output was 3,670 and 3,590 pmol/g-min in the absence and presence of ATP (10^{-5} M), respectively. ATP and related nucleotides reduced the responses of the longitudinal muscle strip to 0.1 Hz stimulation (Fig. 3). A similar

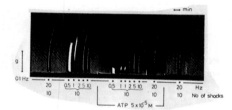

FIG. 3. Inhibitory effect of adenosine triphosphate (ATP) on the neuroeffector transmission of the longitudinal muscle of guinea pig ileum. Note that the higher the stimulation frequency, the less the inhibitory effect of ATP. For experimental conditions see Fig. 1 "Field" stimulation. 10 V/cm, 1 msec. Isometric recording.

reduction was observed when a train of 10 pulses was applied at 0.5, 1.0, and 2.0 Hz stimulation. However, the contraction induced by 10 pulses at 20 Hz was not affected indicating that the sensitivity of the smooth muscle to ACh was not changed. Therefore the inhibitory effect of ATP is presynaptic and may be attributed to a reduction in ACh release. Phentolamine failed to antagonize the inhibitory action of ATP or of related nucleotides.

It is suggested that ATP or related nucleotides, if they are released in the vicinity of the cholinergic nerve terminals, may be inhibitory in action and modulate the release of ACh and gastrointestinal motility.

SITE OF ACTION OF NOREPINEPHRINE

The next question that arises is: where are the α adrenoceptors situated?

Effect on the Nerve Terminals

Paton and Vizi (1969) suggested that the site of action of NE is on the α adrenoceptors situated on the nerve terminals. Since field stimulation reduces ACh release by α-adrenoceptor stimulation, and field stimulation (Fig. 3) excites, besides ganglion cells, the interneuronal (preganglionic) as well as preeffectorial (motor) axons, the effect of NE may be on both the preeffectorial and interneuronal nerve terminals (sites 1 and 2 in Fig. 4). The contractions of the intestine in response to electrical stimulation are due to the ACh released from preeffectorial nerve terminals. Endogenous and exogenous NE can inhibit these contractions. This fact indicates that the effect of NE is on the preeffectorial nerve terminals (site 1). On the other hand, Dawes and Vizi (12) found that NE reduced the release of ACh from the superior cervical ganglion. This fact indicates that NE might also have an action on the nerve-to-nerve synapse, reducing the ACh release from the nerve terminals of interneuronal axons as well.

Christ and Nishi (9) provided electrophysiologic evidence for the presynaptic action of NE in the superior cervical ganglion: while epinephrine reduced the frequency of miniature EPSPs and the quantal contents of EPSPs, the sensitivity of the postsynaptic membrane to ACh was not affected (8). Nishi and North (39) provided electrophysiologic evidence for the presynaptic inhibitory effect of NE in the myenteric plexus. NE (1 to 10

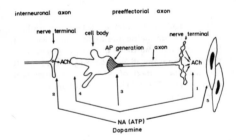

FIG. 4. Scheme of possible sites of action of norepinephrine (NA, noradrenaline).

μM) and epinephrine (0.01 to 0.1 μM) reduced the amplitude of EPSP (North, *personal communication*). In addition, Hirst and McKirdy (24) succeeded in showing electrophysiologically that a preceding train of stimuli applied to periarterial nerves of guinea pig ileum either abolished the synaptic response or reduced the number of EPSPs in the cluster evoked by close transmural stimulation of Auerbach's plexus.

Since NE failed to affect the sensitivity of ganglion cells to iontophoretically administered ACh (40), it can be concluded that the effect of NE in ganglia is on the presynaptic side. All these electrophysiologic findings, in fact, provide evidence for the inhibitory action of NE on interneuronal nerve terminals (site 2). In addition, however, Kosterlitz and Lydon (33), recording impulse transmission from nerve to smooth muscle of the ileum with extracellular electrodes, obtained evidence that epinephrine (0.25 μM) also acts prejunctionally on the preeffectorial nerve terminals.

It has been shown (56) that dopamine reduces the release of ACh. Its effect is probably also on the nerve terminals. Phentolamine prevented its action in a noncompetitive manner (56).

The effect of ATP and related nucleotides is also restricted to the nerve terminals.

Effect on the Cell Body

A possible effect of NE on the postsynaptic side, can be easily excluded by the finding (39) that NE does not affect the sensitivity of ganglion cell to ACh (site 4). The possibility that the site of action of NE is located in a restricted area some distance from the cell soma, where action potentials are generated (site 3), however, cannot be completely excluded. The fact that the release of ACh by field stimulation is inhibited by NE indicates that the site of NE action is located more distal than the cell body or the area where the action potential is generated. Greenhard and McAfee (23) have suggested that dopamine and NE might have an inhibitory effect on the neurochemical transmission of the mammalian superior cervical ganglion by activating dopamine-sensitive adenylate cyclase in the postganglionic neuron and thereby producing hyperpolarization of postganglionic cells.

Effect on the Smooth Muscle

The circulating catecholamines and NE released might have a direct effect on the gastrointestinal smooth muscle (site 5 in Fig. 4). It has been shown (2a) that α adrenoceptors situated on the smooth muscle might be responsible for the increase of potassium conductance which, in fact, results in a hyperpolarization of the membrane.

MORPHOLOGIC EVIDENCE FOR THE INTERACTION BETWEEN CHOLINERGIC AND ADRENERGIC NERVOUS SYSTEM

Functionally all data are consistent with the view that sympathetic nerves situated in the myenteric plexus (17) control the activity of the myenteric plexus which is in fact responsible for the coordinated gastrointestinal motility. Gastrointestinal hormones like gastrin and cholecystokinin are able to excite ganglion cells and cause an increase of ACh release, thereby enhancing gastrointestinal motility. NE released from the noradrenergic nerve terminals is able to reduce ACh release, thereby inhibiting of motility. This conclusion provides full functional corroboration of the morphologic findings of Norberg (41), Jacobowitz (25), Furness (16), Silva et al. (46), and Fehér and Vajda (14) that the chief projection of the adrenergic fiber of the small and large intestine is on the cholinergic ganglion cells of the myenteric plexus and, to a lesser extent, the submucous plexus. The muscle layers themselves are innervated very sparsely, if at all (25,46). Adrenergic terminals are not found in the longitudinal muscle (10,20,21). However, there is no morphologic evidence that adrenergic fibers end synaptically on motor nerve terminals or on preganglionic nerve terminals. The effect of NE on nerve terminals might be explained by diffusion.

SUMMARY

It is suggested that NE released from the sympathetic nerve terminals might bring about relaxation of the gastrointestinal smooth muscle indirectly by removing an excitatory cholinergic tone, which is, in fact, also influenced by gastrointestinal hormones like gastrin and cholecystokinin.

REFERENCES

1. Ahlquist, R. P., and Levy, B. (1959): *J. Pharmacol. Exp. Ther.*, 127:146–149.
2. Baum, T., and Shropshire, A. T. (1971): *Am. J. Physiol.*, 221:1470–1475.
2a. Bülbring, E., and Tomita, T. (1969): *Proc. R. Soc. [Biol. Sci.]*, 172:89–136.
3. Beani, L., Bianchi, C., and Crema, A. (1969): *Br. J. Pharmacol.*, 36:1–17.
4. Bennett, A. J., Eley, K. G., and Scholes, G. B. (1968): *Br. J. Pharmacol.*, 34:630–638.
5. Burnstock, G. (1972): *Pharmacol. Rev.*, 24:509–581.
6. Burnstock, G., Campbell, G., Satchell, D. G., and Smythe, A. (1970): *Br. J. Pharmacol.*, 40:668–688.

7. Crema, A., Frigo, G. M., and Lecchini, S. (1970): *Br. J. Pharmacol.* 39:334–345.
8. Christ, D. D., and Nishi, S. (1969): *Life Sci.*, 8:1235–1238.
9. Christ, D. D., and Nishi, S. (1971): *J. Physiol. (London)*, 213:107–117.
10. Costa, M., and Gabella, G. (1971): *Z. Zellforsch*, 122:357–377.
11. Cowie, A. L., Kosterlitz, H. W., and Watt, A. J. (1968): *Nature*, 220:465–466.
12. Dawes, P. M., and Vizi, E. S. (1973): *Br. J. Pharmacol.*, 48:225–232.
13. Drakontides, Anna B., and Gershon, M. D. (1972): *Br. J. Pharmacol.*, 45:417–434.
14. Fehér, E., and Vajda, J. (1974): *Acta Anat.*, 87:97–109.
15. Furchgott, R. F. (1959): *Pharmacol. Rev.*, 11:429–441.
16. Furness, J. B. (1970): *Histochemie*, 21:295–306.
17. Furness, J. B., and Costa, M. (1971): *Z. Zelforsch.*, 120:346–363.
18. Furness, J. B., and Costa, M. (1973): *Philos. Trans. R. Soc. [Biol. Sci.]*, 265:123–133.
19. Furness, J. B., and Costa, M. (1974): *Rev. Physiol.*, 69:1–51.
20. Gabella, G. (1970): *Experientia*, 26:44–46.
21. Gabella, G. (1971): *Experientia*, 27:280–281.
22. Gershon, M. D. (1967): *J. Physiol. (London)*, 189:317–327.
23. Greenhard, P., and McAfee, D. A. (1972): *Biochem. Soc. Symp.*, 36:87–100.
24. Hirst, G. D. S., and McKirdy, H. C. (1974): *Nature*, 250:430–431.
24a. Illés, P. et al. (1973): *Eur. J. Pharmacol.*, 24:29–36; *Pol. J. Pharmacol. Pharm.*, 26:127–136.
25. Jacobowitz, D. (1965): *J. Pharmacol. Exp. Ther.*, 149:358–364.
26. Jansson, G., and Martinson, J. (1966): *Acta Physiol. Scand.*, 68:184–192.
26a. Juorio, A. V., and Gabella, G. (1974): *J. Neurochem.*, 22:851–858.
27. Kadlec, O., Masek, K., and Seferna, I. (1974): *Br. J. Pharmacol.* 51:565–570.
28. Kazic, T. (1971): *Eur. J. Pharmacol.*, 16:367–373.
29. Kewenter, J. (1965): *Acta Physiol. Scand. Suppl.*, 257:1–68.
30. Knoll, J., and Vizi, E. S. (1970): *Br. J. Pharmacol.*, 40:554–555.
31. Knoll, J., and Vizi, E. S. (1971): *Br. J. Pharmacol.*, 42:263–272.
32. Kosterlitz, H. W., and Waterfield, A. A. (1970): *Br. J. Pharmacol.*, 40:162P, 1970.
33. Kosterlitz, H. W., and Lydon, R. J. (1971): *Br. J. Pharmacol.*, 43:74–85.
34. Kosterlitz, H. W., Lydon, R. J., and Watt, A. J. (1970): *Br. J. Pharmacol.*, 39:398–413.
35. Kuchii, M., Miyahara, J. T., and Shibata, S. (1974): *Br. J. Pharmacol.*, 51:577–583.
36. Lum, B. K. B., Kermani, M. H., and Heilman, R. D. (1966): *J. Pharmacol., Exp. Ther.*, 154:463–471.
37. McDougal, M. D., and West, G. B. (1954): *Br. J. Pharmacol.*, 9:131–137.
38. Neely, J., and Catchpole, B. (1971): *Br. J. Surg.*, 58:21–28.
39. Nishi, S., and North, R. A. (1973): *J. Physiol. (London)*, 231:29–30.
40. Nishi, S., and North, R. A. (1973): *J. Physiol. (London)*, 231:471–491.
41. Norberg, K.-A. (1964): *Int. J. Neuropharmacol.*, 3:379–382.
42. Paton, W. D. M., and Thompson, J. W. (1953): Proc. 19th Int. Physiol. Congr., p. 664.
43. Paton, W. D. M., and Vane, J. R. (1963): *J. Physiol. (London)*, 165:10–46.
44. Paton, W. D. M., and Vizi, E. S. (1969): *Br. J. Pharmacol.*, 35:10–28.
45. Paton, W. D. M., and Vizi, E. S., and Zar, A. M. (1971): *J. Physiol. (London)*, 215:818–848.
46. Silva, D. G., Ross, G., and Osborne, L. M. (1971): *Am. J. Physiol.*, 220:347–352.
47. Petri, G., Szenohradszky, J., and Pórszász-Gibiszer, K. (1971): *Surgery*, 70:359–367.
48. Szabolcsi, I., Vizi, E. S., and Knoll, J. (1974): In: *Symposium on Current Problems in the Pharmacology of Analgesics*, edited by E. S. Visi, pp. 57–64. Publ. House of Hung. Acad. of Sci., Budapest.
49. Del Tacca, M., Soldani, G., Selli, M., and Crema, A. (1970): *Eur. J. Pharmacol.*, 9:80–84.
50. Vizi, E. S. (1968): *Naunyn. Schmiedeberg's Arch. Exp. Pathol. Pharmakol.*, 259:199–200.

51. Vizi, E. S. (1973): *Br. J. Pharmacol.*, 47:465–477.
52. Vizi, E. S. (1974): In: *Neurovegetative Transmission Mechanisms*, edited by B. Csillik and J. Ariens Kappers, *J. Neurol. Transmission Suppl. II*, pp. 61–78.
53. Vizi, E. S., and Knoll, J. (1971): *J. Pharm. Pharmacol.*, 23:918–925.
54. Vizi, E. S., Bertaccini, G., Impicciatore, M., and Knoll, J. (1973): *Gastroenterology*, 64:268–277.
55. Vizi, E. S., Bertaccini, G., Impicciatore, M., and Knoll, J. (1972): *Eur. J. Pharmacol.*, 17:175–178.
56. Vizi, E. S., Rónai, A., and Knoll, J. (1974): *Naunyn-Schmiedeberg's Arch. Exp. Pathol. Pharmakol.*, 284:R89.
57. Vizi, E. S. (1973): *Gastroenterology*, 65:992–994.
58. Weisenthal, L. M., Hug, C. C., Jr., Weisbrodt, N. W., and Bass, P. I. (1971): *J. Pharmacol. Exp. Ther.*, 178:497–508.

Physiology of Smooth Muscle, edited by
E. Bülbring and M. F. Shuba.
Raven Press, New York © 1976.

Inhibition by Acetylcholine of Adrenergic Neurotransmission in Vascular Smooth Muscle

Paul M. Vanhoutte

Department of Internal Medicine, Universitaire Instelling Antwerpen, 2610 Wilrijk, Belgium

The vasodilating properties of choline esters, acetylcholine in particular, are established at least in the intact organism. By contrast, in isolated vascular smooth muscle, the drug has been mostly used as an excitatory agonist. One explanation for the constrictor effects of acetylcholine on vascular smooth muscle is to assume, as suggested by Burn and Rand (1), that acetylcholine liberates norepinephrine from the adrenergic nerve terminals, and thus affects the smooth muscle cells through an indirect sympathomimetic action. However, since acetylcholine causes contraction of noninnervated umbilical veins and of other veins after chemical or surgical denervation and after α-adrenergic blockade, a direct excitatory action of the drug is involved (2–7). The possibility that acetylcholine interferes with adrenergic neurotransmission is suggested by the observations that it depresses the reactivity of isolated arteries to sympathetic stimulation (8–10).

The experiments summarized here provide evidence that, in different blood vessels of the dog, acetylcholine decreases the transmitter release during sympathetic nerve stimulation (11–14).

METHODS

Changes in isometric tension of blood vessel strips (saphenous, mesenteric, femoral, and pulmonary veins; tibial, mesenteric, femoral, and pulmonary arteries) of the dog were recorded. Temperature (37°C), pH (7.4), and PO_2 (140 mmHg) were kept constant throughout the experiment; each preparation was placed at the optimal point of its length–active tension relationship (15). Contractions were obtained with addition of vasoactive agents (norepinephrine, tyramine, KCl) or with electric field stimulation (1 to 10 Hz, 2 msec, 9 V). Contractions of isolated veins of the dog in response to such electric stimulation is abolished by blockade of postganglionic adrenergic transmission with bretylium tosylate or tetrodotoxin, by reserpine pretreatment, by chronic sympathectomy, and by α-adrenergic blockade (2,5,13,16, 17). Thus electric field stimulation causes contraction by release of catecholamines from the adrenergic nerve terminals.

Isolated strips of saphenous and mesenteric veins and of pulmonary arteries were incubated for 4 hr in Krebs-Ringer solution containing 5×10^{-8} g/ml 7-^{3}H-norepinephrine. At the end of the incubation period the preparation was suspended in a moist tunnel-shaped chamber and was superfused at 3 ml/min with aerated Krebs-Ringer solution. The strips were connected to a strain-gauge for continuous isometric tension recording. The superfusate was collected at given intervals (0.5 to 2 min) for direct estimation of the total radioactivity. In some experiments the superfusate for selected 4-min intervals was collected for chromatographic separation of norepinephrine and its metabolites (12,18). For electric stimulation two platinum wires were placed parallel to the strip; both the vessel and the electrodes were continuously superfused. Similar techniques of superfusing preparations previously incubated with tritiated norepinephrine and of following, by radioactivity measurements, the washout of tritiated compounds has been successfully used in different smooth muscles including isolated arteries (19,20) and portal veins (20–22). Stimulation of sympathetic nerves has been shown to increase the ^{3}H-norepinephrine efflux in these preparations (12,18).

RESULTS AND COMMENTS

Cutaneous Veins

Acetylcholine (5×10^{-8} to 5×10^{-7} g/ml) caused contractions of resting saphenous strips, in confirmation of earlier observations (2,4,16,17,23). During the contraction produced by electric stimulation, doses of acetylcholine which did not affect basal tension in resting strips, caused a marked, dose-dependent, and reversible relaxation. By contrast, the same doses of acetylcholine augmented the reaction to exogenous norepinephrine (Fig. 1). When the vein was constricted with an indirectly acting sympathomimetic amine such as tyramine, acetylcholine augmented the contraction in doses which caused relaxation during nerve stimulation. Contractions caused by 25 meq/liter K^+ were also augmented by acetylcholine, but it caused relaxation of strips preconstricted with 30, 35, or 40 meq/liter K^+. After incubation with phentolamine acetylcholine no longer depressed the response to high K^+. Potassium ions cause contraction of cutaneous veins partly through the release of norepinephrine which results from nerve terminal depolarization (18).

These findings demonstrate that acetylcholine has a dual action. At the presynaptic level it inhibits the release of norepinephrine from the tissue stores, leading to relaxation of smooth muscle constricted by the sympathetic impulses or by high K^+; however, acetylcholine apparently cannot prevent the mobilization of norepinephrine by tyramine. At the venous smooth muscle cells, acetylcholine causes contraction. This direct action can be obtained in resting strips but usually with larger doses than those necessary to increase the tension in a strip already contracted by exogenous norepinephrine or tyramine. Both the relaxations during nerve stimulation or exposure

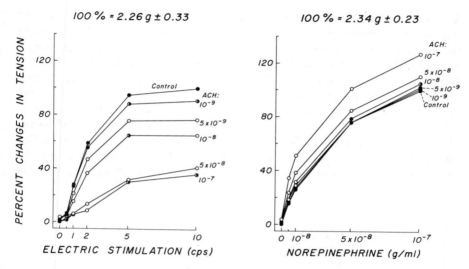

FIG. 1. Effect of increasing doses of acetylcholine on the reactions of dog saphenous vein strips to electric stimulation at five different frequencies and to norepinephrine at four different concentrations. (From Ref. 11 with permission.)

to high K⁺, and the augmented reactions to norepinephrine, are not influenced by nicotinic inhibitors, but are prevented by atropine (Fig. 2). Thus, both must be due to the muscarinic action of acetylcholine (24).

The hypothesis that acetylcholine inhibits the release of the neurotransmitter liberated by the electric impulses or by high K⁺ has been tested in a series of experiments by comparing the effect of acetylcholine on tension changes and on release of ^{3}H-norepinephrine in the same preparation. Acetylcholine caused a slight increase in tension in resting preparations, but did not alter the total ^{3}H-efflux curve. Electric stimulation (2 to 5 Hz) and

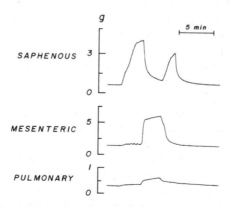

FIG. 2. Effect of atropine (10^{-8} g/ml) on reactions of saphenous (*top*), mesenteric (*middle*), and pulmonary (*bottom*) veins of the same dog to acetylcholine (5×10^{-8} g/ml) during electric stimulation (2 Hz). (Data from Ref. 13 with permission.)

K⁺ (50 meq/liter) augmented the efflux of tritiated compounds into the superfusate. Acetylcholine (10^{-7} g/ml) given during a sustained contraction caused by electric impulses or high K⁺ caused relaxation of the strips, which was paralleled by a decrease in the radioactivity of the superfusate (Fig. 3). With higher concentrations of acetylcholine, the relaxation during electric stimulation was progressively converted to a contraction; however, the radioactivity of the superfusate decreased almost to the resting level. During contractions caused by tyramine (2×10^{-6} g/ml) addition of acetylcholine (10^{-7} g/ml) augmented slightly the contractile response, whereas the radioactivity level of the superfusate was unchanged. The experiments where the superfusate was collected for column chromatographic analysis demonstrated

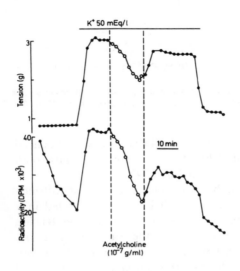

FIG. 3. Changes in isometric tension (*upper*) and in ³H-efflux (*lower*) in a saphenous vein previously incubated with ³H-norepinephrine. During superfusion with solutions containing 50 meq/liter K⁺ acetylcholine causes relaxation and decreases the evoked release of norepinephrine. (From Vanhoutte and Verbeuren, *to be published.*)

that electric stimulation and high K⁺ increased the efflux of both intact norepinephrine and its metabolites. Acetylcholine depressed the output of intact ³H-norepinephrine more than the metabolite fraction. Hence the observed decrease in total radioactivity of the superfusate with acetylcholine reflects an actual decrease in transmitter overflow. These experiments thus provide evidence that acetylcholine interferes with adrenergic neurotransmission by inhibition of norepinephrine release (12,14). Since the acetylcholine inhibition is not blocked by nicotinic antagonists but is abolished by low doses of atropine, and since acetylcholine does not induce a transient increase in release of norepinephrine these experiments do not support the "cholinergic link" hypothesis of Burn and Rand (1) but confirm the presence of muscarinic inhibitory receptors on adrenergic nerve terminals (1,25,26).

The question arises why acetylcholine depressed the increase in norepinephrine release caused by electric impulses or by high K⁺ but not that caused by tyramine. The depressing effect of acetylcholine on norepinephrine release evoked by high K^+ is not affected by tetrodotoxin, excluding an interference with Na^+ movements or spike electrogenesis. A major difference between liberation of norepinephrine from adrenergic nerve endings by tyramine, on the one hand, and by nerve impulses or high K^+, on the other hand, is that the latter are dependent upon the presence of Ca^{++}, while the former is not (27–31). Acetylcholine may depress the evoked release of norepinephrine by decreasing the Ca^{2+}-influx into the nerve endings, thus preventing the release of neurotransmitter associated with the depolarization–repolarization process. The nicotinic action of acetylcholine on adrenergic nerve terminals causes increased membrane activity, which leads to depolarization and eventually depolarization-block of the nerve endings (1, 32–35). Thus the nicotinic action cannot explain the inhibition of adrenergic neurotransmission in the blood vessel wall. It rather seems that the muscarinic action of acetylcholine causes hyperpolarization of the adrenergic nerve endings. This is strongly suggested by the observation that in the cutaneous vein augmentation of K^+ above 50 to 70 meq/liter causes a further release of norepinephrine (14) which is depressed by acetylcholine. The hyperpolarizing muscarinic effect of acetylcholine is well documented in sinoatrial tissue; it has been demonstrated in sympathetic ganglion cells and postulated also at the adrenergic nerve endings of the heart (34,36,37).

Other Blood Vessels

To determine if acetylcholine is a general inhibitor of adrenergic neurotransmission in the vascular system, experiments were performed on different isolated veins and on the corresponding arteries. In resting mesenteric, femoral, and pulmonary veins, acetylcholine (5×10^{-11} to 10^{-6} g/ml) caused an increase in tension; the mesenteric vein was the most sensitive to the direct action of acetylcholine, followed by the pulmonary vein.

In the femoral vein, acetylcholine depressed the reaction to electric impulses but augmented the response to norepinephrine as in the saphenous vein, which demonstrates that also in this preparation acetylcholine inhibits adrenergic neurotransmission. However, in the pulmonary and mesenteric vein strips, acetylcholine augmented the contractions in response to electric impulses, norepinephrine or KCl; these effects of acetylcholine were abolished by atropine (Fig. 2). While these observations do not seem to support the hypothesis that acetylcholine inhibits the release of catecholamines during sympathetic nerve activation in all veins, the strong direct effect of the drug on those particular preparations could mask the inhibitory effect on the sympathetic nerves. To validate this interpretation, experiments were performed on mesenteric veins incubated with tritiated norepinephrine. Although acetylcholine caused further increases in tension during nerve stimulation and high K^+, it markedly reduced the liberation of norepinephrine caused by both procedures (Fig. 4) (13,14).

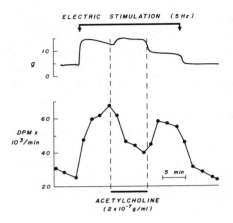

FIG. 4. Isometric tension recording from mesenteric vein strip and radioactivity in superfusate during continuous sampling (2 min/sample). Electric stimulation (2 Hz) caused an increase in tension and in total ^3H efflux; acetylcholine (2×10^{-7} g/ml) augmented the tension but depressed the ^3H efflux (From Ref. 13 with permission.)

In resting isolated arteries (anterior tibial, mesenteric, femoral, and pulmonary), the most common reaction to increasing doses of acetylcholine was a slight relaxation followed by an increase in tension. In all arteries, acetylcholine caused a marked depression of the increase in tension caused by electric stimulation. In most strips, in particular pulmonary and femoral arteries, acetylcholine also depressed the reaction to exogenous catecholamines. When this occurred, the response to electric impulses was more depressed than the response to norepinephrine (13). Atropine abolished both relaxations. From these, and from similar observations in the mesenteric artery of the rat and the ear artery of the rabbit (8–10,38), it can be concluded that acetylcholine inhibits the release of norepinephrine during sympathetic nerve activation of arterial smooth muscle. To strengthen this

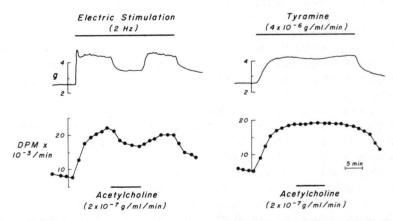

FIG. 5. Comparison of effects of acetylcholine on tension and total ^3H efflux induced in the same pulmonary artery strip by electric stimulation and tyramine. (From Ref. 13 with permission.)

conclusion experiments were performed on pulmonary arteries incubated with ³H-norepinephrine and mounted for superfusion; acetylcholine causes a marked inhibition of the release of adrenergic neurotransmitter by electric impulses or K⁺, but not of that by tyramine (Fig. 5) (13,14).

While the effect of acetylcholine on the adrenergic nerve endings thus seems uniform throughout the vascular system, the direct action on the smooth muscle cells varies. In the isolated veins only contractions were observed. By contrast, in the isolated arteries a transition from relaxation to contraction was obtained. The direct relaxing effect of acetylcholine could be exaggerated by preconstricting the strips with norepinephrine (13). Depressions of the reactions to norepinephrine have already been reported in the aorta and the ear artery of the rabbit and in the mesenteric artery (8–10, 39,40).

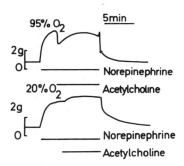

FIG. 6. In the femoral vein, the vasoconstrictor effect of acetylcholine can be changed to relaxation (upper), as illustrated by two preparations from the same donor dog incubated for 4 hr at different PO₂ (From Dalemans and Vanhoutte, to be published.)

Both the inhibitory and the excitatory component are due to a muscarinic action of acetylcholine, since they both are blocked by atropine. One explanation for the biphasic reaction to acetylcholine is that the drug activates two receptors on the smooth muscle cells, one inhibitory and the other excitatory. According to their relative distributions either contractions, relaxations, or biphasic reactions can be obtained. The presence of both inhibitory and excitatory muscarinic receptors has been demonstrated in sympathetic ganglion cells (36,37).

Obviously most venous smooth muscle only has excitatory acetylcholine receptors; arteries possess both types. The inhibitory component has a lower threshold. The different distribution of those two receptors, in conjunction with the inhibitory action of acetylcholine on sympathetic nerve endings may help explain the apparent discrepancy between the *in vivo* vasodilator and the *in vitro* vasoconstrictor properties of acetylcholine. An excitatory reaction can be shifted to an inhibitory one, and vice-versa. Thus, in the femoral vein, incubation in hyperoxic solution unmasks a dilator component of the acetylcholine action (Fig. 6). This suggests that differences in cellular metabolism may explain in part the differences in reactivity between arteries and veins.

In Vivo Experiments

The inhibitory action of acetylcholine on adrenergic neurotransmission was demonstrated in the intact animal. Dogs were anesthetized with chloralose; the left lateral saphenous vein was cannulated at the ankle and perfused at constant flow with the dog's own blood from the terminal aorta. The saphenous venous perfusion pressure was continuously recorded (11, 24). The lumbar sympathetic trunk was stimulated via a platinum bipolar electrode; norepinephrine and acetylcholine were infused at constant rate in the perfusion circuit. In resting conditions, acetylcholine (4×10^{-7} g/ml) caused slight contractions as described by others (41) In every dog, acetylcholine caused a marked inhibition of the response to sympathetic nerve stimulation over a frequency range from 1 to 10 Hz. In the same dogs, during sustained venoconstrictions caused by norepinephrine (2 to 4×10^{-7} g/ml), acetylcholine caused an additional increase in perfusion pressure; in no case was there a venodilatation. In these *in vivo* experiments the electric stimulation was applied directly to the sympathetic nerves. The similarity of the paradoxic effect of acetylcholine on responses to nerve stimulation and exogenous catecholamines with the findings on isolated blood vessels strengthens the conclusion that acetylcholine inhibits the release of norepinephrine.

CONCLUSION

Acetylcholine can inhibit the release of adrenergic transmitter throughout the vascular system. This effect of acetylcholine is due to activation of muscarinic inhibitory receptors on the sympathetic nerve endings in the blood vessel wall; it could explain part of the potent dilator effect of this substance in the intact organism, including the phenomenon of cholinergic neurogenic vasodilation in which acetylcholine is released in the vicinity of the adrenergic vasoconstrictor nerve endings. The present experiments confirm, in venous and arterial smooth muscle the findings of Löffelholz and Muscholl (42,43) on the heart; they are consistent with the view that some of the actions of the cholinergic transmitter are due to inhibition of the sympathetic adrenergic transmission. A similar inhibition pattern has been described for other smooth muscles where exogenous and endogenous catecholamines inhibit the release of acetylcholine by cholinergic motor nerves (44,45). Such local inhibitory mechanisms must allow a very effective control, since on release of the inhibitory agonist the liberation of the excitatory transmitter is automatically interrupted.

REFERENCES

1. Burn, J. H., and Rand, M. J. (1965): *Ann. Rev. Pharmacol.,* 5:163–182.
2. Vanhoutte, P., Clement, D., and Leusen, I. (1967): *Arch. Int. Physiol. Biochem.,* 75:641–657.

3. Ehinger, B., Gennser, G., Ownman, Ch., Persson, H., and Sjöberg, N. O. (1968): *Acta Physiol. Scand.*, 72:15–24.
4. Clement, D., Vanhoutte, P., and Leusen, I. (1969): *Arch. Int. Physiol. Biochem.*, 75:73–87.
5. Vanhoutte, P. and Lorenz, R. R. (1970): *Am. J. Physiol.*, 218:1746–1750.
6. Loh, D. v. (1971): *Pflügers Arch.*, 330:90–98.
7. Altura, B. M., Malaviva, D., Reich, C. F., and Orkin, L. R. (1972): *Am. J. Physiol.*, 222:345–355.
8. Malik, K. U., and Ling, G. M. (1969): *Circ. Res.*, 25:1–9.
9. Rand, M. J., and Varma, B. (1970): *Br. J. Pharmacol.*, 38:758–770.
10. Hume, W. R., De la Lande, I. S., and Waterson, J. G. (1972): *Eur. J. Pharmacol.*, 17:227–233.
11. Vanhoutte, P. M., and Shepherd, J. T. (1973): *Circ. Res.*, 32:259–267.
12. Vanhoutte, P. M., Lorenz, R. R., and Tyce, G. M. (1973): *J. Pharmacol. Exp. Ther.*, 185:386–394.
13. Vanhoutte, P. (1974): *Circ. Res.*, 34:317–326.
14. Vanhoutte, P. M., Verbeuren, T. J. (1974): Presented at the 26th International Congress of Physiological Sciences, New-Dehli.
15. Vanhoutte, P. M., and Leusen, I. (1969): *Pflügers Arch.*, 306:431–353.
16. Vanhoutte, P., and Shepherd, J. T. (1970a): *Am. J. Physiol.*, 218:187–190.
17. Vanhoutte, P., and Shepherd, J. T. (1970b): *Microvasc. Res.*, 2:454–461.
18. Lorenz, R. R., and Vanhoutte, P. M. (1975): *J. Physiol.*, 246:479–500.
19. Su, C., and Bevan, J. A.: *J. Pharmacol. Exp. Ther.*, 172:62–68.
20. Hughes, J., and Roth, R. H. (1971): *Brit. J. Pharmacol.*, 41:239–255.
21. Häggendal, H., Johansson, B., Jonason, J., and Ljung, B. (1970): *Acta Physiol. Scand. Suppl.*, 349:17–32.
22. Häggendal, H., Johansson, B., Jonason, J., and Ljung, B. (1972): *J. Pharm. Pharmacol.*, 24:161–164.
23. Vanhoutte, P. (1970): *C. R. Soc. Biol. (Paris)*, 164:2666–2671.
24. Vanhoutte, P. M., and Shepherd, J. T. (1969): *Circ. Res.*, 25:607–616.
25. Burn, J. H.: (1974): *J. Pharm. Pharmacol.*, 26:212–215.
26. Fozard, J. R., and Muscholl, E. (1974): *J. Pharm. Pharmacol.*, 26:662–664.
27. Hukovic, S., and Muscholl, E.: *Arch. Exp. Pathol. Pharmakol.*, 244:81–96.
28. Lindmar, R., Löffelholz, K., and Muscholl, E. (1967): *Experientia*, 23:933–934.
29. Thoenen, H., Huerlimann, A., and Haefely, W. (1969): *Eur. J. Pharmacol.*, 6:29–37.
30. Blaustein, M. P., Hohnson, E. M., Jr., and Needleman, P. (1972): *Proc. Nat. Acad. Sci. US*, 69:2237–2240.
31. Kirpekar, S. M., Prat, J. C., Puig, M., and Wakade, A. R. (1972): *J. Physiol. (London)*, 221:601–615.
32. Ferry, C. B. (1963): *J. Physiol. (London)*, 167:487–504.
33. Cabrera, R., Torrance, R. W., and Viveros, H. (1966): *Brit. J. Pharmacol.*, 27:51–63.
34. Haeusler, G., Thoenen, H., Haefely, W., and Huerlimann, A. (1968): *Naunyn-Schmiedebergs Arch. Pharmacol.*, 261:389–411.
35. Löffelholz, K. (1970): *Naunyn Schmiedebergs Arch. Pharmacol.*, 267:64–73.
36. Volle, R. L. (1966): *Pharmacol. Rev.*, 18:839–870.
37. Trendelenburg, U. (1967): *Ergeb. Physiol.*, 59:1–85.
38. Steinsland, O. S., Furchgott, R. F., and Kirpekar, S. M. (1973): *J. Pharmacol. Exp. Ther.*, 184:346–356.
39. Jelliffe, R. W. (1962): *J. Pharmacol. Exp. Ther.*, 135:349–353.
40. Graham, J. D. P., Suhaila, A., and Tai, A. (1971): *Br. J. Pharmacol.*, 41:500–506.
41. Rice, A. J., and Long, J. P. (1966): *J. Pharmacol. Exp. Ther.*, 151:423–429.
42. Löffelholz, K., and Muscholl, E. (1969): *Naunyn Schmiedebergs Arch. Pharmacol.*, 265:1–15.
43. Löffelholz, K., and Muscholl, E. (1970): *Naunyn Schmiedebergs Arch. Pharmacol.*, 267:181–184.
44. Vizi, E. S., and Knoll, J. (1971): *J. Pharm. Pharmacol.*, 23:918–925.
45. Taira, N. (1972): *Ann. Rev. Pharmacol.*, 12:197–208.

Physiology of Smooth Muscle, edited by
E. Bülbring and M. F. Shuba.
Raven Press, New York © 1976.

Neural and Hormonal Determinants of Gastric Antral Motility

J. H. Szurszewski

Department of Physiology and Biophysics, Mayo Medical School, Rochester, Minnesota 55900

Gastrin and pentagastrin—a synthetic polypeptide which contains the active fragment of the molecule—stimulate gastric antral motility. Blair et al. (1) found that a histamine-free crude antral extract increased gastrointestinal tone and motility in the anesthetized cat. Gregory and Tracy (2) found in the Heindenhain pouch of the dog that a single intravenous injection of purified gastrin caused a strong contraction which was followed by rhythmic contractions. The stimulatory effect of pentagastrin and gastrin on the motor activity of the antrum has been confirmed in dogs (3–6), humans (7–10), guinea pig ileum (11), rat descending colon (12), the omasum and ruminorecticulum of cattle (13), the fundus of the hamster (12), swine stomach (14), and the opossum stomach and lower esophageal sphincter (15). In dog and human stomach, pentagastrin and gastrin have been found to increase the frequency of contractions when contractions were already present and to induce their occurrence when there were none.

It has been shown in some of the species mentioned above that the increase in motor activity by gastrin and pentagastrin is blocked by atropine indicating that they act indirectly by releasing acetylcholine from the intramural plexuses (11). However, the manner in which pentagastrin and gastrin increase the frequency of contractions and the manner in which acetylcholine produces contractions have not been studied.

The purpose of this paper is to give a short account of an investigation into the basis of the response of the longitudinal muscle of the dog antrum to pentagastrin and acetylcholine. A more detailed account is in preparation.

The results to be described were obtained from isolated longitudinal muscle of the dog antrum. Strips, 900 μm in diameter and 30 mm in length, were placed in a double sucrose-gap apparatus to record changes in membrane potential and isometric tension (16).

The muscle exhibited spontaneous electrical activity which consisted of two components: a fast spike-like potential followed by a plateau-type potential. Spontaneous slow wave complexes (SWC) are shown in Fig. 1a. The configuration of the SWC recorded in this study confirms the observation made with an intracellular electrode in the dog antrum (17) and is similar to slow waves recorded by Papasova et al. (18) in the cat stomach.

The frequency at 37°C, averaged 1.0/min. 10^{-5} M concentrations of atropine and tetrodotoxin did not abolish the SWC. Hence, the SWC is myogenic in origin.

Pentagastrin increased the frequency of the SWC. An example for a high concentration is shown in Fig. 1 and for a lower concentration in Fig. 3. In Fig. 1, the traces are continuous and spontaneous SWCs are shown in panel a. In panel b, pentagastrin was added for only a minute and then washed out. It is clear that its only effect was on the frequency of the SWC which it increased to a maximum of 6.5/min. It had no effect on the resting tone and did not induce phasic contractions. In panel c, 12 min later, the frequency had returned to near the control level. This chronotropic response to penta-

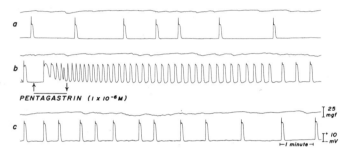

FIG. 1. Effect of a large concentration of pentagastrin on the frequency of the slow wave complex. (a) Control, (b) 10^{-6} M pentagastrin for 1 min, (b) and (c) recovery. Note increase in frequency but no tension change. Traces are continuous.

gastrin was unaffected by atropine or tetrodotoxin alone or in combination. Hence, this action of pentagastrin was on the smooth muscle cell membrane.

Acetylcholine initiated phasic contractions with each SWC. An example, is shown in Fig. 2. The traces are continuous and the dotted line represents the resting membrane potential. Initially, acetylcholine depolarized the membrane, transiently increased the frequency of the SWC, and initiated oscillations which followed each slow-wave complex. The initial depolarization was associated with an increase in tone. Superimposed on the tone which paralleled the depolarization, SWCs produced phasic contractions which continued to increase in amplitude. Although both pentagastrin and acetylcholine increased the frequency of the SWC, the increase during acetylcholine was transient, whereas during pentagastrin, the increased frequency remained constant. Interestingly, the oscillations in Fig. 2 also caused small phasic contractions. The oscillations appeared at a certain threshold for, as the membrane repolarized, the frequency of the oscillations diminished and the smaller ones no longer produced contractions. The responses to acetylcholine were completely blocked by 5×10^{-8} M atropine.

When both acetylcholine and pentagastrin were administered together,

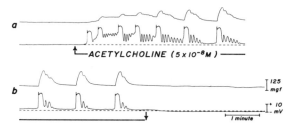

FIG. 2. Effect of acetylcholine on electrical and mechanical activity. Acetylcholine applied at arrow in (a) and washed out at arrow in (b). Dotted line resting membrane potential prior to acetylcholine. Traces are continuous. For further explanation, see text.

phasic contractions occurred at the frequency of the SWC. An example is shown in Fig. 3. In this preparation, there were no spontaneous SWCs. Pentagastrin alone produced SWCs (panel a). Acetylcholine with pentagastrin (panel b) produced SWCs with phasic contraction. When the concentration of acetylcholine was doubled (panel c), the strength of each contraction increased. These data suggest that pentagastrin regulated the frequency of occurrence of SWCs and contractions and acetylcholine determined the strength of contraction.

Figure 4 summarizes my present hypothesis concerning the sites of action of pentagastrin and acetylcholine. There may be two sites of action for pentagastrin: on the postsynaptic smooth muscle cell membrane and also on presynaptic cholinergic nerves. Activation of the receptors on smooth muscle might bring about a metabolic change which acts on the slow-wave generator to increase its frequency of oscillation. The nature of this change is not yet understood. According to the work of others, receptors for pentagastrin may be located presynaptically on cholinergic nerve terminals (2,3) and activation of these receptors may release acetylcholine. The acetylcholine, released

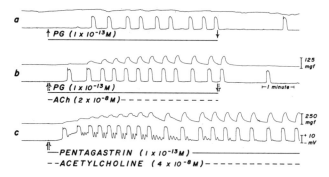

FIG. 3. Response of longitudinal muscle to simultaneous application of pentagastrin and acetylcholine. (a) Effect of pentagastrin, (b) and (c) effect of pentagastrin and acetylcholine. In (c) concentration of acetylcholine doubled and strength of each contraction greater in (c) than in (b).

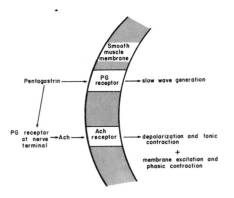

FIG. 4. Proposed sites of action for pentagastrin and acetylcholine. For further explanation, see text.

by pentagastrin, may interact with the muscarinic receptor. This leads to an increase in intracellular ionic calcium which triggers contractions. When both acetylcholine and pentagastrin are present, phasic contractions occur at the frequency of the SWC. Hence, it is reasonable to conclude that following ingestion of a meal, stimulation of cholinergic nerves by gastrin releases acetylcholine which produces phasic contractions in the longitudinal muscle layer, whereas stimulation of the muscle by gastrin leads to an increase in the frequency of contractions.

ACKNOWLEDGMENTS

This investigation was supported by Research Grant AM17238 from the National Institutes of Health, Public Health Service. This work was performed by Dr. Szurszewski during the tenure of an Established Investigatorship of the American Heart Association.

REFERENCES

1. Blair, E. L., Harper, A. A., Lake, H. J., and Reed, D. J. (1961): *J. Physiol. (Lond.)*, 159:72–73P.
2. Gregory, R. A., and Tracy, H. J. (1964): *Gut*, 5:103–117.
3. Jacoby, H. I., and Marshall, C. H. (1969): *Gastroenterology*, 56:80–87.
4. Isenberg, J. I., and Grossman, M. I. (1969): *Gastroenterology*, 56:450–455.
5. Kelly, K. A. (1970): *Am. J. Dig. Dis.*, 15:399–405.
6. Cooke, A. R., Chvasta, T. E., and Weisbrodt, N. W. (1972): *Am. J. Physiol.*, 223:934–938.
7. Bennett, A., Misiewicz, J. J., and Waller, S. L. (1967): *Gut*, 8:470–474.
8. Misiewicz, J. J., Holdstock, D. J., and Waller, S. L. (1967): *Gut*, 8:463–469.
9. Misiewicz, J. J., Waller, S. L., and Holdstock, D. J. (1969): *Gut*, 10:723–729.
10. Cameron, A. J., Phillips, S. F., and Summerskill, W. H. J. (1970): *Gastroenterology*, 59:539–545.
11. Vizi, S. E., Bertaccini, G., Impicciatore, M., and Knoll, J. (1973): *Gastroenterology*, 64:268–277.
12. Mikos, E., and Vane, J. R. (1967): *Nature*, 214:105–107.
13. Ruckebusch, Y. (1971): *Experientia*, 27:10:1185–1186.
14. Roze, C., Couturier, D., Lagneau, P., Gilles, M-R, and Debray, C. (1973): *Rend. Gastro-enterologia*, 5:147.

15. Lipshutz, W., Tuch, A. F., and Cohen, S. (1971): *Gastroenterology,* 61:454–460.
16. Szurszewski, J. H. (1974): In: *Fourth International Symposium on Gastrointestinal Motility,* edited by E. E. Daniel, pp. 409–425. Mitchell Press, Vancouver, British Columbia.
17. Daniel, E. E. (1965): *Gastroenterology,* 49:403–418.
18. Papasova, M. P., Nagai, T., and Prosser, C. L. (1968): *Am. J. Physiol.,* 214:695–702.
19. Nelsen, T. S., and Becker, J. C. (1968): *Am. J. Physiol.,* 214:749–757.
20. Sugawara, K., Isaza, J., and Woodward, E. R. (1969): *Gastroenterology,* 57:649–658.

Physiology of Smooth Muscle, edited by
E. Bülbring and M. F. Shuba.
Raven Press, New York © 1976.

On the Excitatory Action of Prostaglandin E₂ on the Pregnant Mouse Myometrium

H. Suzuki, T. Osa, and H. Kuriyama

Department of Physiology, Faculty of Dentistry, Kyushu University, Fukuoka 812, Japan

Prostaglandins (PGs) were discovered because of their potent biologic actions (23,10; see for reviews 2,12,24,11,19). The actions differ from one series to another not only according to the tissues of the reproductive system and the species but also with the hormonal state of the tissue. For example, in the human myometrium, the frequency and amplitude of spontaneous contractions are usually depressed by PGE_2 but are increased by PGF; on the other hand, the myometria of guinea pig, rat, and rabbit contract in response to both PGE and PGF. In the guinea pig and rat, progesterone added *in vitro* depresses the sensitivity to PGE and PGF (3,9,13,21,18,2,12, 1,19).

The mechanism of the PG action is still incompletely understood. From the biochemical aspect, the PG action is thought to be involved in lipolysis and reesterification in adipose tissue, in platelet aggregation, and in gastric secretion (2). However, the electrophysiologic effects of PG on the activity of smooth muscle have not yet been investigated in detail.

THE EFFECTS OF PGE₂ ON THE MEMBRANE ACTIVITY COMPARED WITH THOSE OF OXYTOCIN

In low concentrations of PGE_2, the spike frequency, the number of spikes in a train discharge, and the number of train discharges were increased (Fig. 1). However, the actual concentration of PGE_2 to produce excitation varied markedly during the progress of gestation.

The effects of PGE_2 and oxytocin on membrane activity were classified into three grades (Fig. 2): *First grade* was an increase in the spike frequency, in the number of spikes in a train discharge, and in the spike frequency of train discharges, without any marked change of the membrane potential; *second grade* was depolarization of the membrane with continuous generation of spikes, eliminating the quiescent periods between train discharges. The *third grade* effect was a depolarization block of spike generation. Figure 1 shows examples of the 1st and 2nd grade effects on the membrane activity of the longitudinal muscle layer induced by 10^{-9} and 10^{-8} g/ml PGE_2. The increased membrane activity was not mediated by the acti-

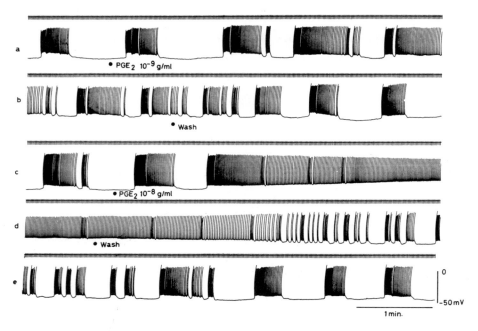

FIG. 1. Effects of PGE₂ on the electrical activity of the longitudinal muscle of myometrium (13th day of gestation). Continuous records taken from the same preparation, but from two different cells are shown in a and b, and in c, d, and e. The drug, at the concentration shown, was added or washed out at the dots.

vation of nervous elements, since tetrodotoxin (10^{-7} g/ml) did not have any effect on the membrane responses (22).

Marked differences between the effects of PGE₂ and oxytocin were observed on the response of the longitudinal muscle of the myometrium at various stages of gestation, i.e., in the early stages of gestation oxytocin showed no excitatory action on the myometrium, and high concentrations of PGE₂ were required to produce an excitatory action. The sensitivity of the muscle membrane to PGE₂ increased at least 100 times between the early and the late stage of gestation; the muscle also became sensitive to oxytocin (15) and, in the late stages of gestation, the effects of PGE₂ and oxytocin on the patterns of membrane activity were similar.

Differences in the effects of the two drugs on the circular muscle of the myometrium were also observed. Train discharges in the longitudinal muscle appeared as bursts on a plateau-like depolarization; but in the circular muscle, the amplitude of this depolarization was much larger and the spikes superimposed on it were often abortive (16). On the 18th day of gestation, only 10^{-6} g/ml PGE₂ was necessary to produce a 2nd grade effect on membrane activity. On the other hand, oxytocin (10^{-3} U/ml) caused depolarization block of the membrane activity (Fig. 3). The low sensitivity of the circular muscle to PGE₂ compared with that of the longitudinal muscle could also

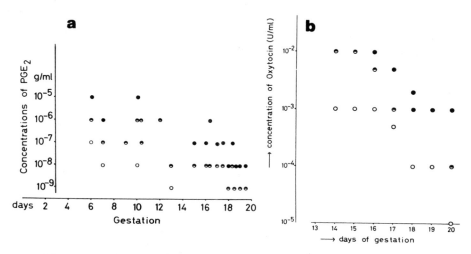

FIG. 2. (a) Effects of prostaglandin E_2 on the electrical activity of the longitudinal muscle of myometrium during the various stages of gestation. ($\bigcirc$) 1st grade response of the PGE_2; ($\ominus$) 2nd grade response of the PGE_2; ($\bullet$) 3rd grade response: depolarization block of spike generation induced by PGE_2. (b) Effects of oxytocin on the electrical activity of the longitudinal muscle of myometrium during the various stages of gestation. Symbols as in (a).

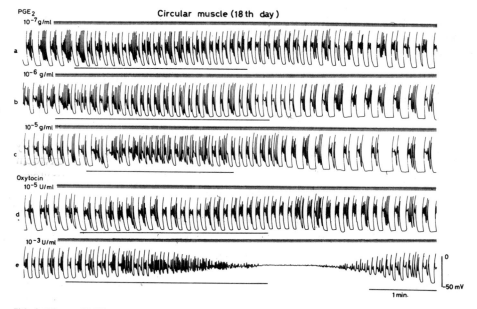

FIG. 3. Effects of PGE_2 or oxytocin on the electrical activity of the circular muscle (myometrium on 18th day of pregnancy at the concentrations shown above each trace). Bars in the figure indicate presence of the drug.

be seen in its mechanical response. With application of PGE_2 (10^{-7} g/ml) the mechanical activity of the longitudinal muscle increased in frequency so that the contractions fused and appeared to be superimposed on an increased muscle tone. However, in the circular muscle, only the frequency of the contractions was increased (Fig. 4).

The effects of PGE_2 and oxytocin on the myometrium also differed in the development of the desensitization. Prolonged treatment of the longitudinal muscle with PGE_2 produced desensitization, i.e., the depolarized membrane gradually repolarized even in the presence of the drug, thus generating spike activity again. In contrast to PGE_2 the myometrium developed very slow

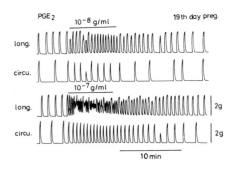

FIG. 4. Effects of PGE_2 on the mechanical activity of the longitudinal and circular muscles of the myometrium (19th day of gestation).

and weak desensitization to oxytocin. Moreover, after desensitization to PGE_2, oxytocin was still able to produce a 3rd grade effect on the myometrium. These results indicate that the receptors to PGE_2 and oxytocin differ.

When the membrane potential was displaced from the depolarized level produced by PGE_2 to the resting membrane potential level, the spike frequency remained still greater than in the control. On the other hand, when the membrane was repolarized to the resting level, after oxytocin had produced a 3rd grade effect, there remained no increased spike frequency. This result suggests that PGE_2 increases the spike frequency mainly due to acceleration of the spike generating mechanism, and oxytocin increases the spike frequency mainly due to depolarization of the membrane.

CHANGES IN THE MEMBRANE RESISTANCE OF THE MYOMETRIUM BY APPLICATION OF PGE_2

When the current–voltage relationships were compared before and during the application of PGE_2, it was found that the amplitude of the electrotonic potential in the presence of PGE_2 was consistently less than that observed in the absence of the drug at the same membrane potential level.

To study the effects of PGE_2 on the membrane resistance at various membrane potential levels, its effects were compared with excess K and depolar-izing current application. Figure 5 shows the relationship between membrane resistance and membrane potential during application of PGE_2, excess K, and electrical displacement of the membrane potential. Even though the membrane potential was displaced with three different procedures, a small depolarization of the membrane consistently increased the resistance and a marked depolarization reduced the membrane resistance. Measurements of the amplitude of the electrotonic potential were not accurate especially with PGE_2 and excess K, since the membrane was always active and the changes in the membrane resistance would cause changes in the length constant. Nevertheless, these results suggest that, when the membrane was depolarized,

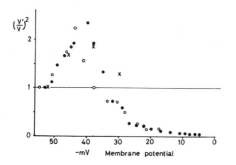

FIG. 5. Effects of displacements of the membrane potential on the membrane resistance (18th day of gestation). Membrane potential was displaced by PGE_2 (O), by excess $[K]_0$ (●), or by current injection (X). Ordinate: Relative change in the membrane resistance expressed as $(v'/v)^2$ by simplification of the cable equation. Abscissa: Membrane potential level ($-$mV).

the pregnant mouse myometrium shows anomalous rectification, i.e., the membrane resistance is increased in spite of the reduction of the membrane potential. Therefore increased membrane resistance during the moderate depolarization of the membrane is not a specific action of PGE_2.

THE IONIC MECHANISM INVOLVED IN THE PGE_2 ACTION

In K-deficient Locke solution (one-tenth the normal concentration of K ion), the membrane was transiently depolarized, and though it gradually repolarized to a level close to the resting membrane potential, spikes appeared as continuous discharge. Application of PGE_2 rapidly produced depolarization of the membrane and blocked spike generation.

In Na-deficient sucrose Locke solution (15 mM Na), the membrane was markedly depolarized. During the course of 1 hr the membrane potential gradually returned to a level near to the resting level. When PGE_2 was applied in Na-deficient solution, the depolarization and increase of the spike frequency were less than in normal solution. In half the normal Na concentration PGE_2 caused no change in membrane potential but the bursts of spikes were accelerated. When the temperature was lowered to 29°C, PGE_2 caused depolarization. When the external Na concentration was reduced further to a quarter of normal, the slow potential was nearly abolished. In this condition PGE$_2$ accelerated the generation of the bursts at 35°C, but depolarization was no longer produced, even if the temperature of the bathing solution was lowered to 29°C (17).

In a solution containing one-quarter Na and one-eighth the normal Ca, spontaneous spike discharge was infrequent. Treatment with PGE$_2$ still produced depolarization which was accompanied by continuous spiking. However, in spite of the presence of PGE$_2$ throughout, the depolarization disappeared and the continuous spiking gradually turned into burst discharges (17).

The effects of PGE$_2$ were further investigated in Ca-free Mg Locke solution. Mg ion was added to the Ca-free solution to prevent the depolarization of the membrane. In this solution the membrane was slightly hyperpolarized and spontaneous spike generation ceased. After 10 min exposure to Ca-free solution, PGE$_2$ caused depolarization which initiated a burst of spikes (Fig. 6). After 20 min exposure to Ca-free solution, PGE$_2$ still depolarized the

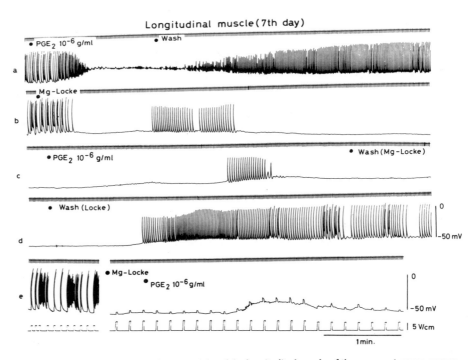

FIG. 6. Effects of PGE$_2$ on the membrane activity of the longitudinal muscle of the pregnant mouse myometrium at the 7th day. a–d and e were obtained from two different preparations. (a) Effect of 10^{-6} g/ml PGE$_2$ in Locke solution; (b) effect of Ca-free Mg Locke solution; (c) effect of 10^{-6} g/ml PGE$_2$ in Ca-free Mg Locke solution; (d) recovery of membrane activity after perfusion with Ca-free Mg Locke solution; (e) effect of 10^{-6} g/ml PGE$_2$ on the membrane potential and outward current pulses in Ca-free Mg Locke solution. Initial part of the record in e is control in Locke solution.

membrane, but outward current pulses of strong intensity could not evoke the spike. With repeated application of PGE_2 the amplitude of the depolarization became gradually less and was finally abolished. This result indicates that PGE_2 may be able to mobilize bound Ca inside the cell.

Further evidence for this possibility of Ca mobilization by PGE_2 was obtained from the effects of PGE_2 on the mechanical response. Figure 7 shows the effects of carbachol, oxytocin, PGE_2, and $PGF_{2\alpha}$ on the K-induced contracture. Isotonic K Locke solution produced a large contracture which gradually declined to a level close to the resting level. When the above agents were applied to the tissue during the tonic response, a small transient mechanical response could be evoked. It has been suggested that carbachol releases sequestered Ca inside the cell and also increases the influx of Ca ion

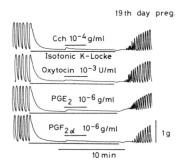

19th day preg.

Cch 10^{-4} g/ml

Isotonic K-Locke

Oxytocin 10^{-3} U/ml

PGE_2 10^{-6} g/ml

$PGF_{2\alpha}$ 10^{-6} g/ml

1 g

10 min

FIG. 7. Effects of carbachol, oxytocin, PGE_2, and $PGF_{2\alpha}$ on the K-induced contracture of the pregnant mouse myometrium (19th day of gestation). Carbachol (CCh) 10^{-4} g/ml, oxytocin 10^{-3} U/ml, PGE_2 10^{-6} g/ml, and $PGF_{2\alpha}$ 10^{-6} were applied during the tonic phase of a K-induced contracture. The presence of isotonic K Locke solution is indicated by the bar below each record. Drugs were applied during the bar above each record.

in an isotonic K solution (20). The small contracture caused by PGE_2 may therefore be due to the same mechanism.

CONCLUSION

PGE_2 has an excitatory action on the mouse myometrium throughout gestation. The action of PGE_2 differed from that of oxytocin in the following ways: (a) PGE_2 had an excitatory action in the early stages of gestation when the uterus is insensitive to oxytocin. (b) PGE_2 had a weaker excitatory action on the circular muscle than on the longitudinal muscle. However, oxytocin has a similar excitatory action on both the longitudinal and circular muscle. (c) Desensitization of the myometrium to PGE_2 appeared more quickly than that to oxytocin. (d) PGE_2 increased the frequency of bursts and the number of spikes in a train, often causing continuous spike discharge, even when the membrane potential was held at the resting level. However, oxytocin increased the frequency of burst discharges which were similar to those observed in normal conditions.

Anomalous rectification of the membrane was observed during depolarization of the membrane by PGE_2, by oxytocin, by excess K, or by current application. This might be due to reduction of K conductance.

It has already been postulated by many investigators that in myometrium, PGE labilizes bound Ca sequestered in the membrane and accelerates the excitation–contraction coupling. Recently, Carsten (4–6) reported that sarcoplasmic reticulum prepared from bovine myometrium released Ca on treatment with PGE_2. Moreover, an increased rate of ^{45}Ca exchange in isolated heart by PGE has been reported (8,7,18,14).

A tentative hypothesis of the mode of action of PGE_2 on the mouse myometrium may be as follows: A low concentration of PGE_2 mobilizes bound Ca at the membrane thus increasing the frequency and number of spikes and depolarizing the membrane. This moves the membrane potential into the range where the K permeability is reduced and anomalous rectification is seen. A higher concentration of PGE_2 increases the Na permeability of the membrane which is also governed by $[Ca]_0$. Presumably, oxytocin has a similar mode of action but ovarian and placental hormones influence the action of oxytocin more than that of PGE_2, as shown by different changes in sensitivity of the myometrium for the two agents during the progress of gestation.

ACKNOWLEDGMENT

We are greatly indebted to Prof. E. Bülbring for helpful discussions and comments in preparing this chapter.

REFERENCES

1. Anderson, R. G. G. (1972): Cyclic AMP and calcium ions in mechanical and metabolic responses of smooth muscles; Influence of some hormones and drugs. *Acta Physiol. Scand.*, Suppl. 382:1–59.
2. Bergström, S., Carlson, L. A., and Weeks, J. R. (1968): The prostaglandins: A family of biologically active lipids. *Pharmacol. Rev.*, 20:1–48.
3. Bergström, S., Dunér, H., von Euler, U. S., Pernow, B., and Sjövall, J. (1959): Observations on the effects of infusion of prostaglandin E_2 in man. *Acta Physiol. Scand.*, 45:133–144.
4. Carsten, M. E. (1973a). Prostaglandins and cellular calcium transport in the pregnant human uterus. *Am. J. Obst. Gynecol.*, 117:824–832.
5. Carsten, M. E. (1973b). Sarcoplasmic reticulum from pregnant bovine uterus. *Gynec. Invest.*, 4:84–94.
6. Carsten, M. E. (1974). Prostaglandins and oxytocin: Their effects on uterine smooth muscle. *Prostaglandins*, 5:33–40.
7. Clegg, P. C., Hall, W. J., and Pickles, V. R. (1966): The action of ketonic prostaglandins on the guinea pig myometrium. *J. Physiol. (Lond.)*, 183:123–144.
8. Coceani, F., and Wolfe, L. S. (1966): On the action of prostaglandin E_1 and prostaglandins from brain on the isolated rat stomach. *Can. J. Physiol. Pharmacol.*, 44:933–950.
9. Eliasson, R. (1959). Studies on prostaglandin. Occurrence, formation and biological actions. *Acta. Physiol. Scand.*, 46:1–73.
10. Goldblatt, M. W. (1935). Properties of human seminal plasma. *J. Physiol. (Lond.)*, 84:208–218.
11. Hinman, J. W. (1972). Prostaglandins. *Ann. Rev. Biochem.*, 41:161–178.
12. Horton, E. W. (1969). Hypotheses on physiological roles of prostaglandins. *Physiol. Rev.*, 49:122–161.

13. Horton, E. W., and Main, I. H. M. (1963). A comparison of the biological activities of four prostaglandins. *Br. J. Pharmacol. Chemother.*, 21:182–189.
14. Klaus, W., and Piccinini, F. (1967): Über die Wirkung von Prostaglandin E_1 auf den Ca-Haushalt isolierter Meerschweinchenherzen. *Experientia*, 23:556–557.
15. Kuriyama, H., and Suzuki, H. (1975): Comparison between prostaglandin E_2 and oxytocin actions on the pregnant mouse myometrium (*in preparation*).
16. Osa, T. (1974): An interaction between the electrical activities of longitudinal and circular smooth muscles of pregnant mouse uterus. *Jap. J. Physiol.*, 24:189–203.
17. Osa, T., Suzuki, H., Katase, T., and Kuriyama, H. (1974): Excitatory action of synthetic prostaglandin E_2 on the electrical activity of pregnant mouse myometrium in relation to temperature changes and external sodium and calcium concentrations. *Jap. J. Physiol.*, 24:233–248.
18. Paton, D. M., and Daniel, E. E. (1967). On the contractile response of the isolated rat uterus to prostaglandin E_1. *Can. J. Physiol. Pharmacol.*, 45:795–804.
19. Pharris, B. B., and Shaw, J. E. (1974): Prostaglandins in reproduction. *Annu. Rev. Physiol.*, 36:391–412.
20. Schild, H. O. (1964). Calcium and the effects of drugs on depolarized smooth muscle. In *Pharmacology of Smooth Muscle*, edited by E. Bülbring, pp. 95–104. Pergamon, Oxford.
21. Sullivan, T. J. (1966). Response of the mammalian uterus to prostaglandins under differing hormonal conditions. *Brit. J. Pharmacol. Chemother.*, 26:678–685.
22. Suzuki, H., and Kuriyama, H. (1975): Effects of prostaglandin E_2 on the electrical property of the pregnant mouse myometrium. *Jap. J. Physiol.*, 25No.2
23. von Euler, U. S. (1934). Zur Kenntnis der Phermakologischen Wirkungen von Nativsekreten und Extrakten männlicher accessorisher Geschlechtsdrüsen. *Arch. Expl. Pathol. Pharmakol.*, 175:78–84.
24. Weeks, J. R. (1972). Prostaglandins. *Ann. Rev. Pharmacol.*, 12:317–336.

Physiology of Smooth Muscle, edited by
E. Bülbring and M. F. Shuba.
Raven Press, New York © 1976.

Effect of Estradiol, Progesterone, and Oxytocin on Smooth Muscle Activity

K. Milenov

Bulgarian Academy of Sciences, Institute of Physiology, Sofia, Bulgaria

Until recently, it has been accepted that estradiol, progesterone, and oxytocin influence specifically the activity of uterine muscle. However, clinical and experimental examinations show that they also participate in the regulation of the activity of other smooth muscle types. In most pregnant women a state of hypotony develops which manifests itself not only in a decrease of spontaneous activity or relaxation of the uterus but also in hypotony of the smooth muscle of the ureters, large bowel (8), and veins (21). The hypotony of the smooth muscle in pregnant women leads to heartburn (11), to prolongation of the emptying time of the stomach (3), and in many cases to obstipation (1).

The cause of this generalized hypotony of smooth muscles is not well established. Some investigators connect it with the hormonal changes accompanying pregnancy. The main role is attributed to progesterone (14,15) the plasma concentration of which increases during pregnancy from 1.5 μg/100 ml peak value in the luteal phase to 14 μg/100 ml peak value during pregnancy (24).

The participation of progesterone and of estrogens in the regulation of extrauterine smooth muscle activity has been demonstrated on the motility of the ureters and gastrointestinal tract. On the basis of hydrourographic studies Hundley et al. (10) demonstrated that estrogens increase and progesterone-like substances decrease the peristaltic activity of ureters. *In vitro,* progesterone (12 to 45 μg/ml) inhibits the spontaneous contractions of human stomach (14) and guinea pig taenia coli (25). Progesterone (extract from corpus luteum) administered i.v. inhibits the motor activity of dog stomach and intestine (19). Progesterone administered per os for long periods to rabbits leads to hypotony of the large bowel (14). Datta et al. (6) compared the intestinal propulsion of mixed cycle mice, pregnant mice, ovariectomized mice pretreated with estradiol or progesterone, or a combination of these two hormones. Intestinal propulsion was diminished in both pregnant and ovariectomized mice. Estradiol alone caused a dose-dependent increase in propulsion in ovariectomized mice.

The participation of oxytocin in the regulation of extrauterine smooth muscle activity has been demonstrated on the blood flow of some veins and on gastrointestinal motility but with different results. Oxytocin injected i.v. increases the blood flow (vasodilator effect) in the human hand (12) and in the kidney and hind leg of rat and dog (18). Oxytocin has a vasoconstrictor effect after sympathectomy of the extremities, and also after pretreatment of the animals with estrogen (7,18).

Konzett et al. (13) found that the synthetic oxytocin (syntocinon) increases ileum peristalsis in rabbits, when administered through a prolonged i.v. infusion, but exerts no constant effect *in vitro* on the spontaneous contractions of intestinal segments. Injected i.v. in anesthetized dogs (17), as well as in conscious dogs (16,23), oxytocin decreases stomach and intestine peristalsis significantly.

We present here data on the effect of estrogen, progesterone, and oxytocin (all synthetically prepared and used clinically) on the bioelectrical and contracile activity of different smooth muscles.

MATERIALS AND METHODS

The experiments were carried out on 15 conscious female dogs 4 of which were ovariectomized. In all dogs silver macroelectrodes were chronically implanted in the muscle wall of the stomach (corpus and antrum), the intestine (jejunum and ileum), and in the uterus (corpus and both horns). Four dogs had a stomach fistula, four had a small pouch from the big curvature according to Heidenhain, and four had an ileum fistula. Every dog was used in the experiments over a period of 5 to 12 months. The electromyograms of the stomach (EGM), intestine (EIM), and uterus (EUM) were recorded in parallel on AC channels of an encephalograph "Schwarzer" with a time constant of 0.1 to 0.03, and the mechanograms by balloon kymographic methods.

The electromyograms of the four ovariectomized dogs were studied before and during a 10-day period of i.m. treatment with estradiol-17β dipropionate (20/μg/kg body weight daily) or progesterone (2 mg/kg body weight daily). In the four not ovariectomized dogs, the electromyograms in anestrus and spontaneous estrus were studied.

The effect of oxytocin was examined on the electromyograms and mechanograms of stomach and intestine, as follows: on eight dogs before and after administration of adrenergic and cholinergic blocking agents (Redergan and Obsidan); on the dogs with Heidenhain's small pouch, as a region free of vagal innervation; *in vitro,* on the spontaneous activity (bioelectrical and mechanical) of muscle strips isolated from the stomach, taenia coli, ureter, and vena portae of guinea pigs (by the method of Golenhofen) (9).

RESULTS

Normal electromyograms of the stomach and intestines of conscious dogs are well described in the literature (2,5,22). They have a regular rhythm consisting of slow potential oscillations called "basic electrical rhythm" (BER) (2). The frequency of this rhythm in our animals was from 4 to 6 cpm for the stomach and 17 to 19 cpm for the intestines (Figs. 1 and 2). It is generally accepted that BER determines the frequency of contractions. Normally every slow potential propagates in aboral direction and is an index of the propagation velocity of the existing impulses in the muscle wall (Fig. 3).

The regular rhythmic pattern in the electromyograms of the stomach was disorganized during estradiol treatment. Figure 1 shows that this effect depended on the dose of estradiol injected. The record in Fig. 1A was obtained

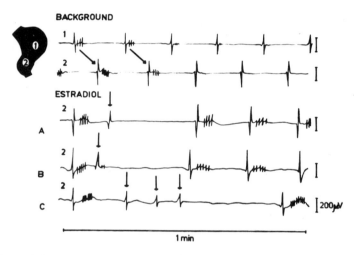

FIG. 1. Background electromyograms of the stomach (EGMG) (leads 1 and 2) of a conscious dog and the changes after estradiol treatment. Appearance of "extra" potentials in antrum: A on the 4th, B on the 6th, and C on the 10th day during i.m. treatment with estradiol.

on the 4th day of treatment; a slow "extra" potential was recorded followed by a prolonged compensatory pause. On the 6th day of estradiol administration (Fig. 1B) an "extra" potential appeared earlier than in Fig. 1A and the compensatory pause became longer. On the 10th day, three "extra" potentials were observed, followed by a compensatory pause (Fig. 1C).

In the electromyograms of the small intestine, estradiol increased, while progesterone decreased the spike potentials with a maximum effect between the 3rd and 7th day of injections. One typical example is shown in Fig. 2. After estradiol each slow potential was accompanied by a group of spikes. The injection of progesterone had the opposite effect; it abolished the spike potentials (Fig. 2).

Simultaneously with these changes estradiol and progesterone influenced the propagation velocity of the BER. Figure 3 shows the propagation velocity of stomach and intestine BER before and during estradiol or progesterone treatment (statistical processing of 40 waves in 4 animals). It is

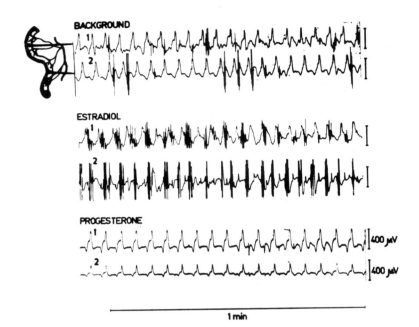

FIG. 2. Background electromyograms of the jejunum (lead 1 and 2) of a conscious dog and the changes during estradiol and progesterone treatment. Estradiol increases the number of groups of spike potentials (recording on the 4th day). Progesterone abolishes all spike potentials and only slow waves are recorded.

evident that estradiol increased, and progesterone decreased the propagation velocity with maximum effect on the 6th day of treatment.

In the electromyograms of the dogs during estrus we observed more groups of spikes and a quicker propagation of BER than in anestrus (in the stomach: 5.55 ± 0.30 mm/sec in anestrus, 6.43 ± 0.18 mm/sec in estrus; and in the jejunum: 132.5 ± 3.30 mm/sec in anestrus, 151.05 ± 4.60 mm/sec in estrus).

The effect of oxytocin on the activity of the gastrointestinal tract was studied in conscious dogs during the first hour after the i.v. injection of the hormone. Syntocinon (Sandoz) and oxytocin (Richter) were used in doses of 0.2 to 0.4 IU/kg body weight. In conscious dogs, after 18 hr starvation, oxytocin (0.4 mg/kg) decreased the tone and abolished peristaltic contractions completely (acinesia) both in stomach and ileum. The recovery of the initial background mechanograms appeared after 8 to 12 min. A similar complete acinesia of stomach and ileum was observed after i.v. injection of atropine sulfate 200 to 300 mg/kg body weight and after norepinephrine or epinephrine 10 mg/kg body weight. The acinesia caused by atropine disappeared after 30 to 40 min, while that caused by catecholamines disappeared after 2 to 5 min.

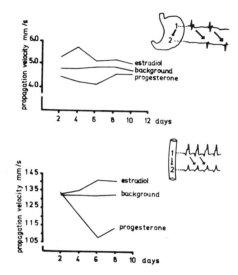

FIG. 3. The propagation velocity of the slow waves in stomach and jejunum before (background) and during estradiol or progesterone treatment. On the ordinate are given the propagation velocity in mm/sec, on the abscissa, the days of treatment. On the right are shown schematically a stomach, a jejunum, the leading electrodes, and the propagating slow waves.

In the electromyograms of stomach and intestine the spike potentials are abolished by oxytocin. A typical example of such an effect is shown in Fig. 4. Whereas the background recording shows groups of spike potentials in the rhythm of some slow waves, all spike potentials disappeared within 8 to 10 min after the administration of oxytocin. At the same time oxytocin caused the opposite effect on the uterus. A continuous discharge of spikes with high amplitude and frequency was recorded (Fig. 4).

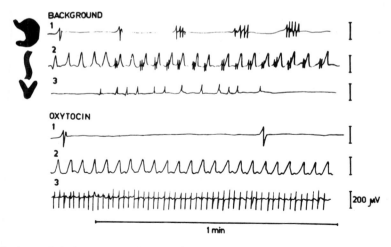

FIG. 4. Background electromyograms of stomach (lead 1), jejunum (lead 2), and uterus (lead 3) of a conscious dog. Oxytocin (0.4 IU/kg body weight) eliminates the spike potentials both in stomach (1) and jejunum (2), but stimulates the spike potentials in the uterus (3). Oxytocin reduces the frequency of BER in the stomach (1) from 4 cpm on background to 2 cpm after oxytocin.

Oxytocin reduced the frequency of BER in stomach two to three times (Fig. 4). The maximum reduction was reached at the 3rd minute, and a gradual recovery occurred from the 3rd to the 12th minute.

Oxytocin increased the propagation velocity of BER in the small intestine by 45% (from 90 to 130 mm/sec).

The effect of oxytocin on the stomach BER requires an intact vagal innervation (Fig. 5). Oxytocin reduced the BER frequency in the stomach (with intact vagal innervation), but at the same time did not reduce the BER frequency in the small pouch (a region free of vagal innervation).

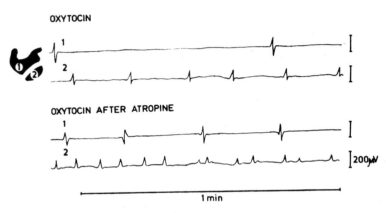

FIG. 5. Effect of oxytocin on the BER of the stomach (lead 1) and on the small Heidenhain pouch (free of vagal innervation, lead 2). Oxytocin (0.4 IU/kg body weight) i.v. reduced the frequency of the stomach BER from 5 to 2 cpm (lead 1) but did not change the frequency of the small pouch BER (lead 2). Oxytocin injected after pretreatment with atropine (300 μg/kg body weight i.v.) had no effect on the BER of the stomach (1) nor of the small pouch (2).

The pharmacologic analysis showed that the effect of oxytocin on BER in the stomach and intestine was not antagonized by adrenergic blocking agents, but was antagonized by cholinergic blocking agents. In animals pretreated with atropine sulfate (200 to 300 mg/kg body weight i.v.), oxytocin did not reduce the frequency of the stomach BER (Fig. 5).

Figure 6 shows the effect of oxytocin on the spontaneous contractile activity of smooth muscle strips isolated from different organs of the guinea pig (stomach, taenia coli, ureter, and vena portae). Oxytocin (30 mIU/ml) inhibited both tone and amplitude of the contractions with maximum effects after 2 min. Larger doses of oxytocin produced a stronger inhibition accompanied by reduction up to complete disappearance of contractions.

The pharmacologic analysis showed that the *in vitro* effect of oxytocin was not antagonized by adrenergic blocking agents (Redergam 3.10^{-6} and

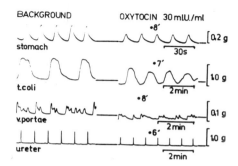

FIG. 6. Background spontaneous activity of muscle strips isolated from stomach, taenia coli, vena portae, and ureter of guinea pig. Oxytocin 30 mIU increases the tone and the amplitude of the contractions.

Obsidan 5.10^{-5}) but was antagonized by cholinergic blocking agents (tetramethylammonium, 10^{-4} mol/ml, and atropine sulfate, 10^{-6} to 10^{-4} mol/ml).

DISCUSSION

These studies indicate that estrogen *in vivo* has a marked excitatory effect and progesterone has a marked inhibitory effect on the spike potentials and the propagation velocity of the BER in the stomach and small intestine. Since the spike potentials trigger the contractions, they are an indirect index of the effect of estrogen and progesterone on the contractions. Thus, our results are in accordance with the results of other workers and support the theory that progesterone participates in the extrauterine smooth muscle hypotony (6,8,14,21,25).

As in the uterus, estrogen increases the excitability of the gastrointestinal smooth muscles; this is accompanied by an increase in the propagation velocity of the BER. The appearance of "extra" potentials in the EGMG of dogs pretreated with large doses of estradiol also indicates a state of high excitability, which allows an "extra" potential to appear as early as during the relative refractory period (Fig. 1A,B,C).

Our results show that synthetic oxytocin inhibits the spike potentials and smooth muscle contractions of the gastrointestinal tract, ureters, and vena portae. The disappearance of the oxytocin effect *in vivo* after 8 to 12 min correlates with the rate of decomposition of hormone (the half-life in plasma is 1 to 4 min) (24). The reduction of the stomach BER in conscious dogs (Figs. 4 and 5) after i.v. injection of oxytocin is an effect for which both n. vagus and cholinergic receptors are required. Oxytocin does not influence the stomach BER in atropinized dogs, nor the BER of the denervated small Heidenhain pouch (Figs. 4 and 5). The antagonism between the effects of atropine and oxytocin on the motility of the ileum in dogs was also observed by Levy (17).

The inhibitory effect of oxytocin on extrauterine smooth muscle activity is also antagonized *in vitro* by cholinergic blocking agents but it is not antagonized by adrenergic blocking agents.

These results indicate that the functions of many extrauterine smooth muscles are affected by ovarian hormones and by oxytocin. This hormonal control may possibly be involved in modulating the function of organs other than the uterus.

REFERENCES

1. Barnes, C. G. (1962): In: *Medical Disorders in Obstetrical Practice,* pp. 103. Blackwell, Oxford.
2. Bass, P., Code, C. F., and Lambert, E. H. (1961): *Am. J. Physiol.,* 201:287.
3. Boyden, E. A., and Rigler, L. G. (1944): *Proc. Soc. Exp. Biol. Med.,* 56:200–201.
4. Csapo, A. I. (1962): *Physiol. Rev.,* 42: Suppl. 5, 7.
5. Daniel, E. E., and Chapman, K. M. (1963): *Am. J. Dig. Dis.,* 8:54.
6. Datta, S., Hey, V. M., and Pleuvry, B. J. (1974): *Pflügers Arch.,* 346:7–95.
7. Deis, R. P., Kitchin, A. H., and Pickford, M. (1963): *J. Physiol. (Lond.),* 166:498–495.
8. Eastman, N. J., and Hellman, L. M. (1961): In: *Williams Obstetrics,* 12th ed., pp. 252–255. Appleton-Century-Crofts, New York.
9. Golenhofen K., v. Loh, D., and Milenov, K. (1970): *Pflügers Arch.,* 315:336–356.
10. Hundley, J. M., Diehl, W. K., and Diggs, E. S. (1942): *Am. J. Obstet. Gynecol.,* 44:858.
11. Hunt, J. N., and Murray, F. A. (1958): *J. Obstet. Gynaecol. Br. Commonw.,* 65:78–83.
12. Kitchin, A. H., Lloyd, S., and Pickford, M. (1959): *Clin. Sci.,* 18:399–408.
13. Konzett, H., Berde, B. B., and Cerletti, A. (1956): *Schweiz. Med. Wochenschr.,* 86:226–229.
14. Kumar, D. (1962): *Am. J. Obstet. Gynecol.,* 84:1300–1304.
15. Kumar, D. (1968): In: *Handbuch der Experimentallen Pharmacology,* Vol. 22, Part 1, p. 530.
16. Karakulina, T. T., and Kusina, M. M. (1972): *Fiziol. Zh. SSSR,* 58:91–95.
17. Levy, B. (1963): *J. Pharmacol. Exp. Ther.,* 140:356–366.
18. Lloyd, S., and Pickford, M. (1962): *J. Physiol. (Lond.),* 163:362–371.
19. Loeper, M., Lemaire, L., and Tauzin J. (1934): *C. R. Soc. Biol. (Paris),* 116:482–484.
20. Marshall, J. M. (1962): *J. M. Physiol. Rev.,* 42, Suppl. 5, 213.
21. McCausland, A. M., Hyman, C., and Winsor, T. (1961): *Am. J. Obstet. Gynecol.,* 81:472–479.
22. McCoy, K., and Bass, P. (1963): *Am. J. Physiol.,* 205:439.
23. Nessen, K. N. (1969): *Fiziol. Zh. SSSR,* 55:486–490.
24. Sawin, C. T. (1972): In: *The Hormones Endocrine Physiology,* Boston.
25. Schatzman H. J. (1966): *Acta Helv. Physiol. Pharmacol.,* 19:106–110.

Physiology of Smooth Muscle, edited by
E. Bülbring and M. F. Shuba.
Raven Press, New York © 1976.

Effect of Histamine, Serotonin, and Other Active Substances on Smooth Muscle Cells of Microvessels

A. M. Chernukh and M. I. Timkina

Institute of Normal and Pathological Physiology, AMS, USSR, Moscow, USSR

A considerable number of studies of the physiology of vascular smooth muscles, especially electrophysiologic studies, have been carried out on large blood vessels. At the same time, the study of smooth muscle cells in the wall of the smallest blood vessels, as structures whose functional state determines to a great extent the capillary blood flow and hence the blood supply of tissues and organs, is of particular interest. Since it is not possible to excise a piece of terminal vessel, the study of the smallest vessels is performed essentially under conditions of vital biomicroscopy mainly on transparent thin organs (mesentery, the web of the bat wing, etc.) (2,4,7).

The first investigations of smooth muscle cells in the wall of terminal microvessels have shown that these cells possess a number of specific properties. For example, the microvascular wall contains fast and slowly contracting smooth muscle cells (5). Moreover, microvessels of different organs and tissues react specifically to vasoactive preparations and physiologically active substances. It is enough to say that histamine, in contrast to its action on large vessels, produces dilatation of microvessels, i.e., relaxation of the smooth muscle cells (4,8). Obviously, such specific reactions of the smooth muscle of terminal vessels may define the functioning of the whole microcirculatory system.

In our laboratory, in the course of some years, functional characteristics of smooth muscle cells in the microvessels of various organs and tissues have been studied in association with the problems of microcirculation. One of the preparations used in our investigation was the mesentery of the rat small intestine with a well-developed vascular network. Treatment of the mesenteric surface with fluorescent stain, orange acridine, and subsequent microscopy in passing light, revealed the structure of the wall of the smallest blood vessels (Fig. 1). The fluorescing nuclei of the smooth muscle cells are oriented in the wall of terminal arterioles in transverse direction and arranged as small groups. Single cellular nuclei may also be noted. In venules, smooth muscle cells are more rarely observed. The transverse location of smooth muscle cells permits them to be identified by vital microscopy in the form of patches on the vascular wall, and allows orientation in electrophysiologic experiments while inserting a microelectrode into the vascular wall.

The microelectrode (diameter of the tip 1 to 2 μm and resistance 10 to 20 MΩ) was inserted into the vascular wall of the mesenteric vessels (diameter from 15 to 80-to-100 μm) with the aid of a specially constructed micromanipulator. For the smaller vessels rigidly fixed microelectrodes were used, and for vessels of 50 to 100 μm diameter showing pronounced contractile activity we used floating microelectrodes. In one series of experiments the tip of the microelectrode was localized by electrophoresis of stain (6).

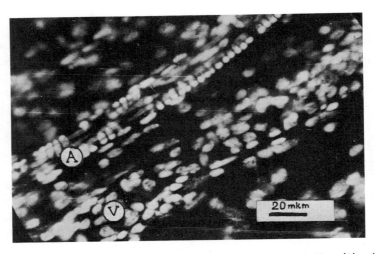

FIG. 1. Fluorescing nuclei of the vascular smooth muscles in the wall of an arteriole 20 μm (A) and a venule 25 μm (V).

There is no doubt that an analysis of any aspect in the study of microvessels is impossible without recording their contractile activity. Actually, fairly good methods exist for measuring the diameter of vessels subjected to microscopic examination, such as the method of image-splitting including TV microscope (1). In our experiments we have used the TV microscope to observe the vasomotor reaction and in some cases combined it with videomagnetophone recording (Fig. 2). This method allowed us to monitor the vasomotor reaction when inserting the microelectrode into the vascular wall. The microelectrode was lowered toward a region where smooth muscle cells had been localized, up to the moment when spontaneous periodic oscillations appeared on the oscillograph screen showing the summed electrical activity of the smooth muscle cells. It should be noted that the position of the microelectrode tip in relation to the smooth muscle cells, the number and the orientation of the cells in the vascular wall were different in different recordings and, as expected, the configuration of the recorded activity was different. In some experiments, after successfully introducing a rather thin microelectrode into a single cell we succeeded in recording the resting po-

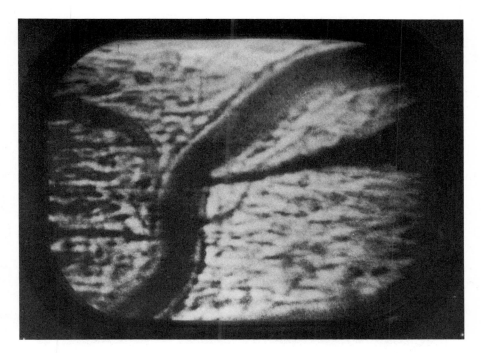

FIG. 2. Metallic microelectrode in the wall of an arteriole. Photograph of TV-microscope screen.

tential and the activity of a single smooth muscle cell (Fig. 3). The configuration and time parameters of the action potentials generated by a single cell in the wall of terminal microvessels were similar to those of smooth muscle cells of larger vessels.

It may be suggested that a certain tone of the microvessels is produced by spontaneous electrical activity of smooth muscle cells in the microvascular wall and that changes in electrical activity are associated with spontaneous vasomotions. Observations made by vital microscopy of the mesentery stained with fluorescent dye have shown that, during spontaneous changes in the vessel diameter, the shape of single smooth muscle cells in the vascular wall can change considerably.

The spontaneous electrical activity of major vessels may be under considerable control by the sympathetic nervous system. However, in the regu-

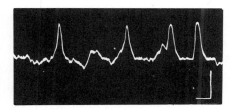

FIG. 3. Electrical activity of a single smooth muscle cell in the wall of an arteriole (20 μm diameter). Calibration: 0.5 mV, 20 msec. Positive deflection upwards.

lation of spontaneous electrical activity and tone of the microvessels, especially the venular parts lacking sympathetic nerve fibers to a considerable extent, the principal role is played by some other still unknown factors. In our work the prerequisites to the study of these factors were the morphologic data of the structure of the nervous and vascular apparatus of the mesentery of rat small intestine (Fig. 4). The extensively branching network of adrenergic nerve fibers innervating the mesentery is intermingled with the network of microvessels, along which run also branches of adrenergic nerve fibers. The preparations show well-distinguishable mast cells with granules,

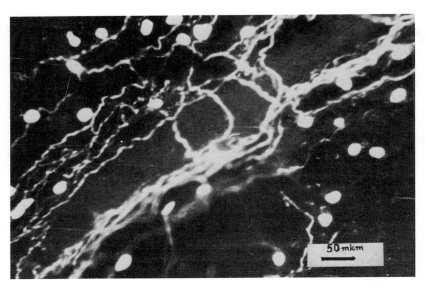

FIG. 4. Distribution of adrenergic nerve terminals in rat mesentery, prepared according to the method of Falck and Owman (1965). For details, see text.

containing, in rats, apart from serotonin, also histamine and heparin. Apparently the smooth muscles of microvessels, branching in the mesentery as well as in other organs and tissues, are under the influence of various local vasoactive substances, for instance those contained in the mast cells and reaching the membranes of smooth muscle cells from the surrounding tissues. These local vasoactive substances are of importance not only for the smooth muscle cells in the wall of the precapillary vessels, but they act also on larger vessels 100 to 200 μm in diameter with well-defined muscle layers.

We have studied the action of histamine, serotonin, bradykinin, and other vasoactive preparations on spontaneous electrical activity recorded from the wall of vessels of different caliber. There is no doubt that the ionophoretic application is the most adequate method of administering a drug to the smooth muscle cells of the microvessels. At the same time, the more widely spread application of a drug like histamine and serotonin to the surface of the mesentery may also be regarded as an adequate method, since the so-called "mediators of inflammation" are synthesized and accumulated in the mast cells of rats and released in their destruction (specific or nonspecific) and spread by diffusion into the mesenteric connective tissue and then to the superficial layer of the microvascular wall.

Electrophysiological experiments have shown that application of histamine (5 to 10 μg in 0.1 ml saline) to the surface of the mesentery of the small

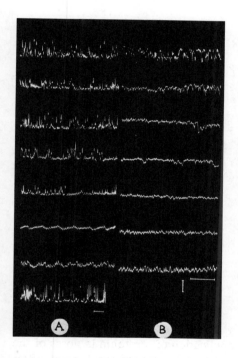

FIG. 5. Electrical activity of the smooth muscle cells in the wall of the terminal mesenteric vessels during application of histamine (5 μg/0.1 ml). A: Upper tracing: activity, recorded in the wall of an arteriole of 20 μm diameter. Every subsequent record 30 sec later. B: The same as A for a venule (25 μm diameter). *Calibration:* 0.5 mV, 100 msec. Positive deflection upward.

intestine of the rat during 20 to 30 sec results in dilatation of the terminal vessels, slowing of blood flow, and, in a number of cases, in the development of stasis. Spontaneous electrical activity is depressed for about 30 to 60 sec following application of histamine (Fig. 5). The duration of the inhibition depends on the extent of the dilatation which, in turn, is due to the dose of the drug. After exhaustion of catecholamine stores, the reaction to histamine is maintained. A depression of spontaneous electrical activity of microvessels, similar to that caused by histamine, is observed on application to the mesenteric surface of preparation 48/80, which leads to degranulation of mast cells and the release of vasoactive substances. Here serotonin is also released and it is difficult to know which of the two substances produce the vasomotor reaction.

When 5 μg serotonin is applied to the mesentery of the small intestine in rat, or to the lucid chamber inserted into the cheek pouch of hamster, the examined vessels react by constriction. In the arterioles the blood flow is only slowed, but in the majority of terminal arterioles, precapillaries, and capillaries stasis develops. Smaller doses of serotonin (0.5 to 1 μg) either have no influence or lead to a hardly detectable vasocontriction. In contrast to the pronounced reaction of the microvessels to histamine, it is difficult to detect changes in electrical activity of microvessels after application of

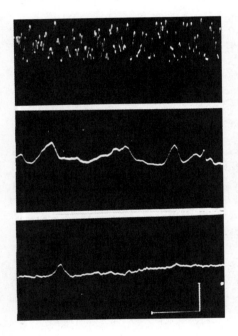

FIG. 6. Electrical activity, recorded in the wall of an arteriole (20 μm) during application of serotonin (5 μg/0.1 ml). *Upper tracing:* Phone activity. *Middle and lower tracings:* 2 and 3 min later. *Calibration:* 1 mV, 100 msec. Positive deflection upward.

small doses of serotonin. Nevertheless, application of serotonin in doses of the order 10 μg in 0.1 ml may lead to a transitory enhancement of impulses recorded from the vascular wall. More frequently, serotonin causes a decrease in amplitude of the spontaneous electrical activity, the record tends to be "flattened," and nonperiodic oscillations are developed (Fig. 6). Restoration of the amplitude of oscillations after application of large doses of serotonin is not observed.

Thus the vasomotor reaction of the terminal vessels of the mesentery of the rat small intestine in response to histamine, preparation 48/80, and large doses of serotonin, is depression of the spontaneous electrical activity and vasodilatation. The constrictor reaction of the microvessels to small doses of serotonin is followed by a hardly detectable transitory rise in amplitude of the electrical activity. Our observations show that the smooth muscle cells of microvessels are rather sensitive to the drugs studied, and suggest that the tone of vascular smooth muscles is greatly influenced by the concentration of these drugs in the perivascular space.

One of the factors determining this concentration may be the activity of sympathetic nerves. In the mesenteric tissue of the rat small intestine, the terminal nerve fibers are intermingled with the mast cells and the vascular network. A similar arrangement, "free nerve endings," has been shown with the electron microscope in the myocardium (3). Evidence showing degranulation of mast cells indicates the possible release of vasoactive substances under the influence of 1-epinephrine.

In trying to analyze this possible interrelation we have studied the electrical activity and vasomotor reaction to epinephrine application. Like serotonin, small doses of epinephrine (up to 5 μg) exert no noticeable influence on the tone of terminal vessels in the mesentery of rat small intestine. Large doses (of the order 10 μg) produce depression of the spontaneous electrical activity, slowing of blood flow, and occasionally stasis and complete vascular dilatation. Recently, we found that stimulation of the splanchnic nerve in rats (which might be compared with epinephrine application) does not lead to a distinct constriction of microvessels with a diameter of less than 20 to 25 μm, although with strong stimulation slowing of blood flow and stasis occurred. The "secondary" dilatatory reaction of the terminal vessels may be due to histamine released from mast cells by adrenergic nervous activity.

During the study of the influence exerted on the smooth muscle cells by "mediators of inflammation" of the type of histamine and serotonin, we have attempted to obtain experimental data on the reaction of microvessels to bradykinin.

Bradykinin is formed in blood plasma during different pathologic reactions. It possesses a pronounced hypotensive effect on intravascular application. However, the data concerning its action on the tone of peripheral vessels are contradictory. We have applied bradykinin to the mesenteric surface as described for histamine and serotonin. We could record no consistent changes of electric and vasomotor activity. Therefore, we have undertaken a series of experiments in which we recorded changes in the outer diameter of microvessels with the method of image splitting.

The results have shown that, depending on the method of application, the reaction of terminal vessels to bradykinin is different and, what is most important, it is different in the first and the subsequent application of the substance. After the intravenous injection of bradykinin, already in the first 30 sec, a dilatation of mesenteric microvessels reaching 10 to 20% is developed, slowing of blood flow, and occasionally stasis. We assume therefore that the superficially applied bradykinin does not reach the membranes of the vascular smooth muscle cells and thus cannot exert a direct influence. Bradykinin may be either destroyed by tissue kininases or, more likely, lead to the release of other biologically active substances which, in turn, conceal the effect of bradykinin itself. It is likely that surface application of bradykinin is not "adequate" since bradykinin is formed mainly in the blood plasma and reaches the membranes of vascular smooth muscle cells from the intravascular space. This may be of little importance for experiments on isolated preparations. But in the analysis of the functional state of the smooth muscle cells of microvessels, a structure involved in the general system of microcirculation, the method of administration of vasoactive substances may play a decisive role.

The example of the influence of bradykinin on microvessels demonstrates that the experimental approach to the study of smooth muscle cells in the wall of terminal vessels is difficult. Electrophysiologic studies are difficult because of the small size of smooth muscle cells. However, vasomotor activity and, what is most essential, the smooth muscle cells in the microvascular wall become well distinguishable with high magnification of the microscope. Nevertheless, the optics necessary to obtain such magnification hamper the manipulations of the microelectrode.

The relatively small number of investigations devoted to the study of the electrophysiology of smooth muscle cells of terminal vessels, may be to a certain extent explained by these circumstances. The present study undoubtedly represents only the first stage in a study of the most general electrical and mechanical processes in smooth muscle cells of microvessels.

REFERENCES

1. Baez, S. (1966): *J. Appl. Physiol.*, 21:299–301.
2. Chambers, R., and Zweifach, B. W. (1944): *Am. J. Anat.*, 75:173–205.
3. Chernukh, A. M., and Alexeyev, O. V. (1974): *Bibl. Anat.*, 11:165–171.
4. Duling, B. R., and Berne, R. M. (1970): *Circ. Res.*, 26:163–170.
5. Funaki, S. (1968): *Proc. Jap. Acad.*, 34:334.
6. Kaneko, A., and Hashimoto, H. (1967): *Vision Res.*, 7:847–851.
7. Nicoll, P. A., and Webb, R. L. (1946): *Ann. N.Y. Acad. Sci.*, 46:697.
8. Schayer, R. W. (1963): *Prog. Allergy*, 7:187–212.

Physiology of Smooth Muscle, edited by
E. Bülbring and M. F. Shuba.
Raven Press, New York © 1976.

Physiological and Biochemical Aspects of the Action of Serotonin on Smooth Muscle Cells of Myometrium

M. D. Kursky, N. S. Baksheev, V. V. Chub, and A. N. Fedorov

A. V. Palladin Institute of Biochemistry, A. Sc. Ukr. SSR, and Department of Obstetrics and Gynecology, No. 1, A. A. Bogomoletz Medical Institute, Kiev, USSR

5-Hydroxytryptamine (serotonin, 5-HT), a mediator in the synaptic transmission of nervous impulses, has a highly specific action on nervous tissue and on smooth muscles. This may be due to the specific molecular organization of the cell membranes and the structure of the 5-HT receptors in these tissues.

In the function of nerve and smooth muscle cells calcium ions play a special role and, as has been indicated by our findings *in vivo* (1–3), Ca exchange increases under the influence of 5-HT.

It is known that a certain concentration of Ca^{2+} is required for the cell function in nerve and muscle. The concentration in solution in the cytoplasm can be a function of the following factors: the presence of agents capable of chelating the ions (proteins, orthophosphates, adenine nucleotides of a number of substrates of the tricarboxylic cycle, etc.); ionic exchange between the cell and the extracellular medium; binding or release of the cytoplasmic ions by mitochondria and endoplasmic reticulum. The latter two functions can be accomplished by way of an active or passive transport of ions. The presence of active mechanisms of Ca^{2+} transport in nerve cells and in sarcoplasmic reticulum of skeletal muscle has been proved and is still being investigated. However, for a long time no one succeeded in isolating functionally active structures from smooth muscles of vertebrates which accumulate calcium. That is why it was believed that the intracellular calcium ion content in smooth muscle cells was regulated by the plasma membrane, i.e., due to Ca exchange between the cell and the extracellular medium.

Recently it has been reported that the mitochondrial and microsomal fractions of myometrium, of the aorta of cows, rabbits, and women, and the intestine of guinea pigs can actively accumulate calcium and this accumulation in one of the subfractions of microsomes is perhaps achieved by a Mg^{2+},Ca^{2+}–adenosinetriphosphatase (ATPase) (4–6).

The aim of our present study was to investigate the mechanism of the action of 5-HT on smooth muscles, and its possible effect on Ca^{2+} transport. We have studied the active accumulation of ^{45}Ca by mitochondrial and microsomal fractions, the ATPase activity, the formation of the intermediate phosphorylic product (IPP) of calcium ATPase, and the effect of serotonin on these processes as well as on the conformation of plasma membranes of myometrium.

METHODS

Spontaneous electrical activity of rat myometrium cells was studied by the double sucrose-gap method (7) and by intracellular recording of electrical potentials (8). The mitochondrial and microsomal fractions were isolated from the myometrium of rabbits and cows by the modified method of Martonosi and Feretos (9).

The activity of ATPases was determined in a medium containing 0.1 KCl, 5 mM MgCl$_2$, 1×10^{-2} to 1×10^{-8} M CaCl$_2$, 50 mM tris HCl buffer pH = 7.4, 2 mM ATP, 0.1 ml suspension of subcellular fractions which had 0.2 mg protein of microsomes and 1 to 2 mg protein of mitochondria, 1×10^{-4} to 1×10^{-7} M serotonin-creatine-sulfate. EGTA (1 mM) was added to the control samples. The total volume of the incubation medium was 1 ml. The incubation time was 10 min. Inorganic phosphorus was determined by the method of Fiske and Subbarow (10).

The absorption of ^{45}Ca by subcellular fractions was determined under conditions similar to the determination of ATPase activity, but the samples contained 1×10^{-4} M CaCl$_2$ as carrier and 30,000 counts min ^{45}Ca per 1 ml. After 10 min incubation the reaction was interrupted by a quick filtration through a micropore filter (0.3 to 0.5 μm) which was twice washed by the cooled solution, replacing the radioactive with the inactive calcium. The filters were dried in air and the radioactivity was calculated by the use of counter YMO-1500M provided with a IIII-15 scaler.

The formation of IPP of the Mg^{2+}-dependent Ca^{2+} activated ATPase of myometrium microsomes was estimated by the method of Jamamoto and Tonomura (11) in a medium containing 0.1 M KCl, 5 mM MgCl$_2$, 50 mM tris-HCl buffer, pH = 7.4, 1×10^{-4} M CaCl$_2$ and 20 μM ^{32}P-labeled ATP (30,000 counts min per 1 ml sample), 0.3 mg protein. Samples without calcium served as controls. Labeled ATP was obtained by the method of Glynn and Chappell (12). The samples were incubated up to 20 sec. Upon completion of the incubation period the samples were filtered through millipore filters (0.5 μm). Radioactivity was estimated on the dry filters with the aid of the YMO-1500M counter. Calcium-45 accumulation by subcellular fractions was carried out by the method of Vasington and Murphy (13).

RESULTS AND DISCUSSION

In normal Krebs solution 5-HT increased myogenic tone, increased the frequency of contractions, and increased the amplitude and frequency of spontaneous spike potentials (Fig. 1,I). 5-HT decreased the membrane potential (Fig. 1,II), increased excitability, and reduced the resistance of the plasma membrane (Fig. 1,III).

In a Ca-free medium 5-HT caused no changes in the membrane potential, permeability of excitability (Fig. 1,IY). These and previously obtained (1–3) biochemical data indicate that 5-HT may increase the membrane permeability to Ca^{2+}, thereby increasing excitability, reducing membrane resistance, and generating spontaneous electrical and contractile activity.

We have shown that in the presence of ATP the mitochondria of horned cattle myometrium accumulate 28% more calcium than the microsomes. The subcellular fractions of myometrium of mother rabbits possess a greater

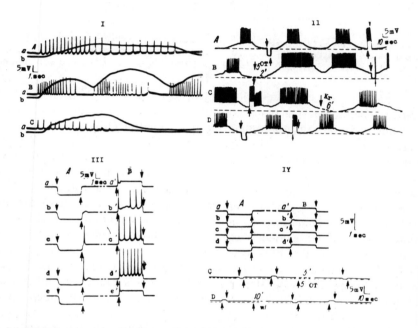

FIG. 1. I: Effect of serotonin on electrical and contractile activity of rat myometrium cells. (A) in Krebs solution; (B) in the presence of serotonin ($10\mu^5$ M) in the Krebs solution; (C) after washing. (a) electrical activity; (b) contractile activity. II: Effect of $10\mu^5$ M serotonin on membrane potential, spontaneous electrical activity, electrotonic potentials, and excitability of smooth muscle cells of myometrium at the beginning of labor. Polarizing current, 0.3 μA. (A) spontaneous electrical activity, membrane potential, and electrotonic potential in Krebs solution; (B) shift of membrane potential, change of spontaneous electrical activity and of electrotonic potential under the influence of serotonin (10^{-5} M) (at arrow marked 5-HT); (C) arrow marked Kr indicates beginning of wash in normal Krebs solution; (D) restoration of membrane potential, spontaneous electrical activity, electrotonic potential, and excitability in Krebs solution. III: Effect of serotonin on excitability and membrane permeability. Current 0.5 μA. (A) anelectrotonic potential; (B) catelectrotonic potential. (a,a″) in normal Krebs solution; (b,b″) 3rd minute; (c,c″) 5th min; (d,d″) 6th to 7th min of action of serotonin (10^{-5} M); (e,e″) 6th minute of washing in Krebs solution. IV: Effect of serotonin (10^{-5} M) in Ca-free solution. Polarizing current 1 μA. (A) anelectrotonic potentials; (B) catelectrotonic potentials. (a,a″) Ca-free solution; (b,b″) 1st to 2nd min, (c,c″) 5th to 6th min, (d,d″) after 5 min washing. (C,D) effect on membrane potential and electrotonic potentials (recording at slow scanning on EPP-09-MI). Arrows: 5-HT, beginning of serotonin application; w, beginning of washing.

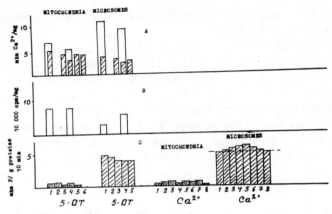

FIG. 2. Effect of serotonin on Ca accumulation (A) on ^{45}Ca release (B) on ATPase activity (C) of subcellular fractions of smooth muscle. Counts per minute on y axis. Concentrations of 5-HT and Ca^{2+} on x axis: (1) control; (2) 10^{-8} M; (3) 10^{-7} M; (4) 10^{-6} M; (5) 10^{-5} M; (6) 10^{-4} M; (7) 10^{-3} M; (8) 10^{-2} M. (☐) rabbit; (▨) cow.

ability to accumulate ^{45}Ca than those of cows, the microsomic fraction accumulating 40% more isotopes than the mitochondria (Fig. 2A).

Calcium ions, in concentrations of 10^{-6} to 10^{-4} M activated ATPase of uterine microsomes (Fig. 2C) in the presence of magnesium ions, inhibiting it at higher concentrations. This may indicate that the accumulation of calcium ions by myometrium microsomes is accomplished by Mg^{2+}, Ca^{2+}–ATPase as was shown by a number of authors for the sarcoplasmatic reticulum of skeletal muscle (11) and for the microsomal fraction of smooth muscles (4–6). In mitochondria calcium ions did not affect the activity of ATPase in the presence of Mg^{2+} ions (Fig. 2C).

5-HT, 10^{-6} M, reduced the accumulation of calcium by uterine microsomes from cows by 24%, having almost no effect on this process in other concentrations. 5-HT also inhibited the accumulation of calcium by microsomes from the myometrium of rabbits (Fig. 2A). Attention is drawn to the fact that 5-HT appreciably affected this process only at 10^{-6} M concentration. In lower and higher concentrations the effect was not so marked. A similar phenomenon was noted by Bloomquist and Curtis (14) on the smooth muscle of insects.

5-HT, 10^{-7} to 10^{-3} M, did not affect the ATPase activity of microsomal and mitochondrial fractions (Fig. 2C). During incubation of these subcellular fractions from rabbit myometrium, "loaded" with radioactive calcium, in a Ca-free medium it was noted that 40% ^{45}Ca is released from mitochondria and 12% from microsomes. Under these conditions 5-HT did not change ^{45}Ca release from mitochondria but increased ^{45}Ca release from microsomal fraction (Fig. 2B).

The formation of IPP in the ATPase reaction was determined with the aid of ^{32}P-ATP. As is seen in Fig. 3, the formation of IPP in smooth muscle was substantially slower than in the skeletal muscle. Thus the maximum of the IPP formation occurred at the 10th second in the skeletal muscle and at the 40th second in smooth muscle. The rate of IPP disintegration for these two types of muscle differed even more, reaching a maximum in 40 to 60 sec in skeletal muscle, whereas in smooth muscle 20% IPP remained in 120 sec. These rates of IPP formation and disintegration correlate well with the rapid contraction cycle in skeletal muscle and the slow cycle in smooth muscle. The evidence perhaps indicates an important role of Mg^{2+}, Ca^{2+}–ATPase in supporting the functional rhythm of the muscles.

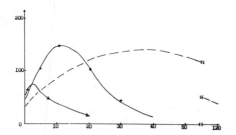

FIG. 3. Dynamics of formation and disintegration of IPP of Mg, Ca–ATPase of microsomes of smooth muscle and of sarcoplasmic reticulum, of skeletal muscles. Time in seconds on x axis. Counts per minute/0.3 mg protein on y axis. (–––) Smooth muscle; (●) skeletal muscle; (▶) binding of ^{45}Ca by skeletal muscle reticulum.

The content of ^{32}P in the proteins of the microsomal fraction was reduced by 5-HT, 10^{-6} M, whereas in concentrations of 10^{-5} to 10^{-3} M, 5-HT caused no change or a slight increase (Fig. 4,A). It may be that serotonin either reduced the rate of IPP formation or accelerated the disintegration.

After incubation of the formed IPP for 20 sec in a medium without ATP-^{32}P we noted that 5-HT, 10^{-6} M, reduced the labeling in microsomes, but it did not change it in higher concentrations (Fig. 4,B). From this observation it may be assumed that low concentrations of this amine increase the rate of disintegration of IPP, whereas high concentrations prevent it. In accordance with the data of Burshtein (15) electron donors, 5-HT being one of them, while not affecting the rate of formation of the enzyme–substrate complex, substantially, sometimes by twice, accelerate its disintegration. Our findings are to a certain degree in line with those data, since 5-HT in 10^{-6} M concentration accelerated IPP disintegration, which is the final stage of the ATPase reaction.

It is known that the formation of IPP Mg^{2+},Ca^{2+}–ATPase is accompanied by binding and transfer of calcium ions and by their accumulation in the microsomal fraction (16). Calcium ions are bound by the sarcoplasmic reticulum of skeletal muscle very quickly, and the binding is already at a maximum in 3 sec (Fig. 3). At this instant the Ca/P ratio amounts to 1.3, i.e., 1 gram-atom of phosphorus binds 1.5 gram-atoms of calcium. The Ca^{2+} binding ability was also found to be present in the microsomal fraction of uterine smooth muscle (Fig. 4C).

Thus we have established that 5-HT has no effect on the activity of Mg^{2+},Ca^{2+}–ATPase, reduces the binding of ^{45}Ca, and stimulates the release of the latter from the microsomal fraction of myometrium. At the same time it increases the Ca^{2+} contents in smooth muscle cells (1–3). Recently, similar results were obtained by other workers on the sarcoplasmic reticulum of smooth muscles of mollusks and on lacrimal glands of insects (14,17).

Considering serotonin as a mediator, we assumed that in the interaction with the receptors of the postsynaptic membrane it might induce structural changes in the membrane, thus facilitating Ca^{2+} penetration into the cell.

It is known that protein molecules which are incorporated in membranes are characterized by a certain alignment and distribution on their surface of polar and nonpolar amino acid side radicals. In accordance with contemporary conceptions, the change in protein conformation is accompanied by a re-

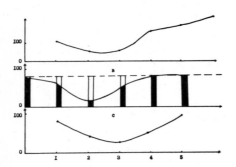

FIG. 4. Effect of serotonin on IPP contents and binding of ^{45}Ca by microsomal fraction of smooth muscle of the uterus. Counts per minute/0.3 mg protein on y axis. A: Inclusion of ^{32}P. B: Disintegration of IPP. C: Binding of ^{45}Ca. 0 = initial contents of IPP (B curve). (1) Controls; (2) effect of serotonin (10^{-6} M); (3) effect of serotonin (10^{-5} M); (4) effect of serotonin (10^{-4} M); (5) effect of serotonin (10^{-3} M).

distribution of a certain quantity of polar and nonpolar groups on the surface of the protein molecule. Information of such changes can be obtained by analyzing the interaction of vital dyes with membrane proteins (18).

We therefore examined the effect of 5-HT and calcium on the interaction of vital dyes [neutral red cation (NR) and photostable direct turquoise anion "K" (PDT)] with the plasma membrane of the smooth muscles of uterus, determining the limiting amount of adsorption centers of the dyes, dissociation constants of the protein–dye complexes, and the binding energy ($-\Delta F$). The coefficients estimated on the basis of the analysis of Langmuir's isotherm are given in Fig. 5. We found that the plasma membrane of cow myometrium had more PDT adsorbent centers than NR adsorbent centers. 5-HT, $1.7.10^{-7}$ M, increased the maximum quantity of NR adsorbent points fourfold, and the quantity of PDT points twofold. During this process the dissociation constant of the protein–dye complexes increased, while the binding energy ($-\Delta F$) decreased by 2.6 Cal/mol. This result indicates that 5-HT may change the interrelation between groups on the plasma membrane surface so that those centers of adsorption which have a smaller binding energy and which were either inaccessible or were screened in the control group, are now accessible. Higher concentrations of 5-HT have less effect on the adsorption of dyes.

1.5 mM Ca²⁺ and 1.5 mM Ca²⁺ + 1.7 × 10⁻⁵ M 5-HT did not affect the limiting quantity of NR adsorption centers on the plasma membrane, though for PDT (which is adsorbed mainly by positively charged amine groups of proteins) the adsorption surface of the plasma membrane changed appre-

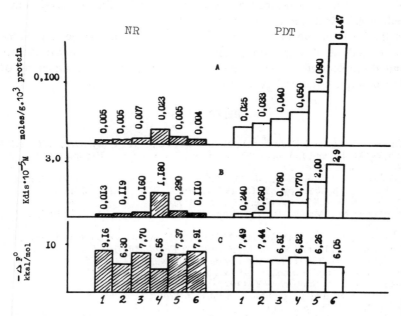

FIG. 5 Effect of serotonin and calcium on thermodynamic parameters of adsorption of vital dyes by the plasma membrane of myometrium. A: Maximum quantity of dye moles bound by 1 kg membrane protein/A_{∞}. B: Dissociation constant (K · 10⁻⁵ M). C: Binding energy (−ΔF° Cal/mol). (▨) NR, neutral red; (☐) PDT, photostable direct turquoise "K". (1) controls; (2) 1.7·10⁻⁵ M 5-HT; (3) 1.7·10⁻⁶ M 5-HT; (4) 1.7·10⁻⁷ M 5-HT; (5) 1.5 mM Ca²⁺; (6) 1.5 mM Ca²⁺ + 1.7·10⁻⁵ M 5-HT.

ciably. Thus 1.5 mM Ca²⁺ increased the maximum quantity of PDT adsorption centers 3.6 times and 1.5 mM Ca²⁺ + 1.7 × 10⁻⁵ M 5-HT, six times.

The results provide evidence for a high sensitivity of the surface of the plasma membranes of uterine smooth muscles to the action of serotonin and Ca²⁺ ions, particularly to their joint action.

The changes in the amount of adsorption centers for the dye may be indicative of structural changes of the plasma membrane which may occur during the interaction with serotonin and Ca²⁺. Perhaps these changes bring about a greater membrane permeability for Ca²⁺. It may be assumed that Ca entry into the cell down the concentration gradient does not create a sufficiently high concentration in the cytoplasm to ensure contraction of the actimyosin system. For this reason another mechanism should operate at the beginning of each cycle of contraction which inhibits active transport and activates the liberation of Ca^{2+} from membrane structures, specifically from sarcoplasmatic reticulum.

The experimental results obtained enable us to think that 5-HT, changing the conformation of the sarcolemma, stimulates a passive supply of Ca^{2+} to the cell. The release of Ca^{2+} from subcellular structures may be caused by 5-HT penetrating into the cell or may be caused by Ca^{2+} passively entering the cell, as is indicated in the literature (19).

On the basis of the data described in this paper it has been possible to devise a method of stimulating contractile activity of the myometrium by combined application of 5-HT and Ca^{2+}, a method which is advantageously employed in obstetrics and gynecology.

REFERENCES

1. Kursky, M. D., and Fedorov, A. N. (1970): *Rep. Acad. Sci. Ukr. SSR,* 8:736.
2. Kursky, M. D., and Baksheev, N. S., Antonenko, S. G., and Fedorov, A. N. (1971): *Ukr. Biokhim. Zh.* 43:219.
3. Baksheev, N. S., Kursky, M. D., Tugay, V. A., Fedorov, A. N., Bondarenko, O. D., and Shchegolkov, A. I. (1972): *Probl. Med. Chem.,* 17:620.
4. Carsten, M. E. (1969): *J. Gen. Physiol.,* 53:414.
5. Boudouin, M., Mayer, P., Farmadjian, S., and Margat, J. C. (1972): *Nature,* 235:336.
6. Godfraind, T., and Verbeke, N. (1973): *Arch. Int. Pharmacodyn. Ther.,* 204:187.
7. Artemenko, D. P., and Shuba, M. F. (1964): *Fiziol. Zh.,* 10:304.
8. Kostyuk, P. G. (1960): *Microelectrode Technique,* K.
9. Martonosi, A., and Feretos, J. (1968): *J. Biol. Chem.,* 243:71.
10. Fiske, C. H., and Subbarow, Y. (1925): *J. Biol. Chem.,* 66, 375.
11. Yamamato, F., and Tonomura, J. (1968): *J. Biol. Chem.,* 64, 137.
12. Glinn, J., and Chappel, I. (1964): *Biochem. J.,* 90:147.
13. Vasington, F., and Murphy, J. V. (1962): *J. Biol. Chem.,* 237:2670.
14. Bloomquist, E., and Curtis, A. B. (1972): *J. Genl. Physiol.,* 59:476.
15. Burshtein, E. A. (1956): *Biophysics of Muscular Contraction,* p. 214. M., Nauka.
16. Weber, A. (1966): *Curr. Top. Bioenergetics,* 1:203.
17. Prince, W. T., Berridge, M. J., and Rasmussen, H. (1972): *Proc. Natl. Acad. Sci. US,* 69:553.
18. Komissarchik, Y. Yu., Levin, S. V., and Passova, R. B. (1971): *Cytology,* 12:286.
19. Ford, L. (1972): *J. Physiol. (London),* 233:21.

Physiology of Smooth Muscle, edited by
E. Bülbringand M. F. Shuba.
Raven Press, New York © 1976.

Electrophysiological Characteristics of Anaphylaxis of Isolated Smooth Muscle

I. S. Guschin

Allergological Research Laboratory of Academy of Medical Sciences of USSR, Moscow, USSR

The involvement of smooth muscle cells in the anaphylactic reaction of isolated smooth muscle organs is accomplished at least in two ways.

As has been postulated in our previous studies, the specific antigen may act directly on the excitable membrane of smooth muscle cells (1–3). This mode of action does not exclude the second principal mechanism which consists in degranulation of tissue mast cells and liberation of biologically active substances which may excite smooth muscle cells.

Recently, it has been shown by our experiments and other investigations that the mode of action of the specific antigen does not consist of an injury of the target cells, but is more likely an activation of the function of these cells (1,3–5).

In this chapter I would like to draw attention not only to smooth muscle cells but also to mast cells which may be closely involved in the regulation of smooth muscle cell function under physiologic and pathologic conditions.

I wish mainly to illustrate with our results an excitatory action of the specific antigen on both cell types, which produces the anaphylactic contraction of smooth muscle organs.

ELECTROPHYSIOLOGICAL CHANGES IN SMOOTH MUSCLE CELLS DURING ANAPHYLACTIC REACTION

Figure 1 illustrates the action of a specific antigen (ovalbumine) on taenia coli from actively sensitized guinea pig. The anaphylactic contracture is preceded by depolarization of the membrane and by increase in the frequency of spontaneous action potentials. The same sequence of events is produced by the exciting mediators, for example, acetylcholine (Fig. 1a). After the anaphylactic reaction and desensitization to the antigen has taken place (Fig. 1c), the excitability of the cells is rapidly restored, as may be tested by repeating the same dose of acetylcholine (Fig. 1d) or by plotting a dose–response curve.

Histologic preparations showed that the taenia coli used in the experiments contained no mast cells (1). Thus the presence of mast cells in smooth muscle is not obligatory for the anaphylactic reaction.

419

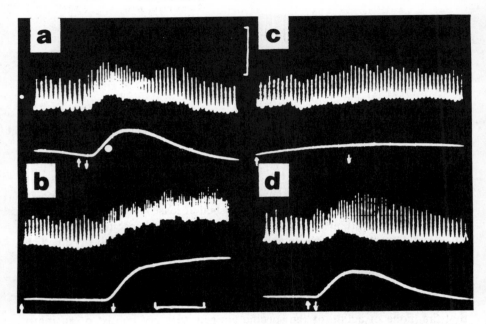

FIG. 1. Electrical and mechanical activity of taenia coli of actively sensitized guinea pig. [sucrose-gap method (6)]. (a, d) Acetylcholine (0.5 μg); (b, c) ovalbumin (200 μg) (c, 10 min after b) *Calibration: horizontal, 10 sec.; vertical, 5 mV and 5 g.* Upward deflection corresponds to depolarization and increase of tension.

It is well known that K-rich solution produces depolarization of the cell membrane (Fig. 2b). Under such conditions propagation of excitation from one cell to another is blocked, and exciting agents produce an increase in tension (Fig. 2c) without visible signs of electrical activity. Depolarized smooth muscle preparations have been widely used for investigating the direct action of different agents on smooth muscle cells (see Refs. 7–11). In our experiments an anaphylactic contracture could be produced in depolarized taenia coli preparations. Figure 3 shows that the anaphylactic contraction of the depolarized smooth muscles was reduced to the same degree as the acetylcholine contraction, which was evoked by the same dose of acetylcholine in normal and K-rich medium.

It may be concluded that electrophysiologic manifestation of the anaphylactic reaction of smooth muscle is similar to those of other excitatory substances. The observation that the anaphylactic reaction can be evoked in polarized and depolarized taenia coli supports the view of a direct action of the specific antigen on the sensitized smooth muscle cells.

MEMBRANE POTENTIAL OF MAST CELLS AFTER ANAPHYLACTIC REACTION

Using intracellular microelectrodes (12) on peritoneal rat mast cells it was shown that the membrane potential of mast cells, pretreated with specific

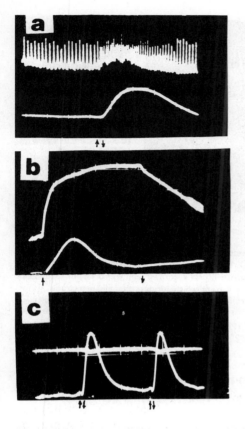

FIG. 2. Reactions of taenia coli to acetylcholine in normal and K-rich media. (a) Acetylcholine (1 μg) in normal medium; (b) effect of K-rich (150 mM) solution; (c) acetylcholine (0.1 mg) in K-rich medium. *Calibration: horizontal,* 10 sec (a), 24 sec (b and c); *vertical,* 10 mV and 4 g (a); 20 mV and 10 g (b); 10 mV and 1 g (c). Upward deflection corresponds to depolarization and increase of tension.

antigen and washed free of the antigen, did not differ from the membrane potential of untreated cells (Fig. 4; Table 1).

The anaphylactic reaction of mast cells was evaluated by the percentage of histamine release. Histamine was measured by a modification of the spectro-

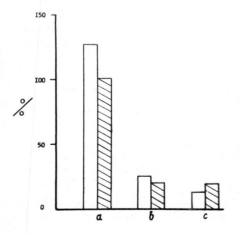

FIG. 3. Contractions of taenia coli (open columns) and ileum (shaded columns) of actively sensitized guinea pigs in K-rich medium. (a) Anaphylactic contraction (200 μg/ml ovalbumin) in normal medium; (b) contraction to acetylcholine (1 μg/ml) in K-rich medium; (c) anaphylactic contraction (200 μg/ml ovalbumin) in K-rich medium. All contractions are evaluated as % of the acetylcholine contraction in normal medium.

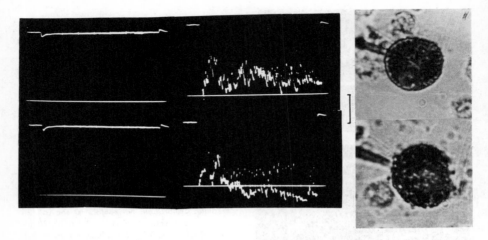

FIG. 4. An example of registration of membrane potential of untreated mast cells (*top*) and mast cells previously exposed to the specific antigen (*bottom*). The mast cells were obtained from the peritoneal cavity of actively sensitized (to ovalbumin) rats. *Left:* artifacts recorded with extracellular location of microelectrodes during the switching on and off the current which triggers the microelectrode driver. *Right:* traces at the moment of inserting the microelectrode into the cell. Microelectrode resistance (filled with 3 M KCl)—150 MΩ. Duration of the push of microelectrode, 1 sec. Downward deflection corresponds to electronegativity. *Calibration:* 10 mV. On the microphotograph: untreated mast cell (*top*) and mast cell previously exposed to the antigen (*bottom*) at the moment of inserting of microelectrodes (×400).

fluorometric assay (4). Anaphylactic histamine release was $23 \pm 0.67\%$ and coincided with $68.2 \pm 1.9\%$ of morphologically changed mast cells.

K-rich (127 mM) solution caused slight depolarization of the cell membrane (about 25% of the initial value of membrane potential). But this depolarization failed to cause histamine release (Table 2) and did not prevent histamine release by chemical histamine liberators (5 mM ATP) or by the specific antigen (Table 2).

These findings confirm the idea that the anaphylactic reaction is not followed by disruption of the cell membrane (3,4), such as may be seen after mechanical trauma by repeated pricking of the cells with micropipets of comparatively large tip diameter or after treatment with Triton X-100 (4). Both procedures were followed by progressive decrease and disappearance of the membrane potential.

TABLE 1. *Membrane potential of untreated mast cells and of mast cells pretreated with the specific antigen*

Mast cells	Membrane potential (mV)	Histamine release (%)
Untreated	15.42 ± 0.48 (n = 71)	$1.87 \pm 0.24^*$ (N = 10)
Exposed to antigen	16.48 ± 0.47 (n = 80)	23.8 ± 0.67 (N = 10)

$\bar{X} \pm$ SE; n: number of cells; N: number of cell samples.
* Spontaneous release.

TABLE 2. *Histamine release from mast cells by liberators in different media*

Media	Releasing agents	Histamine release (%) ($\bar{X} \pm SE$)	N
Normal	—	1.74 ± 0.13*	30
	ATP (5 mM)	61.6 ± 5.9	6
	antigen	24.38 ± 1.15	28
K-rich	—	2.17 ± 0.3*	9
(127 mM)	ATP (5 mM)	67.5 ± 6.8	7
	antigen	29.78 ± 3.84	9

* Spontaneous release.

Careful examination of ultrastructural changes of mast cells during the anaphylactic reaction also shows that this reaction cannot be a result of cell disruption (13). It resembles more the morphology of secretion processes. A similar process is also suggested by the role of divalent cations in secretion and in the anaphylactic release of histamine (14). In both cases the reactions are inhibited in the absence of Ca ions and by manganese. The anaphylactic release of histamine from mast cells was reduced by 81.21 ± 6.8% in Ca-free solution, and by Mn (0.5 mM) by 61.96 ± 2.2%.

INFLUENCE OF ELEVATION OF CELL CYCLIC 3′,5′-ADENOSINE MONOPHOSPHATE CONTENT ON ANAPHYLACTIC HISTAMINE RELEASE

Many investigators are currently studying the role of cyclic 3′,5′-adenosine monophosphate (cAMP) in secretion of mediators and hormones. This substance also plays an important role in the anaphylactic histamine release from different tissues and cells including mast cells (15–18).

We have tried drugs, which modulate cAMP content, on the anaphylactic histamine release from mast cells.

Papaverine has a high potency inhibiting phosphodiesterase activity. However, papaverine only slightly increased cAMP content in mast cells and did not influence the anaphylactic histamine release (in the presence of 10 mM glucose).

On the other hand, prostaglandin E_1 (PGE$_1$) produced a significant increase of cAMP content and this effect was potentiated by papaverine (Fig. 5). PGE$_1$ (0.14 mM) alone reduced the anaphylactic histamine release up to 55% of the control. Papaverine (0.1 mM) augmented the inhibitory effect of PGE$_1$, so that the anaphylactic histamine release in the presence of a combination of papaverine with PGE$_1$ was reduced to 39% of the control.

All these data obtained on mast cells, including the maintenance of a normal membrane potential in antigen-treated cells, the role of divalent cations, and the role of cAMP in anaphylactic histamine release show that the trig-

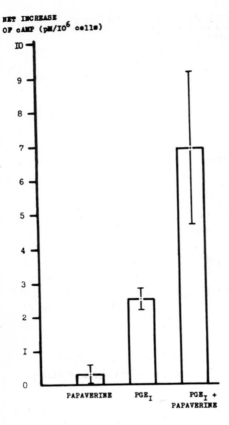

NET INCREASE
OF cAMP (pM/IO⁶ cells)

FIG. 5. Cyclic 3',5'-adenosine monophosphate content in rat mast cells pretreated with papaverine and PGE₁. Concentrated rat mast cell suspensions (0.5 to 1 × 10⁶ cells per sample) were preincubated for 5 min without or with papaverine (0.1 mM) and PGE₁ (0.14 mM). The reaction was stopped by chilling the tubes in ice-water and the cells were deproteinized with 0.3N perchloric acid. The supernatants were chromatographed on columns of Dowex 50 W X 4 (200 to 400 mesh) in hydrogen form and cAMP was eluted with 5 ml distilled water. cAMP was determined by the method of Brown et al. (see ref. 21) using bovine adrenal cAMP-binding protein. Results are expressed as the net increase of cAMP content (pM/10 cells) and are given as mean values ± SE (quadruplicate determinations from five cell pools). Basal cAMP level was 0.95 ± 0.27 pM/10 cells.

gering mechanism of anaphylactic histamine release resembles a secretion process more than cell injury.

Taking into account the electrophysiological features of the anaphylactic reaction of smooth muscle cells, one can reach a more general conclusion: i.e., the initiation of the anaphylactic reaction of target cells is brought about by excitation of these cells by the antigen–antibody complex.

REFERENCES

1. Guschin, I. S. (1966): *Bull. Exp. Biol. (USSR)*, 12:25.
2. Guschin, I. S. (1967): *Pathol. Physiol. Exp. Ther. (USSR)*, 2:45.
3. Guschin, I. S. (1973): *Anaphylaxis of Smooth Muscle and Heart.* Medicina, Moscow.
4. Guschin, I. S., Orlov, S. M., and Czju, N. L. (1974): *Allerg. Immunopathol.*, 2:69.
5. Stanworth, D. R. (1973): *Immediate Hypersensitivity. The Molecular Basis of Allergic Response*, p. 290 (*Frontiers of Biology*, Vol. 28). North-Holland, Amsterdam, London, New York.
6. Guschin, I. S. (1966): *Bull. Exp. Biol. (USSR)*, N10: 120
7. Edman, D. H. L., and Schild, H. O. (1961): *Nature*, 190:74.

8. Evans, D. H. L., Schild, H. O., and Thesleff, S. (1958): *J. Physiol. (Lond.)*, 143: 474.
9. Sperelakis, N. (1962): *Am. J. Physiol.*, 202:731.
10. Waugh, W. H. (1962): *Circ. Res.*, 2:264.
11. Burnstock, G., Holman, M. E., and Prosser, C. L. (1963): *Physiol. Rev.*, 43:482.
12. Guschin, I. S., and Sverdlov, Yu. S., (1973): *Bull. Exp. Biol. (USSR)*, 4:121.
13. Anderson, P., Slorach, A., and Uvnäs, B. (1973): *Acta Physiol. Scand.*, 88:359.
14. Foreman, J. C., and Mongar, J. L. (1972): *J. Physiol. (Lond.)*, 224:753.
15. Bourne, H. R., Lichtenstein, L. M., and Melmon, K. L., (1972): *J. Immunol.*, 108:695.
16. Kaliner, M., and Austen, K. F. (1973): *J. Exp. Med.*, 138:1077.
17. Kaliner, M., and Austen, K. F. (1974): *J. Immunol.*, 112:664.
18. Lichtenstein, L. M., DeBernardo, R., and Sjoersma, A. (1971): *Biochem. Pharmacol.*, 20:2287.
19. Rutten, W. J., Schoot, B. M., and De Pont J. J. H. H. M. (1973): *Biochim. Biophys. Acta*, 313:378.
20. Loten, E. G., and Sneyd, J. G. T. (1970), *Biochem. J.*, 120:187.
21. Brown, B. L., Albano, J. D. M., Ekins, R. P., Sgherz, A. M., and Tampion, W. (1971): *Biochem. J.*, 121:561.

Subject Index

  replacement of, 181
  slow potential dependent on, 210, 214
Sodium nitroprusside (NP), 197, 204, 205
Sodium-potassium exchange
  measurement of, in vascular smooth muscle, 20-21
  mechanisms of, 18
Sodium-potassium dependent current, in myometrium of guinea pig, 78-79, 81
Sodium-potassium flux measurement, for determination of membrane potential in vascular smooth muscle, 33-36
Sodium-potassium pump, 7, 8, 103, 104, 109